EcoWISE

Innovative Approaches to Socio-Ecological Sustainability

EcoWISE (Ecological wisdom inspired science and engineering) series aims to publish authored or edited volumes that (1) offer novel perspectives and insightful reviews, through the lens of ecological wisdom, on emerging or enduring topics pertaining to urban socio-ecological sustainability research, planning, design, and management; (2) showcase exemplary scientific and engineering projects, and policy instruments that, as manifestations of ecological wisdom, provide lasting benefits to urban socio-ecological systems across all temporal and spatial scales; or (3), ideally, coalesce (1) and (2) under a cohesive overarching framework. The series provides a forum, the first of its kind, for the broad international community of scholars and practitioners in urban socio-ecological systems research, planning, design, and management.

Books in the EcoWISE series will cover, but not be limited to, the following topical areas:

- Transdisciplinary studies of ecological wisdom, *ecophronetic* practice research, complex adaptive systems, traditional ecological knowledge (TEK), urban resilience, global climate change, urbanization, phronetic social science, and sustainability science;
- Ecological sciences for and practices in ecosystem rehabilitation, habitat reconstruction, landscape restoration, sustainable agroecology, permaculture, architecture, landscape and urban planning and design, resource conservation and management, disaster management, urban flood control and management, and low impact development;
- Benchmarks for sustainable landscape and urban planning and design (e.g., the Sustainable SITES Initiative [SITES], Leadership in Energy & Environmental Design [LEED]);
- Engineering science, technology and policy for environmental restoration (e.g., river, lake, hazardous waste sites, brown fields), low CO_2 emission, wastewater and soil treatment, renewable energy, urban green infrastructure and building materials.

More information about this series at http://www.springer.com/series/15217

Varenyam Achal · Abhijit Mukherjee
Editors

Ecological Wisdom Inspired Restoration Engineering

Springer

Editors
Varenyam Achal
Guangdong Technion-Israel Institute of Technology
Shantou, China

Abhijit Mukherjee
Department of Civil Engineering
Curtin University
Perth, WA, Australia

ISSN 2367-0029 ISSN 2367-0037 (electronic)
EcoWISE
ISBN 978-981-13-0148-3 ISBN 978-981-13-0149-0 (eBook)
https://doi.org/10.1007/978-981-13-0149-0

Library of Congress Control Number: 2018947271

This Springer imprint is published by the registered company Springer Nature Singapore Pte Ltd.
The registered company address is: 152 Beach Road, #21-01/04 Gateway East, Singapore 189721, Singapore

Preface

Ecological wisdom inspired science and engineering (EcoWISE) offers novel perspectives and insightful research and reviews, through the lens of ecological wisdom, on emerging or enduring topics pertaining to urban socio-ecological sustainability research, planning, design, and management. It was first proposed during the *1st International Symposium on Ecological Wisdom for Urban Sustainability* on October 17–18, 2014, in Chongqing, China, by Prof. Wei-Ning Xiang, and reiterated during *2nd International Symposium on Ecological Wisdom for Urban Sustainability*, November 17–20, 2016, held at the University of Texas at Austin. Following this conference, "Ecological Wisdom Inspired Restoration Engineering" took shape of a book and attracted contributions from authors from all around the world.

Ecological Wisdom Inspired Restoration Engineering is one of first two edited books in the EcoWISE book series. It is focused on learning and adapting EcoWISE in restoration engineering through theories, hypotheses, policies, practical understandings, and case studies. Understanding nature's processes is a prerequisite for prudent human actions to sustain the healthy functioning of a habitable earth. The book aims to (1) provide a guide for the readers seeking to understand and build sustainable urban socio-ecological systems by restoration technologies coming from ecological wisdom and (2) explore ecological wisdom principles from various perspectives pertaining to ecological restoration.

There are 14 chapters in this book. Chapter "Development of Environmentally Sustainable Materials" examines environmentally sustainable materials from a standpoint that might be said to be deeply ecological as it advocates for the increased use of sustainability metrics in assessing evaluations of materials' greenness. Construction industry is facing big problem in achieving sustainability. Approaching from the ecological wisdom perspective, microbial strategies are providing promising strategy to achieve green building materials. Popularly investigated strategies are listed as biocementation, biomasonry, biorepair, and bioconsolidation (Chapter "Overlooked Strategies in Exploitation of Microorganisms in the Field of Building Materials"). In Chapter "Microbially Induced Calcite Precipitation (MICP) for Soil Stabilization," biogrout is explained as environmentally friendly alternate to chemically based grouting materials in improving the engineering properties of soils. Chapter "Utilization of Microbially Induced Calcite Precipitation for Sand Solidification Using *Pararhodobacter* sp." shows how a bacterium, *Pararhodobacter* sp. contributed to the application of a new technique for sand improvement using biostimulation, while Chapter "Effect of Plant-Derived Urease-Induced Carbonate Formation on the Strength Enhancement of Sandy Soil" presents the research on plant-derived urease using crude extract of watermelon (*Citrullus lanatus*) seeds in improving sand specimens with potential in controlling soil liquefaction.

In order to save carbon emission and reduce environmental impact, making an effective use of straw fibers in concrete is no doubt a significant step toward low carbon and sustainable construction in the era of rapid economic growth and industrialization in China. It is explained in Chapter "Material Properties of Agriculture Straw Fibre-Reinforced Concrete", a chapter that focused on advance knowledge on the mechanical characteristics of agricultural straw fiber reinforced concrete (ASFRC) in improving its compressive and tensile performance. Aiming for a clean future through smart technology in construction, Chapter "Agro-Industrial Discards and Invasive Weed-Based Lignocelluloses as Green Building Materials: A Pertinent Review" delineates the viability potential and hurdles in the path of using agro-industrial discards and invasive weed-based lignocelluloses in building materials.

There are various environmental and social problems such as pollution, difficulties in food and water supply, poverty or homelessness, faced by modern city, where specific ecosystem services offer structural solutions. Structural integrating ecosystem services in the built-up urban space can solve major urban environmental and social problems to improve urban sustainability and revitalize degraded urban areas. Chapter "Integration of Ecosystem Services in the Structure of the City is Essential for Urban Sustainability" shows that nature-based solution management differs from technological management and provides ecosystem services in any restoration project. Chapter "Integrated Blue and Green Corridor Restoration in Strasbourg: Green Toads, Citizens, and Long-Term Issues" provides perspectives on several ecosystem services, including supporting services (habitat for amphibians), regulating services (water quality enhancement), and cultural services (urban landscape greening) from ecological, engineering, and sociological points of view. Further, brownfields are challenging problem, especially in industrial and post-industrial cities in many countries. In Chapter "The Role of Ecological Wisdom in Brownfields Redevelopment in China," ecological wisdom is synthesized and integrated from other countries where brownfield redevelopment is studied and applied to deepen the current understanding of brownfield redevelopment. Chapter "Ecological Wisdom-Inspired Remediation Technology for Aquaculture Water Quality Improvement in Ecological Agricultural Park" investigates ecological wisdom-inspired remediation technology for the improvement of aquaculture water quality in ecological agricultural park of Shanghai, China. In addition, wetlands are significant for the development of resources and environmental protection in the city of Shanghai and discussed in Chapter

"Wetlands Restoration Engineering in the Metropolitan Area." The management of wastewater being generated from domestic, industrial, and agricultural sources is also a big issue worldwide. Chapter "Modern and Emerging Methods of Wastewater Treatment" employs the principle of ecological wisdom in designing new treatment strategies and also to strengthen the available natural treatment methods in order to achieve the goal of sustainable water development. The concept of ecological wisdom also guides us to look for alternative to improve crop production and soil health to enhance sustainable agricultural production (Chapter "The Role of Microbes to Improve Crop Productivity and Soil Health").

I wish to express my appreciations to the multidisciplinary team of authors for wonderful scientific contribution to this book. We are very grateful to Prof. Wei-Ning Xiang (the Founding Editor in Chief of the Springer Nature EcoWISE book series) for his insightful comments and for accepting this book to be a part of the EcoWISE book series. Thanks to Ms. Xiaoli Pei from Springer Nature and her production team for supporting us constantly during the editorial process. It is intended that this book will serve as a useful resource for environmental engineers, biotechnologists, civil engineers, researchers, and graduate students in these areas. On behalf of all the authors, I firmly believe that the knowledge and wisdom in this book will greatly enhance the EcoWISE enterprise's effort to resolve various environmental issues under the beacon of ecological wisdom.

We hope you like reading this book.

Shanghai, China Varenyam Achal

Contents

Prologue

Ecological Wisdom: Genesis, Conceptualization, and Defining Characteristics

EcoWISE

Ecological practice is the action and process that humans involuntarily engage themselves in the aim to bring about a secure and harmonious socio-ecological condition that serves human beings' basic need for survival and flourishing. It is the most fundamental and arguably primordial practice. *Homo sapiens* has been engaging in over thousands of years of coevolution with nature and falls into one or any combination of the following categories—ecological planning, design, construction, restoration, and management.

From ecological practice, humans acquired a distinctive master skill par excellence, *ecological wisdom*, that enables them to address and act well on intractable socio-ecological issues that are crucial to their survival and flourishing. While manifesting itself in a myriad of ecological and landscape projects and public policy instruments that has been beneficent to both humans and other residents on the earth, this invaluable intellectual asset of ecological wisdom continues to evolve in the contemporary society of unprecedented socio-ecological transformations, inspiring advancement in modern science, and stimulating technological and engineering innovations for the greater good. Ecological wisdom inspired science and engineering (EcoWISE, for brevity) is therefore the emerging transdisciplinary field of scholarly inquiry that seeks novel insights, deliberately through the lens of ecological wisdom, into contemporary socio-ecological issues, and aims to develop innovative, prudent, and efficacious scientific and engineering solutions. The Springer Nature EcoWISE book series provides a forum, the first of its kind, for the international community of scholars and practitioners to collectively advance this worthy enterprise.

The book series aims to publish authored or edited volumes that (1) offer novel perspectives and insightful reviews, through the lens of ecological wisdom, on emerging or enduring topics pertaining to ecological practice and research; (2) showcase exemplary scientific and engineering projects, and policy instruments that, as manifestations of ecological wisdom, provide lasting benefits to socio-ecological systems across all temporal and spatial scales; or (3) ideally, coalesce (1) and (2) under a cohesive overarching framework. The series is intended to serve the broad international community of scholars and practitioners in socio-ecological practice and research.

Integral to EcoWISE are the questions pertaining to the genesis, conceptualization, and defining characteristics of ecological wisdom: What is it? Where does it come from? What defining characteristics does it have? In the following sections, I shall explore these three questions.[1]

Ecological Wisdom

There are three ways in which the scholarly construct of ecological wisdom is defined (for recent and succinct reviews of various definitions, see Liao and Chan 2016, pp. 111–112; Wang et al. 2016). As described chronologically below, they derive from different etymologies and reflect varied intellectual traditions.

Ecological Wisdom as an Ethical Belief: *Ecosophy*

In a 1973 essay on the main characteristics of the deep ecology movement, Norwegian ecological philosopher Arne Naess coined the term *ecosophy*, by combining the ancient Greek words *ecos* (household place) and *sophia* (theoretical wisdom), to represent an individual's own personal "philosophy of ecological harmony or equilibrium (between human and nature—the author)" (Naess 1973, p. 99). Despite his intention to use this term "to mean ecological wisdom or wisdom of place" (Drengson and Devall 2010, p. 55), no formal articulation was made until 16 years later. In a 1989 essay entitled *From ecology to ecosophy, from science to wisdom*, he inaugurated the fused nexus of ecological wisdom and *ecosophy* with a strong proclamation that for humans "to live on Earth enjoying and respecting the full richness and diversity of life-forms of the ecosphere, … [e]co-wisdom (ecosophy) is needed" (Naess 1989, p. 185).

Along with this *ecosophical* line of reasoning, there was a strikingly parallel development in a noncognate context and with no direct intellectual contact. In a

[1]Drawing primarily on Chinese and English literature owing to my linguistic capabilities, this synthesis is inevitably limited in its scope and thus subject to expansion.

1996 seminal Chinese book *On ecological wisdom* (《生态智慧论》, *sheng tai zhi hui lun*), Chinese philosopher Zhengrong She (佘正荣) coined the term 生态智慧 (*sheng tai zhi hui*), in a way similar, yet unrelated, to Naess', by combining the Chinese words 生态 (ecological) and 智慧 (wisdom). He defined ecological wisdom as *ecohumanism* with the following proclamation (She 1996, pp. 3–4)[2]: "At the transitional juncture from the industrial to ecological civilization, human beings must supplant the anthropocentric humanism with ecohumanism. A tripartite worldview that blends seamlessly ecological sciences, ecological ethics, and ecological aesthetics, ecohumanism is the ecological wisdom human beings need, and can guide the contemporary human beings through the jungle of industrial civilization toward the bright future of ecological civilization."

Acknowledging that "ecological wisdom is the wisdom for living and survival that is rooted in and developed through the primordial process of human adaptation to the environment" (She 1996, p. 2),[3] he posited that the ecological philosophies (i.e., *ecosophies*, as defined by Naess) of some of the greatest thinkers in human history, including those of Laozi, Aldo Leopold, Aurelio Peccei, Holmes Roston III, Arnold Joseph Toynbee, and Zhuangzi, are but archived individual convictions drawing on collective *ecosophical* beliefs (Ibid. p. 3).[4]

This collective perspective of *ecosophy* deviates from the "whole personal view" (Drengson and Devall 2010, p. 56) of Naess'. According to Canadian philosopher Alan Drengson and American sociologist Bill Devall, Naess believes that "[s]ince there is an abundance of individuals, languages, cultures, and religions, there will be an abundance of *ecosophies*." (Ibid.) To differentiate, "[e]ach person's *ecosophy* can be given a unique name, possibly for the place they live, or for something to which they feel strongly connected." Exemplifying this individual's personal view are Naess' "*Ecosophy* T" (Drengson and Devall 2010, pp. 56–57) and Chinese ecological aesthetician Xiangzhan Cheng's "*Ecosophy* C" (2013).

Ecological Wisdom as a Dual Ability: *Ecophronesis*

In a 2017 article entitled *Ecological philosophy and ecological wisdom*, Chinese ecological philosopher Feng Lu (卢风) defined ecological wisdom as the dual human ability to make ethically and politically sound judgment and to take ensuing

[2]"人类在从工业文明向生态文明转变的历史关头,必须超越人类中心主义的价值观,形成一种使生态规律、 生态伦理和生态美感有机统一的新的价值观。这就是生态人文主义的价值观。生态人文主义是当代人类所需要的生态智慧, 它将引导人类安全地走向未来的生态文明。"

[3]"生存智慧来源于生物对环境的适应, 因而生存智慧实质上就是生态智慧。对环境的适应是一切智慧最原始和最深刻的根源。"

[4]"生态哲学给人类提供了一些深刻的生存智慧。但是这并不是说, 在现今的生态哲学中已经达到了尽善尽美的生存智慧, 也不是说在生态哲学出现之前就没有产生过相当深刻的生态智慧。事实上, 在东方古代的文化传统中就产生过非常深刻的生态直觉 (智慧—作者), 这些生态直觉 (智慧) 对于当代人类生态观的发展和完善具有十分重要的价值。"

prudent actions in particular circumstances of ecological practice (Lu 2017, p. 278; p. 285).[5] This human ability approach to wisdom definition has its intellectual root in Aristotelian conception of *phronesis* (i.e., practical wisdom; for a recent and succinct review of Aristotelian *phronesis*, see Xiang 2016, pp. 54–55). It is in fact the philosophical underpinning of a 2016 essay on ecological practical wisdom by American geographer and planning scholar Wei-Ning Xiang (Xiang 2016). In a way similar to that employed by Naess and She, Xiang coined the term *ecophronesis*, by combining two ancient Greek words *ecos* and *phronesis*, to represent ecological practical wisdom which he defined as "the master skill par excellence of moral improvisation to make, and act well upon, right choices in any given circumstance of ecological practice" (Xiang 2016, p. 55). Here, Xiang noted the term *skill* is used as an uncountable mass noun synonymous with the term *ability* (as in "the skill") [Ibid.].

Despite the nascent coinage of *ecophronesis*, both the term and the *ecophronetic* line of reasoning it represents are indeed, according to Xiang, an *ex post* recognition of and a revered tribute to an outstanding group of human beings throughout history (Xiang 2016, p. 59). *Ecophronimoi* are the people of ecological practical wisdom whose master skill par excellence of *ecophronesis* enabled them to be successful in challenging circumstances of ecological practice (Ibid.). Among the prominent *ecophronimoi* are the Chinese ecological planner and engineer Li Bing (480 BC–221 BC) and his colleagues of many generations who collectively designed, built, and sustained the Dujiangyan irrigation system (256 BC to present) in Sichuan, China (Needham et al. 1971, p. 288; Xiang 2014, pp. 65–66), and the American ecological planner and educator Ian McHarg (1920–2001) and his colleagues who planned and developed the town of the Woodlands in Texas, the USA, in the 1970s (McHarg 1996, pp. 256–264; Xiang 2017a; Yang and Li 2016). Their *ecophronetic* practices of stellar quality have brought lasting benefits to the people and other living communities in the areas the projects serve, and clearly achieved the paramount level of "doing real and permanent good in this world" (Xiang 2014, p. 65).

Ecological Wisdom as a Cohesive Whole of *Ecophronesis* and *Ecosophy*

In his 2016 essay on *ecophronesis*, Xiang made the observation that not only are *ecophronesis* and *ecosophy* so profoundly linked, but the connection between them is indeed integral to *ecophronesis*. He noted that in the instances of prudent and successful ecological practice throughout human history, like those of aforementioned Dujiangyan irrigation system and the Woodlands, *ecophronimoi*'s mastery and execution of improvisational skill were mindfully bound by a moral covenant with nature, and inspired and informed by the human beings' enlightened

[5] "生态智慧是在生态学和生态哲学指引下养成的判断能力、直觉能力和生命境界 (涵盖德行)。生态智慧与人的生命和实践'不可须臾离'" (卢风, 2017, p. 285).

self-interest (Xiang 2016, pp. 57–58).[6] This union of improvisational ability and moral commitment is what American planning scholar John Forester calls "moral improvisation" (Forester 1999, pp. 224–241). It is with this very master skill par excellence of moral improvisation that *ecophronimoi* became capable of being "doubly responsible" (Nussbaum 1990, p. 94) in any particular instance of ecological practice—honoring commitments and upholding principles, on the one hand, and attending specific circumstantial particulars, on the other (Xiang 2016, p. 58).

This observation corroborates Xiang's argument that as an *ex post* and long overdue recognition of a reverable human virtue in ecological practice, the scholarly construct of ecological wisdom is incomplete and unbalanced in the absence of either *ecosophy* or *ecophronesis* (Xiang 2016, p. 58). It provides support for his proposal, as depicted symbolically in Eq. (1)[7] (Xiang 2017b), that both *ecosophy* and *ecophronesis* should be juxtaposed at the core of ecological wisdom (Xiang 2016, p. 53).

$$\text{Ecological wisdom} = \text{ecosophy} + \text{ecophronesis} \tag{1}$$

This *ecophronetic* line of reasoning for "the *ecophronesis-ecosophy* nexus of ecological wisdom" (Xiang 2016, p. 58) finds supporting arguments in Naess' work on *ecosophy*. In the 1989 essay aforementioned, Naess argued that the ethical belief of *ecosophy* is a source of inspiration and guidance for both action and research. "[N]ot a philosophy in the academic sense" (Naess 1989, p. 187), he wrote, "[a]n articulated *ecosophy* includes an attempt to outline *how to inhabit the Earth* conserving her long range, full richness and diversity of life as a value in itself (Ibid. p. 186)." As such, "[w]ithin the framework of Ecosophy research enters primarily as 'action research'" that is "subordinated to practical policies," (Ibid. p. 188), and aimed at "the derivation of particular prescriptions (that are) adapted to particular situation." (Ibid. p. 187) The practical orientation of *ecosophy* and contextual characterization of *ecosophical* research Naess articulated here manifest in his own work on Ecosophy T and the Apron Diagram, and are readily evident throughout his later writings (for a succinct review, see Drengson and Devall 2010).

---

[6]Human beings' enlightened self-interest is a term used in environmental virtue ethics that serves the same *ecosophical* function as Naess' *ecosophy* does—it is an ethical belief of the ecological harmony between human and nature (Cafaro 2001, pp. 3–5). According to Xiang (2017, p. 56), under the premise that there exists a relationship of human-nature reciprocity, "it states plainly that it is in human beings' self-interest—ethical, moral as well as material—to respect and appreciate the intrinsic value of all living and non-living beings on the earth; and that human beings' own flourishing, at individual and collective levels, should be conceived and pursued in ways that both sustain and depend on the flourishing of the entire 'more-than-human whole' of which humans are part." As "such nonanthropocentrism is a part of wisdom" (Cafaro 2001, p. 15) that is widely shared by people from around the world and across generations, including Naess (see his 1986 essay, p. 72), I use it here as a collective *ecosophy*.

[7]In delivering this keynote speech in Chinese, Xiang presented the equation as 生态智慧=生态哲思+生态实践智慧. A copy of the PowerPoint presentation is available from the author upon request.

An Embracing Definition of Ecological Wisdom

At the point where the *ecosophical* and *ecophronetic* lines of reasoning converges, Xiang posited in a 2017 speech (2017b), emerges a definition of ecological wisdom that embraces *ecophronesis* and *ecosophy* into a cohesive whole. One such definition he initially presented (Ibid.) is further elaborated below.

> Ecological wisdom is the master skill par excellence of moral improvisation for and from ecological practice; it enables a person, a community, or an organization to make ethical judgment and take circumspect actions in particular circumstances of ecological practice; it is a cohesive whole of the *ecosophical* belief in the relationship of human-nature reciprocity and the *ecophronetic* ability to make, and act well upon, contextually and ethically right choices.

This definition highlights two defining characteristics of ecological wisdom—the ability in ecological practice to achieve the ideal of the unity of moral knowledge and virtuous action, and the ability to conduct preeminent ecological practice research.

Ecological Wisdom as the Ability to Achieve the Unity of Moral Knowledge and Virtuous Action

Fine hundred years ago, Chinese Neo-Confucian philosopher Wang Yangming (王阳明, 1472–1529) coined the term *the unity of knowledge and action* (知行合一, *zhī xíng hé yī*) to designate a state of moral ideal that he believes "exists for all (humans)" (Ching 1976, p .68) and can be achieved through and in practice (Ibid. p. 72). In this state of moral ideal, ethical knowledge (*i.e.*, knowledge of the good) and virtuous action (*i.e.*, action to do the good) are only two words describing the same one effort; as such, one acts spontaneously yet virtuously upon deep moral convictions (Ibid. pp. 68–69).[8] Similar ideas are also found in Aristotle's thinking over two millennia ago. "For Aristotle," wrote Canadian political scientist David Tabachnick, "to be 'ethical' was more than simply knowing right from wrong, but also meant the capacity to act upon that (moral—author) knowledge." (Tabachnick 2013, p. 32).

Ecological wisdom as defined above enables a person, a community, or an organization to achieve Wang's ideal state of the unity of moral knowledge and virtuous action (for brevity, thereafter, *the unity of knowledge and action*), and to meet the Aristotelian ethical standard. As a master skill par excellence of moral improvisation, it activates and amplifies the action-guiding function of *ecosophical*

[8]It should be noted that, according to Julia Ching, a Canadian philosopher and a word leading Wang Yangming scholar, for Wang Yangming, "...just as true knowledge is always knowledge of virtue, true action should always be virtuous action. 'The unity of knowledge and action' is primarily a moral ideal rather than a principle of epistemology." (Ching 1976, p. 66) Unfortunately, by many with good intentions it has been mistaken as a principle of epistemology (Dong 2013).

belief such that the ethical knowledge of the good serves as a moral benchmark for one's sound judgment and virtuous action in particular circumstances of ecological practice (Xiang 2016, p. 56). The ensuing outcomes, in the form of ecological plans, designs, construction and restoration projects, and management policies, are thus simply tangible manifestations of the knowledge of the good, and exemplified by, among others, the aforementioned Dujiangyan irrigation system and the Woodlands. This process of activating and realizing one's *ecosophical* belief is analogous to, if not the same as, *zhì liáng zhī* (致良知)—extending and realizing one's innate conscience (i.e., knowledge of the good) through virtuous actions in practice—a process that, according to Wang Yangming, leads to *the unity of knowledge and action.*[9]

In a 2003 essay, Chinese philosopher Mingying Deng proposed the concept of *eco-conscience* (生态良知, shēng tài *liáng zhī*) and defined it as a coalesced nexus of "the consciousness of being part of a more-than-human whole; the sense of moral goodness of one's own conduct, intentions, or character; and a feeling of ethical obligation to do right or be good in the best interest of the more-than-human whole." (Deng 2003, p. 86).[10] *Eco-conscience* such defined is comparable to the *ecosophy* component of ecological wisdom (section "Ecological Wisdom as a Cohesive Whole of *Ecophronesis* and *Ecosophy*") with a shared belief in the relationship of human-nature reciprocity. A subtle difference is that *eco-conscience*, or conscience by and large, is often regarded as an innate quality of every human being [i.e., "the innate knowledge of the good" (Zhang 2017, p. 341)], while *ecosophy* is not reportedly so.[11] The difference can nevertheless be omitted here and now since even Wang Yangming himself makes no distinction between conscience and moral knowledge in his conception of *zhì liáng zhī* (Ching 1976, p. 67). As such, it suffices to say that ecological wisdom, through activating, extending, and realizing *eco-conscience* (*zhì* shēng tài *liáng zhī*, 致生态良知) or *ecosophical* belief grounded in eco-conscience, is capable of empowering a person, a community, or an organization to achieve Wang's state of moral ideal of *the unity of knowledge and action* in ecological practice and thus to become ethical by the Aristotelian standard.

---

[9] "致吾心良知之天理于事事物物，则事事物物皆得其理矣。致吾心致良知者,致知也。事事物物皆得其理者，格物也。是合心与理为一者也。" (王阳明《王阳明全集》卷二《传习录中.答顾东桥书》, 上海古籍出版社, 1992).

[10] "(生态良知) 是指人类自觉地把自己作为生物共同体的一员，把自身的活动纳入生物共同体的整体活动， 并在此基础上形成的一种维持生物共同体和谐发展的深刻的责任感以及对自身行为的生态意义的自我评价能力。" (邓名瑛, Deng 2003, p. 86).

[11] More investigation is requested into the relationships between eco-conscience and *ecosophy*. In the writings on *ecosophy*, authors (Naess, Drengson, Deval, and Cheng, among others) predominantly treated *ecosophy* as a belief of environmental ethics with no articulation to eco-conscience. In a 2017 essay, on the other hand, Chinese ecological philosopher Xuezhi Zhang posited that one's achievement of the ideal moral state of *the unity of human and nature* is grounded in eco-conscience and speculated whether eco-conscience could integrate environmental ethics (2017, p. 342). However, no rigorous investigation into the relationships has been found in the literature.

Ecological Wisdom as the Ability to Do Preeminent Ecological Practice Research

In addition to activating the action-guiding function of *ecosophical* belief or *eco-conscience*, *ecophronesis* in the scholarly construct of ecological wisdom is capable of empowering scholar-practitioners and practitioners to do outstanding research for ecological practice.

Scholar-practitioners are scholars who are engaged in use-inspired basic research for practice (i.e., *practice research*) in Pasteur's quadrant and dedicated to generating *new* knowledge that is *useful* to practitioners (Xiang 2017, pp. 2243–2244). Common to all scholar-practitioners who have done outstanding ecological practice research, like McHarg, is their *ecophronetic* way of conducting practice research (Ibid. p. 2245). Wrote Xiang (Ibid. unless essential, citations in the original text are omitted for brevity), with *ecophronesis*,

> scholar-practitioners like McHarg became much capable of generating *new* knowledge that is *useful to the real*: not only did they advance scholarly rigorous—thorough and valid—knowledge that was also *immediately relevant*, *actionable*, and *potentially efficacious* to the real-world practitioners who were in specific knowledge needs under particular circumstances of ecological practice; but they also produced high caliber scholarship that is enlightening to scholars and practitioners from around the world and across generations who have interest in ecological practice research. Furthermore, because *ecophronesis* embraces inherently a transdisciplinary research capability in socio-ecological systems, *ecophronetic* scholar-practitioners like McHarg were immune from the pathogenic influence of 'ivory tower syndrome' (Toffel 2016, p. 1494). They became readily capable of bridging the arguably unbridgeable gap between scientific rigor and practical relevance, and taming the seemingly intractable problems of 'knowledge production' and 'knowledge transfer' (Sandberg and Tsoukas 2011, p. 338), all of which have been and remain to be persistent concerns in both circles of ecological practice and science in the modern-day world. As such, their *ecophronetic* way of conducting practice research manifested itself in a myriad of ecological projects and public policy instrument that has been providing lasting ecosystem services benefits to the human beings across generations.

It is noteworthy that the empowerment of *ecophronesis* equally benefits many practitioners who are engaged in pure applied ecological research that is motivated solely by the applied goals of problem-solving in practice (Xiang 2017, pp. 2242–2243). Exemplifying these *ecophronetic* practitioners are aforementioned Li Bing and his colleagues of many generations. Without seeking a scholarly understanding of the encountered phenomena through the scientific lens, they were enabled to conduct in an *ecophronetic* way preeminent research that contributed to the very success of their ecological practice of stellar quality.

Role Models and the Community of Scholar-Practitioners

One premise underlying *ecological wisdom inspired science and engineering* (EcoWISE), the overarching concept of this book series, is that science and engineering need to and should be inspired by ecological wisdom. I hope that the preceding sections on the genesis, conceptualization, and defining characteristics of ecological wisdom have corroborated the premise to be just and appropriate. With regard to the subsequent question of *how* science and engineering should be inspired by ecological wisdom to serve the community of more-than-human whole on the earth, one way forward would be for us to emulate the role models of *ecophronetic* scholar-practitioners, whom the last two sections were dedicated to.

Ecological wisdom is not an abstract concept in the scholarly papers, and it is on clear display in the well-lived and fully realized lives of many practitioners and scholar-practitioners who have done preeminent ecological practice and research, and achieved the ideal moral state of *the unity of knowledge and action*. "A good example is the best sermon," to follow the example of these outstanding human beings, *ecophronimoi*, is both fitting and indeed rewarding. As a student of McHarg's in the 1980s, for example, not only did I witness that the unity of moral knowledge and virtuous action was like second nature to him, but I can also testify that like many of his students, my academic aspiration has been ever since inspired and professional path illuminated by his role model as an *ecophronetic* scholar-practitioner. With a gentle caveat that this way of inquiry for EcoWISE aims to examine and advocate *ecophronetic* practice research as a distinctive mode of practice research drawing upon the experience and examples of *ecophronetic* scholar-practitioners, rather than promoting the individuals themselves, I trust that the EcoWISE book series will become a celebrated venue for the building of a strong community of e*cophronetic* scholar-practitioners around the world and am confident that it will serve the community well in their pursuit of exemplary ecological wisdom-inspired ecological practice research and outstanding *ecophronetic* scholarship.

Holly Hill Farm, Charlotte, NC, USA
March 2018

Wei-Ning Xiang
University of North Carolina at Charlotte
Charlotte, USA

and

Tongji University
Shanghai, China

References

Cheng XZ (2013) Aesthetic engagement, Ecosophy C, and ecological appreciation. Contemporary Aesthetics 11. http://quod.lib.umich.edu/c/ca/7523862.0011.009/–aesthetic-engagement-ecosophy-c-and-ecological-appreciation?rgn=main;view=fulltext

Ching J (1976) To acquire wisdom: the way of Wang Yang-ming. Columbia University Press, New York

Deng MY (2003) On eco-conscience (论生态良知). Ethics research (《伦理学研究》) 2003(2) 86–89 (In Chinese)

Dong P (2013) The unity of knowledge and action: the magnificent state of life ("知行合一":无限阔大的生命境界). A public lecture at Zhejiang University, Hangzhou, China, November 27, 2013. Zhejiang University News, http://www.news.zju.edu.cn/2013/1127/c763a63860/page.htm. Accessed on 26 February 2018

Drengson A, Devall B (2010) The deep ecology movement: origins, development and future prospects. The Trumpeter 26(2):48–69

Forester J (1999) The deliberative practitioner: encouraging participatory planning processes. The MIT Press, Boston, MA, USA

Liao KH, Chan JKH (2016) What is ecological wisdom and how does it relate to ecological knowledge? Landscape and Urban Planning 155:111–113

Lu F (2017) Ecological philosophy and ecological wisdom (《生态哲学与生态智慧》). In: Lu F, Cao MQ (eds) Ecological philosophy: the pulsing spirit of the Anthropocence (《生态哲学: 新时代的时代精神》). Chinese Social Sciences Press, Beijing (中国社会科学出版社, 北京), pp 277–286

McHarg IL (1996) A quest for life: an autobiography. Wiley, New York

Naess A (1973) The shallow and the deep, long-range ecology movement. A summary. Inquiry 16:1, 95–100

Naess A (1986) Intrinsic value: will the defenders of nature please rise. In: Reed P, Rothenberg D (eds) Wisdom in the open air: the Norwegian roots of deep ecology, 1992. The University of Minnesota Press, Minneapolis, MN, USA, pp 70–82

Naess A (1989) From ecology to ecosophy, from science to wisdom. World Futures 27, 185–190.

Sandberg J, Tsoukas H (2011) Grasping the logic of practice: theorizing through practical rationality. Acad Manag Rev 36(2):338–360

She ZR (1996) On ecological wisdom (《生态智慧论》, *sheng tai zhi hui lun*). Chinese Social Sciences Press, Beijing (中国社会科学出版社, 北京)

Tabachnick DE (2013) The great reversal: how we let technology take control of the planet. University of Toronto Press, Toronto

Toffel MW (2016) Enhancing the practical relevance of research. Prod Operations Manag 25 (9):1493–1505

Wang X, Palazzo D, Carper M (2016) Ecological wisdom as an emerging field of scholarly inquiry in urban planning and design. Landscape and Urban Plan 155:100–107

Xiang W-N (2014) Doing real and permanent good in landscape and urban planning: ecological wisdom for urban sustainability. Landscape and Urban Plan 121, 65–69

Xiang W-N. (2016) *Ecophronesis*: the ecological practical wisdom for and from ecological practice. Landscape and Urban Plan 155:53–60

Xiang W-N (2017a) Pasteur's quadrant: an appealing *ecophronetic* alternative to the prevalent Bohr's quadrant in ecosystem services research. Landscape Ecol 32(12), 2241–2247, https://doi.org/10.1007/s10980-017-0583-y

Xiang W-N (2017b) Ecological wisdom inspired ecological practice is integral to the building of ecologically civilized society (生态文明社会的构建需要生态智慧引导下的审慎的城乡生态实践). A keynote speech at the International forum on sustainable solutions for new era of ecological civilization, Tsinghua University, Beijing, China, December, 2017

Zhang X (2017) Ecological consciousness and the conscience in the writings of Wang Yangming. In: Yao X (ed) Reconceptualizing confucian philosophy in the 21st century. Singapore, Springer Nature

Development of Environmentally Sustainable Materials

Mona Wells

Abstract This chapter examines environmentally sustainable materials from a standpoint that might be said to be deeply ecological as it advocates for the increased use of sustainability metrics in assessing evaluations of materials' "greenness." As such, the discussion begins by examining different paradigms of environmental sustainability, and in the context of lead use and contamination, a current environmental issue. The use of sustainability metrics is discussed, segueing into identification of the two major drivers of environmental damage (fossil fuel consumption and agriculture). Subsequently, examples of interesting environmental materials are discussed: One section is devoted to materials that increase efficiency of energy use, and another section discusses the reduction of damage from agriculture via materials engineering that will enable less land use and thereby promote ecological recovery. A final section on materials development comes full circle in considering the possibility of peak metals and innovative electronic technologies that may someday be extensible to reducing the needed circuits in buildings. In closing, the fundamental tension between technology and materials' consumption is considered in the context of Jevon's paradox as a cautionary note regarding the development of sustainable materials and the need for strong sustainability and deep ecological wisdom.

Keywords Sustainable materials · Ecological wisdom · Ecological recovery Land use · Jevon's paradox

1 What is a Sustainable Material?

Consider the case of lead. Lead is the most prevalent pollutant in the world today (McCartor and Becker 2010), and, according to the United States Environmental Protection Agency (US EPA 2000), there is "no demonstrated safe concentration of lead in blood." It has long been known that lead is particularly dangerous for

M. Wells (✉)
National Institute of Water and Atmospheric Research, Dunedin, New Zealand
e-mail: mona.wells@niwa.co.nz

V. Achal and A. Mukherjee (eds.), *Ecological Wisdom Inspired Restoration Engineering*, EcoWISE, https://doi.org/10.1007/978-981-13-0149-0_1

children (US EPA 2000); however, now it is increasingly apparent that childhood lead exposure conveys lifelong effects (Brubaker et al. 2009; Cecil et al. 2008). By far, the major world producer of lead is China (ILA 2017a), where in all but 3 of 34 provincial-level administrative units, the blood lead level of children exceeds the 100 μg/L "level of concern," and, by land area, in nearly half of the country over 40% of children have blood lead levels exceeding this (Ye et al. 2007). Despite these facts, despite lead's being a non-renewable resource, despite the production and use of lead's having increased exponentially over recent decades, and despite the laws of thermodynamics (which dictate that no recycling process could ever be 100% efficient), the International Lead Association claims to support a sustainable lead industry (ILA 2017b) and lead producers call for the use of more lead–acid batteries in hybrids as "supporting sustainability" (Prweb 2017).

In his book Beyond Growth: The Economics of Sustainable Development, Daly (1996) notes that

> One way to render any concept innocuous is to expand its meaning to include everything. By 1991 the phrase 'sustainable development' had acquired such cachet that everything had to be sustainable, and the relatively clear notion of environmental sustainability of the economic subsystem was buried under 'helpful' extensions such as social sustainability, political sustainability, financial sustainability, cultural sustainability, and on and on. Any definition that excludes nothing is a worthless definition… which is why
>
> *we define sustainability as the ability to continue a defined behavior indefinitely.*

In other words, the term sustainability has been largely co-opted to include things that are *not*. According to a recent North American Product Survey, over 95% of the products reviewed made claims that violated one or more of the Seven Sins of Greenwashing (defined as disinformation disseminated by an organization so as to present an environmentally responsible public image, Terrachoice 2010). Our point of departure, therefore, is to first delineate what is meant herein by sustainability, and secondly to consider ways in which claims of sustainability might constructively be assessed.

Perhaps the most well-known conception of sustainability is the so-called triple bottom line (TBL, Elkington 1998), wherein something might only be considered sustainable if involving harmony between environmental, social, and economic considerations (also known as the three pillars of sustainability). In the TBL view of sustainability (Fig. 1, left), the three pillars of support are all of equal importance, though it should be remembered that TBL is referred to as an accounting framework, which is not a priori a particularly envirocentric construct. TBL, by extension, has led to the conceptualization in Environmental Economics that different types of capital (environmental, social, economic) are substitutable; hence, social and economic factors might readily outweigh environmental concerns. This has been referred to as weak sustainability (Fig. 1, center). Countering weak sustainability, the field of Ecological Economics has given us strong sustainability (Fig. 1, right, Daly 1995; Daly and Cobb 1989). Strong sustainability is based on the laws of thermodynamics, wherein growth is constrained by the size and resources of the earth and no productive matter and energy change is possible without an irreversible entropic degradation process that generates waste, and while it is possible to reduce the amount of waste by increasing efficiency, there are nonetheless insurmountable entropic limits to efficiency gains (Daly 2007). Strong sustainability recognizes that economic capital is derived from social capital (i.e., economy is a social construct) and that in turn social and economic capital are derived from environmental capital. Both TBL and weak sustainability represent a fundamentally anthropocentric view, whereas strong sustainability recognizes that humans need the earth, and not vice versa, and strong sustainability may therefore be said to be more closely aligned with deep ecology and ecological wisdom. Whether there are any actual "Laws" of economics might be said to be debatable (Karabell 2013). The laws of thermodynamics have been extensively tested for over a century by the scientific method, and this article will adopt the perspective that sustainability is only valid in having scientific underpinnings, i.e., in referencing strong sustainability.

2 Measuring the Sustainability of Materials

In order to know if a material is environmentally sustainable or not, it is important to have some way to measure sustainability—development of sustainable materials, therefore, must occur in tandem with development of better evidence-based approaches to quantification of sustainability. Sustainability metrics and sustainability development indices (SDIs) are tools that measure the benefits achieved through the implementation of sustainability/sustainable practices. It is important to keep in mind that sustainability metrics crosscut all dimensions (e.g., economic, social, environmental, and others) and paradigms of sustainability (e.g., weak and strong), as well as different views of development (e.g., whether or not globalization and/or urbanization are desirable). Most metrics encounter difficulties with transparency in the need for weighting, which comes into play with multi-dimensional SDIs, particularly in the context of weak sustainability. The purpose of this section is not to provide a comprehensive overview of sustainability metrics and SDIs, but rather to raise awareness for readers who are interested in sustainable materials about the most well-developed metrics and indices used in determining the *environmental sustainability* of materials.

It is helpful to distinguish metrics and indicators in terms of generality and specificity. Four indicators that might be referred to as general, i.e., ecosystem—scale metrics, include the Environmental Vulnerability Index (EVI), the Environmental Performance Index (EPI), the Living Planet Index (LPI), and Ecological Footprint Accounting (EFA). The EVI (SOPAC 2005) is calculated based on 50 environmental indicators of three aspects of environmental vulnerability: risks to the environment (natural and anthropogenic), environmental resilience, and ecosystem integrity (the health or condition of the environment as a result of past impacts). The Environmental Performance Index (EPI) is a method of quantifying and numerically marking policy-based environmental performance and

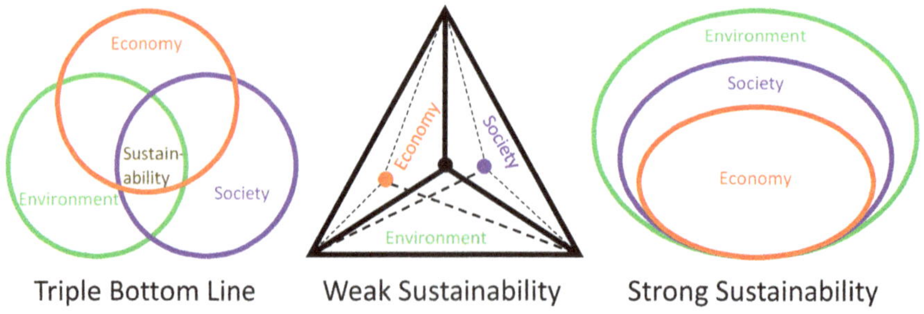

Fig. 1 Graphical depictions of conceptualizations of sustainability, adapted from Wu (2013)

is calculated based on 25 indicators of two overall environmental quantities: environmental health (human) and ecosystem vitality (Esty et al. 2008; Hsu et al. 2016). The LPI (Loh et al. 2005; McRae et al. 2017; WWF 2016) is global, calculated based on the Living Planet Database using over 14,000 population time series, and overall species trends are aggregated for the terrestrial, marine, and freshwater systems. The three system indices are averaged to produce the global LPI. EFA (GFN 2017; Rees and Wackernagel 2013) is based on the biological concept of carrying capacity and tracks the amount of land and/or water needed to produce the resources needed to support and absorb waste produced by any given population. Results are typically expressed in terms of how many planets are needed to support a given region, which is shockingly high in some developed nations. All four of these indicators are potentially useful in providing states or businesses with information to undertake self-assessment and policy refinement regarding their own environmental damage and/or vulnerability, particularly taking regional considerations into account, but from the standpoint of sustainable materials, these indicators are not practically useful to evaluate a product or process.

We distinguish indicators that we refer to as specific as being those indicators that are practically useful to evaluate a specific product or process. Examples of these include exergy analysis (EXA), embodied energy, emergy analysis (EMA), and life cycle assessment (LCA). Exergy is a thermodynamic term that reflects the maximum useful work possible during a process that brings a system into equilibrium with a heat reservoir. In practical terms, exergy may be thought of as the energy that is available for use. Exergy is eliminated for irreversible processes in proportion to the entropy of the system and its surroundings. Exergy analysis may be used to evaluate the impact of human activities on the environment (Apaiah et al. 2006; Hajjaji et al. 2012; Kanoglu et al. 2009); however, a flaw is that this analysis requires assumptions about thermodynamic reference states that are not testable assumptions.

Embodied energy includes the primary and secondary energy use for the production of materials, including process (Stein et al. 1981). Primary energy includes sources such as fossil fuel, and secondary energy is, for example, electricity.

Work of the biogeosphere that may be required (e.g., space heating for a building) in the future is not explicitly included; hence, the embodied energy is not able to offer information on a given product or process over from a life cycle perspective.

Emergy sums of all the available energy of one kind required directly and indirectly for the production of a product or service, and this does reference the biogeosphere (e.g., often expressed in terms of solar radiation). Emergy has thus been described as the total environmental support (rather than a measure of actual energy content) of a product or process (Odum 1996; Odum and Odum 1980; Scienceman 1987).

LCA is a tool to assess the environmental impacts and resources over a product or process lifetime, typically "from cradle to grave" (ISO 2006a, b). LCA identifies flows into (e.g., resources use) and emissions from (e.g., pollutants) a product or process, and then uses these flows to assess impacts across a range of potential impact categories, such as land use, ecotoxicity, etc. While the process was historically laborious and is subject to known limitations, the advent of new software and ever-increasing amounts of software-compatible data have brought this technique into the mainstream. Perhaps no example is as illustrative of the potential materiality and transparency of LCA as its role in the French Grenelle II laws (Cros et al. 2010).

A number of other indicators in addition to those described herein, particularly as pertains to building and construction materials, exist. While the use of sustainability metrics and indicators represent a quantum leap forward, most of the indicators in common use might be said to be incomplete, particularly in respect of neglecting environmental indirect costs of human activity, as discussed for instance by the Millennium Ecosystem Assessment. In the next sections of this chapter, we will examine particular areas wherein development of sustainable materials might be said to have the greatest cost–benefit potential, and the different categories of sustainable material will be evaluated according to material evaluation of sustainability where possible.

3 What to Do First?

Being able to measure sustainability may to some extent address issues of whether a product, process, or material is sustainable or not; it is not simply enough to know this. In addition to Greenwashing, the urgency of the need to transition to sustainability has arguably generated many "solutions looking for a problem." The United Nations (UN) has sensibly called for a "What do I do first?" approach, i.e., which problem is most important in order to determine how to make a meaningful contribution to sustainability? The UN Environment Programme (UNEP) International Panel for Sustainable Resource Management has assessed the best available science, and on a global level, to identify the primary drivers of environmental damage, which perforce arise from an ever-growing human population and concomitant ever-increasing level of production and consumption. The UNEP report *Assessing the Environmental Impacts of Consumption and Production: Priority Products and Materials* (UNEP 2010) reviews "all the available science and conclude that two broad areas are currently having a disproportionately high impact on people and the planet's life support systems—these are energy in the form of fossil fuels (including for agriculture and alternative energy products such as solar panels) and agriculture, especially the raising of livestock for meat and dairy products."[1] These areas of impact are not limited to but largely center upon biodiversity loss, climate change, and freshwater resources depletion. As such, the focus on sustainable materials should be informed by these two key drivers of fossil fuel use and agriculture.

4 Energy Use and Building Insulation

A group of scientists at Cambridge University recently conducted an analysis and found that "73% of global energy use could be saved by practically achievable design changes to passive systems" (Cullen et al. 2011). Of the three possible areas that they identified for energy savings, buildings accounted for over half of the energy used and the greatest potential energy savings of 80%. Insulation, for the buildings themselves, but also for appliances and services in buildings, accounted for the greatest proportion of possible energy savings. The technology to produce increasingly effective insulation has developed in a manner that might be described as meteoric in the last two to three decades, and three examples of advanced insulation materials discussed herein include aerogels, vacuum, and photonic crystals.

Aerogels are solids with a porosity of greater than 50%, a density in the range of 1–150 kg/m^3 and are typically 90–99.8% air by volume (Cuce et al. 2014; Schiavoni et al. 2016). Silica is the most common aerogel substrate; however, aerogels can also be based on alumina, lanthanide and transition metal oxides, metal chalcogenides, organic and inorganic polymers, and carbon (Sadineni et al. 2011). Production involves drying a gel at supercritical temperature, and aerogels may be produced as either granular materials or monoliths. Aerogels are best known for having low thermal conductivity, in the range of 0.012–0.020 W/mK (see Fig. 2, right), some of the lowest values known, however also have excellent fire resistance (Fig. 2, left), are (as monoliths) excellent vapor barriers, have excellent resistance to direct sunlight, have a high service temperature, excellent durability, excellent sound resistance, and may also serve as effective infiltration barriers (Cuce et al. 2014). Thus, in most ways, aerogels outperform other conventional insulation materials and most other currently existing advanced insulation materials. While aerogel insulation materials are currently available commercially, the barrier to

[1]Note—LCA-based evidence was the predominant basis for formulation of conclusions and the highest available standards for quality of evidence were used.

uptake thus far is cost; however, use of aerogel insulation is increasing exponentially and cost may be usefully reduced as manufacturing and production improve and sales volumes increase.

Another characteristic of aerogels that holds promise for emerging applications relates to aerogel optical properties. Some silica aerogels have high transmittance of visible light, with, for instance, a 10-mm-thick panel having a transmittance of 88% overall (and transmittance largely attenuated in the ultraviolet or UV portion of the spectrum, Sadineni et al. 2011). It is expected that modifications in the composition of gels used in production of aerogels will enable tuning of optical properties in future. In addition to transmittance, greater control on the aerogel pore size might enable tuning of reflectance/scattering characteristics of aerogel insulation. Granular aerogel has been used inside the cavity of double-glazed windows and polycarbonate construction panels for windows that weigh less than 20% of the equivalent glass unit and have 200 times more impact strength (Baetens et al. 2011; Sadineni et al. 2011; Schiavoni et al. 2016). Such windows are increasingly used in both vertical and roof-lighting applications to supplement ambient lighting from windows.

Potential health hazards associated with aerogel-based insulation are largely associated with dust, much as with many other forms of insulation, and are amendable therefore to standard operational controls in handling and installation (Schiavoni et al. 2016). Solvents and energy used in production of aerogels are inherently environment-unfriendly, and probably because the commercial use of these materials is relatively new, there are few detailed sustainability metrics. Embodied energy has been investigated and one study found that, on a per weight basis, a number of conventional materials have lower embodied energy than silica aerogel (Schiavoni et al. 2016). On a per area basis, compared to EPS, XPS, glass wool, cork, foam glass, PUR and PIR, only glass wool had a lower embodied energy than silica aerogel. However, the most appropriate metric is instead the amount of material used to achieve a given level of insulation (akin to the functional unit in LCA) and on this basis aerogels require a factor of ~2–3 times less material (Schiavoni et al. 2016). Embodied energy is a relatively limited measure of sustainability, and hopefully more detailed analysis of environmental impacts will be forthcoming soon. With respect to aerogel glazing/window replacement, two LCA studies have been thus far been performed: one detailed, one streamlined, and both focusing on climate change as the primary impact category of interest (Dowson et al. 2012; Lolli and Andresen 2016). The detailed study found emissions savings of up to 9%, as compared to triple-glazing with argon gas. The streamlined study found that aerogel windows "paid" for themselves in CO_2-e emissions reduction after two years. Based on the information of environmental performance available thus far, future development of improved aerogels appears promising, but particularly in respect of the potential optical properties.

By definition, since a vacuum contains no atoms or molecules, thermal conduction and convection are not possible; ergo the thermal conductivity of a perfect and infinite vacuum is zero. This physical fact forms the basis for the development of vacuum insulation panels (VIPs), which are constructed of an evacuated, open-porous material that is enveloped into a multilayer film. The common core

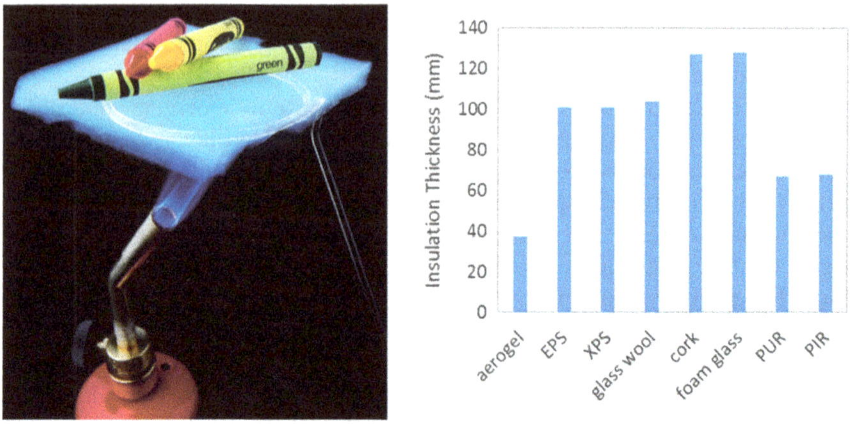

Fig. 2 Photograph of aerogel by courtesy of the US National Aeronautics and Space Administration (left) and comparison of the thickness of various materials needed to achieve a thermal insulation U-value of 0.3 W/m^2K (right) illustrating that the aerogel material is a far superior insulator to conventional insulation materials (data from Cuce et al. 2014). EPS, XPS, PUR, and PIR are expanded polystyrene, extruded polystyrene, polyurethane, and polyisocyanurate, respectively

material is fumed silica; however, aerogels, EPS and PUR, and fiberglass are also in use (Alotaibi and Riffat 2014). A vacuum is imposed on the voids of the core material, and thus, a barrier is required to preserve the integrity of the vacuum during handling, and VIPs are typically enclosed in one or more thicknesses of metal film. Getters or desiccants are also incorporated to absorb water vapor or other gases that might penetrate the barrier. In consequence of the need to preserve vacuum, VIP performance is dependent on the performance of each of its parts, both singly and integrated (Alotaibi and Riffat 2014; Kalnæs and Jelle 2014).

As might be expected in consequence of part of the VIP material being under vacuum, the thermal conductivity of VIPs is very low, on average some 17 times lower than that of conventional glass fiber batts (Baetens et al. 2010). The range of thermal conductivity for VIPs suitable for use in buildings has been cited to be 0.0035–0.008 W/mK, with VIPs suitable for use in appliances being even lower (Kalnæs and Jelle 2014). Thus, the thermal conductivity of VIPs is about half to a third that of aerogels, on average, and VIPs are generally recognized as having the overall best thermal performance of any single insulation material in commercial use for buildings (Alotaibi and Riffat 2014; Jelle 2011). This also conveys a simultaneous savings in space needed to accommodate insulation. On the negative side, one review describes VIPs as suffering from the four cardinal weaknesses of fragility, perforation vulnerability, increasing thermal conductivity during time and lack of building site adaption cutting (Jelle 2011). Additionally, this insulation technology is still relatively costly (Baetens et al. 2010; Jelle 2011).

VIPs have been sufficiently adopted for use in Switzerland and Germany that the Institute of Energy at the University of Applied Sciences, Basel, Switzerland, has conducted analysis of environmental performance to see whether more energy is used in production than is actually saved and what the life cycle disposition of ecological damage is (Schonhardt et al. 2003). In the Swiss study, one VIP material was compared with glass wool and EPS. Results indicate that the embodied energy in the VIP material was the greatest, with the embodied energy of EPS and glass wool being 90% and <50% of the VIP, respectively, for the same amount of insulation over a given area. In terms of ecological damage, the LCA result indicates glass wool is least environmentally damaging, followed by VIPs and EPS. Glass wool outperformed VIPs in most damage categories, including (in descending order of importance) respiration hazard, release of carcinogenic substances, climate change, and land use. Glass wool was also best for resource use, with VIPs being worst, with major VIP damage categories being (in descending order of importance) emissions to air, generation of radioactive waste, landfill waste, and emissions to water. Most of the ecological damage originates from energy use. Overall, while the attractive thermal conductivity of VIPs drives their current use, practical issues entail that development of alternatives, particularly in the area of nanomaterials, is an active area of research. The LCA results indicate that much of the environmental damage comes from energy use in production and raises the question of whether improvements in energy savings from improved building insulation should be substituted for increased energy use in production of improved building insulation.

The quest for new materials that will have the same or better thermal performance as VIPs, without the disadvantages of VIPs, continues apace. Examples of materials that have gained attention and research effort include vacuum insulation materials (VIMs), dynamic insulation materials (DIMs), and nano-insulation materials (NIMs) (Baetens et al. 2010; Jelle 2011). VIMs differ from VIPs in that the base material is homogeneous and has a closed pore structure, thus eliminating problems with punctures and being able to cut the material for use in construction. DIMs are materials that exhibit changing thermal conductivity in a dynamic and/or controllable fashion. This field might be regarded as exploratory. NIMs are based on nanomaterials, and the literature in this area is vast. One form of NIM that is of particular interest is photonic crystals. Photonic crystals consist of periodic nanostructures, the arrangement of which affects how light travels through them and enables blocking of specific kinds of radiation, including thermal or infrared radiation. Recently, a team of researchers at Stanford University found that 1 μm thick layers of such nanostructures separated by 90 μm gaps of vacuum would theoretically reduce the thermal conductance of the material to about half that of a pure vacuum across the same thickness (Lau et al. 2009). In theory, the ability to tailor the material characteristics would mean that, as well as blocking light, such material might be used to capture heat energy for solar-thermal applications. While research into this sort of material is in its infancy, at least one patent application has already been filed for the use of photonic crystals as thermal insulation (Sterzel et al. 2008).

While this area of materials science is no doubt exciting, both in terms of progress in fundamental knowledge and in the development of new and ever more exotic technologies, it is easy to become distracted from the eventual purpose.

Many of the new materials in development require inputs of toxic chemicals and/or large quantities of energy, and very few schemes for new materials development begin with life cycle thinking built into research from the outset. This represents both a severe deficit in current approaches to sustainable materials and a golden opportunity for innovation looking forward.

5 Land Use and Decreasing the Damage from Animal Industries

According to the UNEP report (UNEP 2010) on the environmental impacts of consumption and production, which found that agriculture is one of the two main sources of environmental damage, major areas of agricultural damage include habitat change, climate change, resource use (freshwater), and toxic emissions. The report notes that

> Substantial habitat losses have arisen due to increased demand for land for agriculture and grazing, and significant declines in game and fish populations have resulted from over-harvesting.

The report concludes that

> Impacts from agriculture are expected to increase substantially due to population growth, increasing consumption of animal products. Unlike fossil fuels, it is difficult to look for alternatives: people have to eat. A substantial reduction of impacts would only be possible with a substantial worldwide diet change, away from animal products.

While much emphasis has been placed on urbanization in recent years, particularly in countries such as China, the UNEP report finds that the major damage incurred from habitat change globally results from land transformation (80%, the majority of which is for agriculture), whereas only 12% of the habitat change damage results from land occupation. The report points out that the most logical solution to the problem is a global shift to a plant-based diet. The impacts from animal agricultural may be described in terms of simple math rather than calculus: Animal production has low efficiency for biomass conversion, with animals consuming more freshwater than grain plants per food calorie produced (i.e., 10,000–20,000 kg of freshwater consumed per kg of beef produced versus 1000 kg of freshwater per kg of wheat produced, Chapagain and Hoekstra, 2003), generating a large amount of manure, and releasing a large quantity of strong greenhouse gasses (GHGs) (Chen and Zhang 2015). In factory farm production, animals are raised in very close proximity, both necessitating the use of antibiotics and leading to antibiotic resistance, increasing risks of transfer of infectious diseases such as avian flu, swine flu, and the hepatitis E virus to humans (Chen and Zhang 2015). While damage from animal industries may be clear, the appeal of the UNEP's recommendation is less so. In many countries, vegetarianism could be said to be on the rise; however, there are no reliable statistics on the number of people following a plant-based or vegan diet globally, and the incidence is presumed to be small.

In light of the environmental damage of animal industries, both terrestrial and marine, scientists have begun working on a new type of material that is at once a material in the sense that it is first engineered and then produced, but is also a foodstuff and shows promise as a meat replacement: in vitro meat, also known as cultured meat, synthetic meat, and clean meat. Production of in vitro meat is a form of advanced tissue engineering and involves extraction of a small quantity of cells from a farm animal, transferring them into a suitable liquid growth medium and subsequent culturing on scaffolds in a bioreactor until sufficient proliferation occurs for harvest. In vitro meat is thus practically animal-free meat (Bhat et al. 2015; Chen and Zhang 2015).

In vitro meat has long been conceived and at an early date was investigated by US NASA for long space voyages (Bhat et al. 2015; Pandurangan and Kim 2015). While it is now a reality, commercialization is yet in its infancy. The world's first laboratory-grown burger, cultured in a laboratory by scientists from Maastricht University, Netherlands, was cooked and eaten at a news conference in London in 2013 (BBC 2013). The burger was prepared by a famous chef and tasted by a food researcher who commented overall positively and stated that even in a blind trial she would have identified it as meat and not a meat substitute. At least three startup firms for production of laboratory meat, chicken, and duck now exist (Memphis Meats, Mosa Meat and Supermeat; see Sect. 8). The lead scientist whose laboratory synthesized the first burger, Mark Post, has commented that the cost of his 2013 burger, €250,000, had decreased to €8.00 by 2015 (ABC Australia 2015). Post estimates that his product will be economically competitive in approximately a decade, an optimism echoed by other sources (ABC Australia 2015; Chiles 2013).

Because it is engineered, in vitro meat has the potential to be healthier than conventional meat; for instance, fat content and the amount of saturated fat might be controlled and beneficial additives might be used (Bhat et al. 2015), similar to the supplementation of milk with vitamin D and flour with riboflavin. Because in vitro meat is produced in a laboratory, it is amenable to the application of Good Manufacturing Practices, which admits the possibility of reducing the incidence of food-borne diseases and it would also be much easier to control the risk of exposure to other hazards associated with conventional meat production (pesticides, arsenic, dioxins), hence to produce chemically safe meat (Bhat et al. 2015; Chen and Zhang 2015). Laboratory-based meat production would be subject to a number of efficiencies that are not practically feasible with conventional meat production. The use of a simple culture and growth under continuously optimized conditions offers potential production turn around times of weeks instead of months or years, and nutrients may be more efficiently used since the process of growth and maintenance of bones, respiratory system, digestive system, skin, nervous system, etc., is not required for in vitro meat (Bhat and Fayaz 2011). While many people are comfortable with the animal suffering that accompanies conventional meat production (as evinced by sales), in vitro meat will also facilitate the reduction of the currently estimated 70 billion farm animals (often raised under brutal conditions,

CIWF 2013; Solotaroff 2013) and many trillions of fish harvested each year globally (Mood and Brooke 2010).

Because of the well-known environmental damage of secondary production and overharvesting, the environmental benefits of in vitro meat (and other future in vitro animal products) are anticipated to be substantial based on first principles and thermodynamics. A recent study used LCA to assess potential environmental impacts of large-scale in vitro meat production using cyanobacteria hydrolysate as the primary nutrient and energy source growth (Tuomisto and Teixeira de Mattos 2011). The results are shown in Fig. 3; for the four impact categories shown, energy, GHGs, land, and water, in all cases the life cycle environmental impacts of in vitro meat are lowest except for the life cycle energy associated with poultry. As beef consistently has the highest environmental impact, it is possible to assess cumulative impact as compared to beef by normalizing each category to beef and summing. The life cycle cumulative impacts of in vitro meat range from approximately half that of poultry to one-tenth that of beef. It would be of interest to know if in vitro meat might currently be cost competitive with conventional animal products if full environmental direct and indirect costs were factored into product cost.

The findings of Tuomisto and Teixeira de Mattos are consistent with the findings of Mattick et al. (2015), though Mattick et al. found an even greater benefit of in vitro meat production as compared to conventional meat in terms of GHG savings. Sun et al. (2015) have examined the potential benefits of in vitro meat production in China, concluding that the overall impact of replacing livestock products with in vitro meat would be beneficial for China's environment. These authors further concluded that in vitro meat would be a logical mode of production to improved food security in China because of limits of agricultural land and that a switch to in vitro meat would entail that less land would be needed to produce a given amount of protein and energy.

Advances in in vitro meat production are occurring so rapidly that the needs and challenges looking forward are also changing very fast. Aside from technical challenges, it is well accepted that one significant challenge will be developing a market, even if in vitro meat becomes cost competitive. A recent study involving 410 participants identified three clusters of participants, two of which would be or might be amendable to change, and one of which, representing almost 50% of participants, were described by the hallmark of having "a pattern of attachment to meat and unwillingness to change habits" (Graça et al. 2015). Overall, these findings might be said to be positive–if $\sim$50% of meat eaters would be willing to switch to in vitro meat, this might cause a culture change for the remainder. One challenge with in vitro meat, at the intersection of engineering and consumer expectation, involves taste (Pandurangan and Kim 2015). While the initial taste trial mentioned above was promising, taste is highly specific and dependent upon growth medium and method and therefore may create particular quality control challenges. Additional challenges are involved with growth media, namely that complex media, compared to strictly pure chemical media, offer more opportunities for tailoring taste and providing an all-in-one nutrient and energy source,

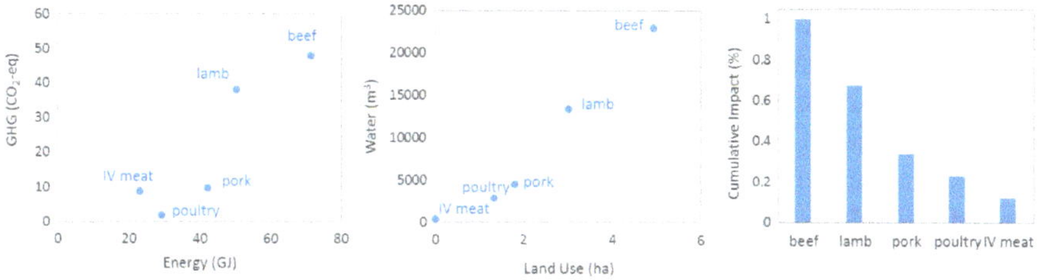

Fig. 3 Comparison of life cycle impacts of beef, lamb, pork, poultry, and in vitro meat according to energy, GHG, land use, and water use impact categories (left and center), as well as cumulative impacts for the four categories normalized to beef (data from Tuomisto and Teixeira de Mattos 2011)

but convey more complex needs to be well characterized to ensure food safety implications might be addressed. A number of other engineering and hygienic challenges will need to be addressed to improve production to a fully commercial state (Pandurangan and Kim 2015). Once arriving at a commercially feasible state, additional market challenges will exist for in vitro meat, notable of which is the lobbying capacity of the meat industry. It may be that specifically because of the environmental damage of conventional animal agriculture, the meat industry has the most to gain in supporting in vitro meat and simultaneously managing environmental risk (Bonny et al. 2015).

While this section has focused on in vitro meat, it is worth noting that the further development and commercialization of the tissue engineering techniques used to produce in vitro meat are also potentially beneficial to the rapid expansion of biomaterials outside of the present conventional sphere of medical applications. An example of an offshoot of in vitro meat is Modern Meadows (2017) a startup company focused on the production of laboratory-grown leather. These two new applications may represent the tip of the iceberg for future materials based on tissue engineering.

6 Sustainable Lead Redux—Can We Replace Metals?

While the major gains in environmental sustainability are no doubt to be had in energy conservation and reduction of impacts from agriculture, there are nonetheless many other areas where the need for improvement is critical. To return to our point of departure, wherein we considered the prevalence of lead as an environmental pollutant and human toxicant, the causative factors in this circumstance should also be considered. As with every other resource, human use of metals has caused metal remobilization in the earth's crust at an unprecedented rate. While metal deposits were concentrated in the form of ores over geological time scales of millions of years, humans remobilize, and via pollution, randomize the distribution of metals, "using them up" in an unsustainable manner (Cohen 2007).

This is true of all metals, though some metals may prove more problematic than others. Thus, for instance, photonic crystal technology may show great promise, but many such structures contain metals in their structural matrix. Another example is solar cells, common designs for which utilize gallium and indium. One expert, René Kleijn of Leiden University is of the opinion that reserves "would not allow a substantial contribution of these cells" to the future supply of solar electricity (Cohen 2007).

Eventually, no matter how efficient metals recovery and recycling are, humans must find a replacement for the large amounts of metal that we use. As one author points out, the circumstance is entirely logical—while humans rely exclusively on electrons as dominant charge carriers, and hence, we need conductors to facilitate that work, biological systems rarely use electrons, instead uses ions and molecules (Simon et al. 2016). As the topic of how to replace metals, and work being undertaken to that end, is vast, only a few representative examples will be given here. By far the most progress has been made in the area of bioelectronics, which increasingly seeks to assemble flexible circuits that are capable of interfacing with humans.

Sustainable and/or biodegradable electronics substrates have been a focus in recent years. Paper in particular has been investigated due to its cost-effectiveness and has been used for the fabrication of rechargeable batteries for energy storage applications, in combination with electronically conductive polymers such as polypyrrole, polyaniline, and polythiophene, which can be easily oxidized and reduced at very high rates and thus might have potential in higher power (Nyholm et al. 2011). Another interesting application of paper in electronics has been the creation of solar cell structures integrated on a paper-based module (Barr et al. 2011). Dielectrics and semiconductors are other examples of important materials in electronics that have high, i.e. environmentally negative, production intensity. In this area, one very interesting development has been the development of a genetically engineered quartz-binding peptide sequence for the precise control of threshold voltages in key electronic components (Dezieck et al. 2010). This result has generated much excitement as it opens the door for binding nature-inspired organic semiconductor small molecules (for example, indigo, Tyrian purple), in order to fuse dielectric and semiconductor layers in one single molecule for high-performance electronics. Indigo and other indigoids like Tyrian purple are serving as the basis for a new generation of building blocks for hydrogen-bonded semiconductors that may be deposited on surfaces such as vacuum processed polyethylene (Glowacki et al. 2013; Kanbur et al. 2012). Recent work in this area suggests that there might be a wide scope to design new nature-inspired molecules with interesting semiconducting properties (Berry et al. 2002).

While the challenges of finding a replacement for using electrons as charge carriers are enormous, so too is the amount of work currently underway in the area of organic electronics. While this work is far from providing an answer to replacing conducting metals for high-load electrics in buildings, there is more latitude for nearer term application in the area of Smart Buildings, particularly as human–electronics interfaces mature. Someday soon, people may be optimizing energy use in their houses and apartments using their organic electronic wrist interface rather than their smartphone, and, like the telegraph, smartphones themselves may be obsolete. These new technologies may be alluring, mesmerizing even, but relatively few have been subjected to evidentiary methods of sustainability assessment, and as such may not be, really, green.

7 Looking Forward: "Sustainable Materials" and Consumption

While there is no doubt of the need for development of more sophisticated sustainable materials in the face of global environmental problems, technology, in the form of sustainable materials or other, is often used as a misdirection to the main problem, which is that of a growing world population and increasing consumption. In 1865, the economist William Stanley Jevons documented how technological improvements that increased the efficiency of coal-use led to the increased consumption of coal in a wide range of industries. Further studies by Jevon gave birth to Jevon's paradox, which entails that technologically mediated increases in the efficiency with which a resource is used causes the rate of resource consumption to increase as a result of increasing demand, and Jevon's paradox is now one of the most widely known axioms of environmental economics (Polimeni et al. 2008). The common incidence of examples that fall within the paradox is commonly ignored, and for obvious reasons. So, for instance, it is considered far preferable to increase the energy efficiency of homes than to use less energy, and while, according to UNEP, adopting a plant-based diet may be a simple and healthy expedient to reducing the environmental impacts of agriculture, few people in practice are willing to do so. Without a change in prevailing attitudes to consumption, technology may be more of a hindrance than a help to the causes of sustainability.

References

Memphis Meats http://www.memphismeats.com/. Accessed 30 Apr 2017

Mosa Meat https://culturedbeef.org. Accessed 30 Apr 2017

Supermeat http://supermeat.com/meat.html. Accessed 30 Apr 2017

Alotaibi SS, Riffat S (2014) Vacuum insulated panels for sustainable buildings: a review of research and applications. Int J Energ Res 38:1–19

Apaiah R, Linnemann AR, van der Kooi HJ (2006) Exergy analysis: a tool to study the sustainability of food supply chains. Food Res Int 39:1–11

ABC Australia (2015) Mark Post of Maastricht University in the Netherlands has developed synthetic beef patties. http://www.abc.net.au/news/rural/2015-03-27/mark-post-synthetic-beef/6352860. Accessed 30 Apr 2017

Baetens R, Jelle BP, Thue JV et al (2010) Vacuum insulation panels for building applications: a review and beyond. Energ Build 42:147–172
Baetens R, Jelle BP, Gustavsen A (2011) Aerogel insulation for building applications: a state-of-the-art review. Energ Buildings 43:761–769
Barr MC, Rowehl JA, Lunt RR et al (2011) Direct monolithic integration of organic photovoltaic circuits on unmodified paper. Adv Mater 23:3500–3505
BBC–British Broadcasting Company (2013) World's first lab-grown burger is eaten in London. http://www.bbc.com/news/science-environment-23576143. Accessed 30 Apr 2017
McCartor A, Becker, D (2010) World's worst pollution problems. Blacksmith Institute Report, in Cooperation with Green Cross/Switzerland, Blacksmith Institute, New York
Berry A, Dodge TC, Pepsin M et al (2002) Application of metabolic engineering to improve both the production and use of biotech indigo. J Ind Microbiol Biotechnol 28:127–133
Bhat ZF, Fayaz H (2011) Prospectus of cultured meat—advancing meat alternatives. J Food Sci Technol 48:125–140
Bhat ZF, Kumar S, Fayaz H (2015) In vitro meat production: challenges and benefits over conventional meat production. J Integr Agric 14:241–248
Bonny SPF, Gardner GE, Pethick DW et al (2015) What is artificial meat and what does it mean for the future of the meat industry? J Integr Agric 14:255–263
Brubaker CJ, Schmithorst VJ, Haynes EN et al (2009) Altered Myelination and axonal integrity in adults with childhood lead exposure: a diffusion tensor imaging study. Neurotoxicology 30:867–875
Cecil KM, Brubaker CJ, Adler CM et al (2008) Decreased brain volume in adults with childhood lead exposure. PLOS Med 5:e112
Chapagain AK, Hoekstra AK (2003) Virtual water trade. In Hoekstra AK (ed) Proceedings of the international expert meeting on virtual water trade. IHE Delft, The Netherlands
Chen H-G, Zhang Y-HP (2015) New biorefineries and sustainable agriculture: Increased food, biofuels, and ecosystem security. Renew Sust Energ Rev 47:117–132
Chiles R (2013) If they come, we will build it: in vitro meat and the discursive struggle over future agrofood expectations. Agric Hum Values 30:511–523
CIWF–Compassion in World Farming (2013) Stategic Plan 2013–2017. CIWF Godalming, Surrey, United Kingdom
Cohen D (2007) Earth's natural wealth: an audit. New Sci 2605:34–41
Cros C, Fourdrin E, Réthoré O (2010) The French initiative on environmental information of mass market products. Int J Life Cycle Assess 15:537–539
Cuce E, Cuce PM, Wood CJ et al (2014) Toward aerogel based thermal superinsulation in buildings: a comprehensive review. Renew Sust Energ Rev 34:273–299
Cullen JM, Allwood JM, Borgstein EH (2011) Reducing energy demand: what are the practical limits? Environ Sci Technol 45:1711–1718
Daly HE (1995) On Wilfred Beckerman's critique of sustainable development. Environ Values 4:49–55
Daly HE (1996) Beyond growth: the economics of sustainable development. Beacon Press, Boston
Daly HE (2007) Ecological economics and sustainable development: selected essays of Herman Daly. Edward Elgar, Cheltenham, UK
Daly HE, Cobb JC (1989) For the common good: redirecting the economy toward community, the environment, and a sustainable future. Beacon Press, Boston
Dezieck A, Acton O, Leong K et al (2010) Threshold Voltage control in organic thin film transistors with dielectric layer modified by a genetically engineered polypeptide. Appl Phys Lett 97:013307
Dowson M, Grogan M, Birks T et al (2012) Streamlined life cycle assessment of transparent silica aerogel made by supercritical drying. Appl Energ 97:396–404
Elkington J (1998) Cannibals with forks: the triple bottom line of 21st century business. New Society Publishers, Gabriola Island, BC
Esty DC, Kim C, Srebotnjak T et al (2008) Environmental performance index. Yale Center for Environmental Law and Policy, New Haven, Connecticut, USA

GFN–Global Footprint Network (2017) Ecological footprint. http://www.footprintnetwork.org/our-work/ecological-footprint. Accessed 19 Apr 2017

Glowacki ED, Irimia-Vladu M, Kaltenbrunner M et al (2013) Hydrogen-bonded semiconducting pigments for air-stable field-effect transistors. Adv Mater 25:1563–1569

Graça J, Oliveira A, Calheiros MM (2015) Meat, beyond the plate. Data-driven hypotheses for understanding consumer willingness to adopt a more plant-based diet. Appetite 90:80–90

Hajjaji N, Pons M-N, Houas A et al (2012) Exergy analysis: an efficient tool for understanding and improving hydrogen production via the steam methane reforming process. Energ Policy 42:392–399

Hsu A et al (2016) 2016 Environmental performance index. Yale University. www.epi.yale.edu. Accessed 19 Apr 2017

ILA–International Lead Association (2017a) Lead production and statistics. http://www.ila-lead.org/lead-facts/lead-production–statistics. Accessed 19 Apr 2017

ILA–International Lead Association (2017b) Lead production and statistics. http://www.ila-lead.org. Accessed 19 Apr 2017

ISO–International Organisation for Standardisation (2006a) 14040 (2006): environmental management–life cycle assessment–principles and framework. ISO, Geneve

ISO–International Organisation for Standardisation (2006b) ISO 14044 (2006): Environmental management–life cycle assessment–requirements and guidelines. ISO, Geneve

Jelle BP (2011) Traditional, state-of-the-art and future thermal building insulation materials and solutions–properties, requirements and possibilities. Energ Build 43:2549–2563

Kalnæs SE, Jelle BP (2014) Vacuum insulation panel products: a state-of-the-art review and future research pathways. Appl Energ 116:355–375

Kanbur M, Irimia-Vladu ED, Glowacki G et al (2012) Vacuum-processed polyethylene as a dielectric for low operating voltage organic field effect transistors. Org Electron 13:919–924

Kanoglu M, Dincer I, Cengel YA (2009) Exergy for better environment and sustainability. Environ Dev Sustain 11:971–988

Karabell Z (2013) The 'Laws of Economics' don't exist. The Atlantic. https://www.theatlantic.com/business/archive/2013/04/the-laws-of-economics-dont-exist/274901. Accessed 19 Apr 2017

Lau WT, Shen JT, Fan S (2009) Universal features of coherent photonic thermal conductance in multilayer photonic band gap structures. Phys Rev B 80:155135

Loh J, Green RE, Ricketts T et al (2005) The Living Planet Index: using species population time series to track trends in biodiversity. Phil Trans R Soc B 360:289–295

Lolli N, Andresen I (2016) Aerogel versus argon insulation in windows: a greenhouse gas emissions analysis. Build Environ 101:64e76

Mattick C, Landis AE, Allenby BR (2015) A case for systemic environmental analysis of cultured meat. J Integr Agric 14:249–254

McRae L, Deinet S, Freeman R (2017) The diversity-weighted living planet index: controlling for taxonomic bias in a global biodiversity indicator. PLoS One 12:e0169156

Modern Meadows (2017) Leather reimagined. http://www.modernmeadow.com. Accessed 30 Apr 2017

Mood A, Brooke P (2010) Estimating the number of fish caught in global fishing each year, based on FAO data. http://fishcount.org.uk/published/std/fishcountstudy.pdf and http://fishcount.org.uk/published/std/fishcountappendixa.pdf. Accessed 30 Apr 2017

Nyholm L, Nyström GV, Mihranyan A et al (2011) Toward flexible polymer and paper-based energy storage devices. Adv Mater 23:3751–3769

Odum HT (1996) Environmental accounting: emergy and environmental decision making. Chichester Wiley, New York

Odum EC, Odum HT (1980) Energy systems and environmental education. In: Bakshi TS, Naveh Z (eds) Environmental education: principles, methods and applications. Plenum Press, New York, pp 213–231

Pandurangan M, Kim DH (2015) A novel approach for in vitro meat production. Appl Microbiol Biotechnol 99:5391–5395

Polimeni K, Mayumi K, Giampietro M et al (2008) The Jevons Paradox and the myth of resource efficiency improvements. Earthscan, London

Prweb (2017) Lead–acid batteries deserve bigger role in hybrid vehicles, Doe Run CEO tells international lead conference. http://www.prweb.com/releases/lead-acid-batteries/doe-run-company/prweb1549684.htm. Accessed 19 Apr 2017

Rees WE, Wackernagel M (2013) The shoe fits, but the footprint is larger than earth. In: Mace GM (ed) PLoS Biology, vol 11, p e1001701

Sadineni SB, Madala S, Boehm RF (2011) Passive building energy savings: a review of building envelope components. Renew Sust Energ Rev 15:3617–3631

Schiavoni S, D'Alessandro F, Bianchi F et al (2016) Insulation materials for the building sector: a review and comparative analysis. Renew Sust Energ Rev 62:988–1011

Schonhardt U et al (2003) Oekobilanz eines Vakuum-Isolations-Panels (VIP). University of Applied Sciences, Institute of Energy, Basel, German

Scienceman DM (1987) Energy and Emergy. In: Pillet G, Murota T (eds) Environmental economics: the analysis of a major interface. R. Leimgruber, Geneva, pp 257–276

Simon DT, Gabrielsson EO, Tybrandt K et al (2016) Organic bioelectronics: bridging the signaling gap between biology and technology. Chem Rev 116:13009–13041

Solotaroff P (2013) In the belly of the beast. http://www.rollingstone.com/feature/belly-beast-meat-factory-farms-animal-activists. Accessed 30 Apr 2017

SOPAC–South Pacific Applied Geoscience Commission (2005) Building resilience in SIDS. The environmental vulnerability index (EVI) 2005. SOPAC Technical Report, Suva, Fiji Islands

Stein R, Stein C, Buckley M, et al (1981) Handbook of energy use for building construction. DOE/CS/20220-1, US Department of Energy, Washington, DC

Sterzel HJ, Schauernheim D, Ellerstadt KK (2008) Photonic crystals for thermal insulation US Patent Application 2008/0233391, 25 Sep 2008

Sun Z-C, Yu Q-L, Lin Han (2015) The environmental prospects of cultured meat in China. J Integr Agric 14:234–240

Terrachoice (2010) The sins of greenwashing: a report on environmental claims made in the North American consumer market. Underwriters Laboratory, Ottawa

Tuomisto HL, Teixeira de Mattos MJ (2011) Environmental impacts of cultured meat production. Environ Sci Technol 45:6117–6123

UNEP–United Nations Environment Programme (2010) Assessing the environmental impacts of consumption and production: priority products and materials. A report of the working group on the environmental impacts of products and materials to the international panel for sustainable resource management. UNEP Geneva

US EPA–United States Environmental Protection Agency (2000) America's children and the environment: a first view of available measures. US EPA, Washington EPA 240-R-00-006

Wu J (2013) Landscape sustainability science: ecosystem services and human well-being in changing landscapes. Landscape Ecol 28:999–1023

WWF–World Wildlife Fund (2016) Living planet report 2016: risk and resilience in a new era. WWW International. http://wwf.panda.org/about_our_earth/all_publications/lpr_2016. Accessed 19 Apr 2017

Ye X, Fu H, Guidotti T (2007) Environmental exposure and children's health in china. Arch Environ Occup Health 62:61–73

Overlooked Strategies in Exploitation of Microorganisms in the Field of Building Materials

Yusuf Cagatay Ersan

Abstract Resource efficiency reports released in the last decade point out construction industry as one of the key sectors that needs improvement in terms of ecological sensitivity. Being aware of this unfavorable reputation of construction industry, researchers embarked on replacing the ongoing conventional methods with more sustainable and environmentally friendly ones. One of the approaches for the latter is incorporating microorganisms into construction industry. Popularly investigated strategies can be listed as biocementation, biomasonry, biorepair, and bioconsolidation. Most of these processes are the outcome of a single approach, namely microbial-induced calcium carbonate precipitation (MICP) which was mostly investigated by means of axenic cultures and through one single microbial process, ureolysis. The state of the art about the latter is close to saturation. Moreover, approaching from the ecological wisdom perspective it can be said that some promising microbial strategies to achieve green building materials were overlooked and drawing attention to these strategies became necessary. This review study reveals the overlooked promising microbial strategies in the field of construction biotechnology. The context mainly discusses the potential of five overlooked microbial strategies: (i) heterotrophic and autotrophic MICP pathways, (ii) microbial strategies for surface treatment, (iii) microbial-induced corrosion inhibition, (iv) microbial sequestration of greenhouse gases, and (v) microbial-produced polymers, for their application in the field of construction materials. Further suggestions aim to integrate the microbial resource management approach and non-axenic cultures into the relevant fields of research for the development of environmentally friendly building materials.

Keywords Self-healing concrete · Bio-based materials · Resource efficiency
Construction biotechnology · Microbial induced calcite precipitation

Y. C. Ersan (✉)
Department of Civil Engineering, Abdullah Gul University, 38080 Kayseri, Turkey
e-mail: yusuf.ersan@agu.edu.tr

V. Achal and A. Mukherjee (eds.), *Ecological Wisdom Inspired Restoration Engineering*, EcoWISE, https://doi.org/10.1007/978-981-13-0149-0_2

1 Introduction

The world's population continues to grow while the amount of resources (energy, raw material, water) deplete. Justice should be prioritized in dealing with the resource limitation problems, since resources are vital for current and future generations. Considering the growing anthropogenic effects and increasing demand on resources, it became essential to focus on effective and sustainable use of resources. The increasing public awareness on the concept of "sustainability" and relevant legal adjustments motivated industries to modify their conventional processes into more sustainable ones. Starting from the food industry and agricultural activities, the sustainability concept moved toward energy sector and for the last 20 years it has been integrated into the research fields of construction.

Construction industry is the leading industry at global scale considering the net materials addition to stock per annum (Matthews et al. 2000). Therefore, it is also one of the largest users of resources (energy, raw material, water) and results in formidable pollution at the end (Horvath 2004). The complete environmental assessment of construction industry is complicated, since the extent of the components (raw materials), facilities (processing, manufacturing, construction, operation, maintenance), and infrastructure (buildings, transportation elements, governmental installations) is immense. It is known that buildings constitute the major portion of the construction industry, and thus, the severe impact of the construction industry can be envisioned around them. According to the recently released reports, 77%, 75%, and 70% of the construction activity were for buildings in Europe, USA, and China, respectively (The European Cement Association 2011; FMI 2016; EU SME Centre 2015). Buildings are thought to be responsible for 40% of the global energy and material flow and 50% of the global water use (United Nations Environment Programme 2009). It was also estimated that the buildings contribute to 23% of the global air pollution, 30% of the greenhouse gas (GHG) emission, 40% of the global wastewater production, and 40% of the global solid waste production (United Nations Environment Programme 2009; Willmott Dixon 2010).

Considering the volume-based consumption, the major construction materials can be listed as crushed rock, aggregates (coarse and fine), cement, cement concrete, asphalt concrete, timber products, clay brick, concrete block, steel, aluminum, and others. In EU-27 countries, the amount of concrete, aggregates, and bricks used in buildings (construction, renovation, etc.) between 2006 and 2010 correspond to 90% (by weight) of the total construction material consumption (Fig. 1) (Herczeg et al. 2014). Among them, concrete production is responsible for 5% of the anthropogenic CO_2 footprint and around 80% of this CO_2 emission is due to the cement production process (NRMCA 2012; Crow 2008; Carbon Dioxide Information Analysis Center 2016). In 2016, 4.2 billion Mt of global cement production was reported and the trend is upward (Fig. 2). According to the data provided in global carbon budget, the global CO_2 production was 39.5 billion Mt in

2013, and solely cement production was responsible for around 2 billion Mt of it (Carbon Dioxide Information Analysis Center 2016).

Having a clear picture, official authorities realized that it was necessary to tackle the environmental impact of building sector with a greater sensitivity. United Nations prioritized tackling with the emissions from building sector in the framework of United Nations Environment Programme (UNEP) to reach 2020 and 2050 targets for global GHG emission (United Nations Environment Programme 2009). Moreover, under the European Commission's EU 2020 strategy, the Roadmap to Resource Efficient Europe (RERM) was revised and building sector was named among the three key sectors to tackle (Herczeg et al. 2014). The four milestones to achieve more environmentally friendly buildings were listed as (i) increasing energy efficiency, (ii) decreasing the amount of raw materials used, (iii) increasing cost-efficient renovation rate, and (iv) recycling at least 70% of the non-hazardous construction and demolition waste (Herczeg et al. 2014).

As research, industry, and official authorities are driving forces of each other, defined milestones increased the popularity of research that is based on ecological wisdom. Ecological wisdom suggests the effective use of biological nature and the profound ecological knowledge that has been developed to date, to solve issues of the today's world and the near future (Liao and Chan 2016). Intensive integration of the ecological wisdom into research has led to a number of novel research areas aiming to replace the ongoing conventional methods with more sustainable and environmentally friendly ones. For instance, one of the innovative approaches was the introduction of the microbiology to the field of construction to enhance durability and sustainability of the materials. Due to their major role in biodeterioration, microbes were unfavorable in the field of construction and building materials (Wei et al. 2013); yet, a fresh introduction in an engineered way created a promising harmony. Owing to their consistent, precise, controllable, and nonstop metabolic activities, microorganisms offered sustainable alternatives to binders, material additives, repair, and renovation methods in the field of construction (Achal and

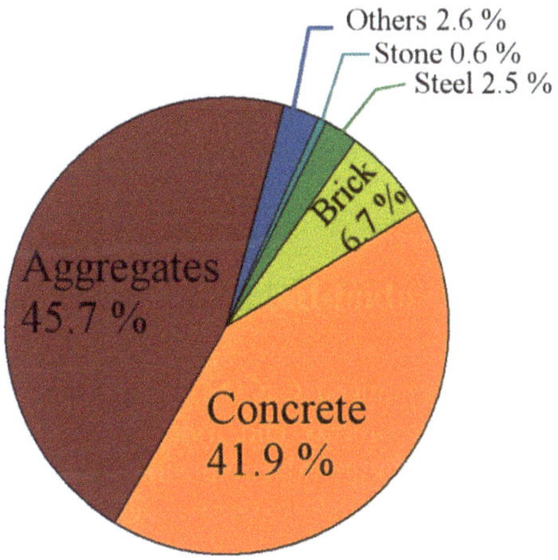

Fig. 1 Percentage of the construction materials used for buildings based on the data collected in EU-27 countries between 2006 and 2010. (Redrafted after Herczeg et al. 2014)

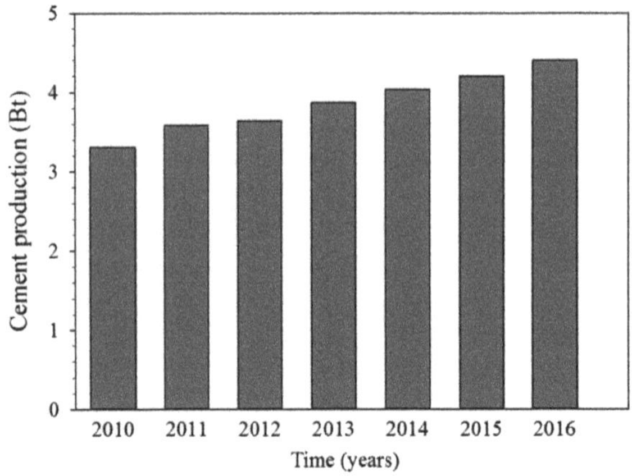

Fig. 2 Global cement production in years (data acquired from Portland Cement Association Report Portland Cement Association 2014)

Mukherjee 2015; De Muynck et al. 2010; Wiktor and Jonkers 2011; Stabnikov et al. 2015) and revealed great potential to contribute to solving problems such as greenhouse gas emission, gray cities, and construction wastes (Grabiec et al. 2012; Ganendra et al. 2014; Manso and Aguado 2016). A biogenic process known as microbial-induced calcium carbonate precipitation (MICP) was one of the primarily investigated microbial strategies in the field of construction. MICP showed promising initial results, and its applications were diversified in the field of construction and building materials. As a result, concepts such as biocementation, bioconsolidation, and biorepair arose (Achal and Mukherjee 2015; De Muynck et al. 2010). As MICP concept gained popularity in the field of construction and building materials, several other microbial strategies were overlooked. The popularity of MICP concept created a bandwagon effect, and unfortunately, recent studies on MICP applications in construction materials started to lose their conceptual diversity. The main reason of such saturation can be attributed to the limited awareness of the variety offered by microorganisms. Therefore, it became necessary to draw attention to the overlooked MICP pathways, their possible applications, and various other overlooked microbial strategies that would potentially improve the sustainability of construction materials. This study sets a course for further MICP research and discovers the overlooked microbial strategies in the field of construction materials.

2 Microbial-Induced Carbonate Precipitation (MICP)

Microbial-induced $CaCO_3$ precipitation was first reported in the late nineteenth century by Murray and Irvine (1889/1890) and Steinmann (1899/1901) in concurrence with urea decomposition by the marine microbiota (Ehrlich and Newman 2008). The first extensive evidence for the interaction between $CaCO_3$

precipitation and bacteria was revealed by Nadson (Ehrlich and Newman 2008; Nadson 1903, 1928).

Ion activity product plays a major role on chemical precipitation of compounds in aqueous solutions. When the solution is saturated with the products' ions, no more dissolution of the compounds is possible and thus precipitation of the compound is inevitable. $CaCO_3$ can be considered as a slightly soluble compound in water as its solubility constant (K_{sp}) is 4.8×10^{-9} at 25 °C. Therefore, increasing CO_3^{2-} concentration in the presence of Ca^{2+} leads to $CaCO_3$ precipitation. The former process is where the bacteria are intensely involved in. Most of the bacteria are capable of inducing $CaCO_3$ precipitation if the conditions allow (Ehrlich and Newman 2008). Simply, bacteria produce CO_2 through specific metabolic activities and contribute to CO_3^{2-} concentration in an aqueous environment (Eq. 1). In the presence of Ca^{2+}, this contribution is followed by the precipitation of $CaCO_3$ crystals. However, the role of bacteria in MICP is not limited to CO_2 production.

$$CO_{2(g)} \leftrightarrow CO_{2(aq)} + H_2O \leftrightarrow H_2CO_3 \leftrightarrow H^+ + HCO_3^- \leftrightarrow 2H^+ + CO_3^{2-} \quad (1)$$

H_2CO_3 represents both the dissolved carbon dioxide and the carbonic acid. At 25 °C and 1 atm, the pK_H value of $CO_{2(aq)}$ is 1.46 and the pK_a values for carbonic acid and bicarbonate are 6.35 and 10.32, respectively.

The four key parameters driving MICP are listed as the presence of (i) Ca^{2+}, (ii) dissolved inorganic carbon (DIC), (iii) alkalinity, (iv) nucleation sites (Hammes and Verstraete 2002). Bacteria play an active role in three of the mentioned parameters. In addition to CO_2 production, certain metabolic pathways lead to the production of alkalinity which shifts the carbonate balance more toward the CO_3^{2-} (Eq. 1). Moreover, during the precipitation process bacteria can serve as nucleation sites by concentrating Ca^{2+} ions around the negatively charged cell membrane. Urea hydrolysis has been the most prominent pathway in studies investigating MICP in construction materials (Fig. 3).

Fig. 3 Ureolysis-based MICP research in the field of construction materials outnumbers the studies conducted based on alternative MICP pathways

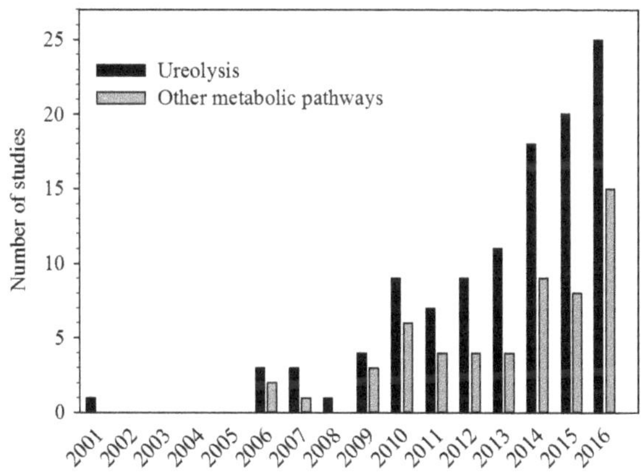

As shown in Fig. 3, the amount of studies investigating MICP through alternative pathways has been in minority, and thus, the information about the MICP applications based on alternative metabolisms is scarce in the field. It is known that besides ureolysis, aerobic or anaerobic oxidation of organic carbons (aerobic or anaerobic respiration), aerobic or anaerobic oxidation of organic nitrogen compounds, sulfate reduction, methane oxidation, iron reduction, hydrolysis of carbon dioxide, and photosynthesis can be mentioned among the metabolic activities that can induce $CaCO_3$ precipitation in the presence of Ca^{2+}. There are significant amount of detailed review studies available which are oriented around the application of ureolysis-based MICP in the field of construction materials (Achal and Mukherjee 2015; De Muynck et al. 2010). Therefore, in this review study, MICP applications in the field were discussed by slightly touching on ureolysis-based studies and focusing more on the potential of the alternative MICP pathways. From the aforementioned pathways, aerobic respiration, nitrate reduction, methane oxidation, and carbon dioxide hydrolysis were evaluated for their potential to achieve more sustainable construction materials. Prior to discussion on their potential in the field of construction materials, it is essential to understand the MICP mechanisms of these pathways.

3 Alternative MICP Pathways and Their Working Principles

3.1 Aerobic Oxidation of Organic Carbon

Aerobic oxidation of organic carbon (aerobic respiration) is a common process among the most bacterial species for energy production and cell synthesis. The process leads to formation of carbon dioxide as shown in Fig. 4. When aerobic respiration takes place under alkaline conditions, dissolved CO_2 acts as a weak acid and reacts with free OH^- ions to form CO_3^{2-} ions (Eq. 1). In the presence of Ca^{2+}, these CO_3^{2-} ions precipitate in the form of $CaCO_3$ (Eqs. 2, 3) (Tziviloglou et al. 2016).

$$Ca(C_3H_5O_3)_2 + 6O_2 \rightarrow Ca^{2+} + 4CO_2 + 2HCO_3^- + 4H_2O \tag{2}$$

$$4CO_2 + 2HCO_3^- + 6Ca(OH)_2 \rightarrow 6CaCO_3 + 6H_2O + 2OH^- \tag{3}$$

Since cementitious composites are mostly considered among the highly alkali materials, MICP through aerobic oxidation of organic carbon can be mentioned among the alternative pathways for their improvement. However, it should be noted that the necessity of external alkalinity for formation of carbonate ions makes the process inefficient for use in carbonated concrete, bricks, stone, and aggregates.

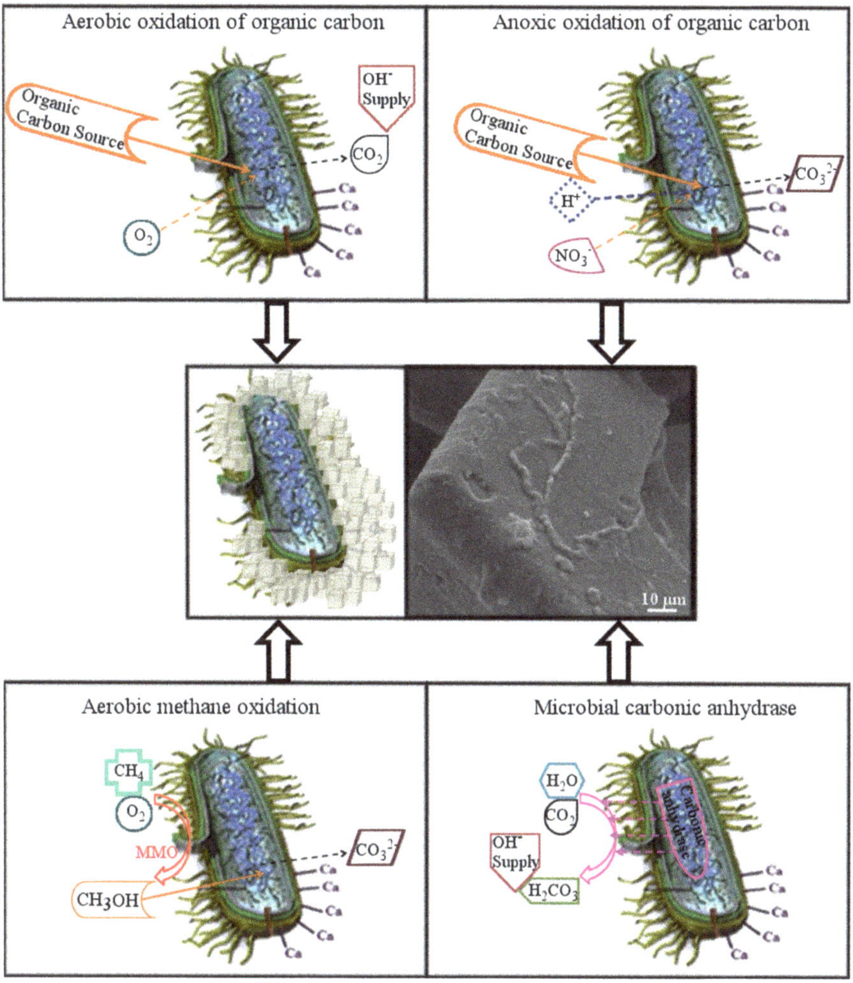

Fig. 4 Schematic overview of the MICP occurring around the cell membrane as a result of different microbial activities (*MMO*: Methane Mono-Oxygenase enzyme) (Redrafted after De Muynck et al. 2008)

3.2 Anoxic Oxidation of Organic Carbon

In nature, oxygen is the most abundantly used electron acceptor for organic carbon oxidation; yet, certain kind of bacteria can use alternative electron acceptors for growth and energy production. Nitrate (NO_3^-), sulfate (SO_4^{2-}), and manganese dioxide (MnO_2) can be listed among the inorganic compounds that can serve as electron acceptor for bacteria. Oxidation of organic carbon through microbial reduction of NO_3^- precedes other anoxic pathways due to its high standard free energy (Tziviloglou et al. 2016) (Eq. 4). Therefore, nitrate reduction can be considered as the primary MICP pathway in the absence of oxygen. Reaction that leads to MICP through nitrate reduction is given in Eq. (4).

$$5Ca(HCOO)_2 + 4NO_3^- \rightarrow 2N_2 + 6HCO_3^- + 4CaCO_3 + 2H_2O + Ca^{2+} \quad (4)$$

$$6Ca(OH)_2 + 6HCO_3^- \leftrightarrow 6CaCO_3 + 6H_2O + 6OH^- \quad (5)$$

Nitrate-reducing bacteria produce alkalinity during the process. Therefore, under alkaline conditions (i.e. applications in cementitious materials) their $CaCO_3$ precipitation yield becomes higher than the MICP yield of the bacteria that are solely supplemented by the external alkalinity (Eqs. 2–5). Being independent from the external alkalinity, nitrate reduction pathway has great potential to be an alternative to ureolysis which is responsible for the production of toxic by-products (ammonia) and unpleasant odor.

3.3 Methane Oxidation

Microbial methane oxidation to CO_2 is mostly conducted in marine and freshwater sediments by certain microbial communities so-called methane-oxidizing bacteria (MOB). The process can occur under both aerobic and anoxic conditions. Under aerobic conditions, the process starts with the conversion of methane to methanol at the cell membrane with the aid of oxygen and methane mono-oxygenase (MMO) enzyme (Eq. 6 and Fig. 4). The produced methanol is the sole carbon source for further cell synthesis and energy production. Initially, the produced methanol is converted to formaldehyde in periplasm and further converted to formate through enzymatic processes inside the cell. Dissolved formate is in equilibrium with formic acid, and MOB oxidizes formic acid to CO_2 by using formate dehydrogenase enzyme (Eqs. 7–10). Under alkaline conditions, the produced CO_2 turns into CO_3^{2-} and precipitates as $CaCO_3$ in the presence of adequate amount of Ca^{2+} (Eq. 11) (Ganendra et al. 2014; Ettwig et al. 2010). By supplying additional formate to aerobic MOB, it is also possible to increase pH of the environment (Eq. 9) and thus induce $CaCO_3$ precipitation even in the absence of external alkalinity (Ganendra et al. 2015).

$$CH_4 + O_2 \xrightarrow[NADH+H^+ \rightarrow NAD^+]{Methane\ Mono-Oxygenase} CH_3OH + H_2O \quad (6)$$

$$CH_3OH \xrightarrow[PQQ \rightarrow PQQH_2]{Methanol\ Dehydrogenase} CHOH \quad (7)$$

$$CHOH + H_2O \xrightarrow[NAD^+ \rightarrow NADH+H^+]{Formaldehyde\ Dehydrogenase} HCOO^- + H^+ \quad (8)$$

$$HCOO^- + H_2O \leftrightarrow HCOOH + OH^- \tag{9}$$

$$HCOOH \xrightarrow[NAD^+ \rightarrow NADH + H^+]{Formate\ Dehydrogenase} CO_2 \tag{10}$$

$$Ca^{2+} + CO_2 + 2OH^- \leftrightarrow CaCO_3 + H_2O \tag{11}$$

3.4 *Carbon Dioxide Hydrolysis*

Carbon dioxide in the atmosphere is in equilibrium with the carbonic acid (H_2CO_3) in aquatic environment. Carbonic acid is formed through the reaction of CO_2 with water and is in equilibrium with bicarbonate (Eq. 12). Bicarbonate is also in equilibrium with the carbonate ions, and the equilibrium is controlled by the pH of the environment. When an adequate amount of Ca^{2+} ions and alkalinity are present, atmospheric CO_2 can be captured in aquatic environments in the form of $CaCO_3$ (Eq. 13). However, the complete reaction rate is limited with the conversion of CO_2 to bicarbonate. A microbial enzyme called carbonic anhydrase catalyzes the carbonic acid formation, and thus, presence of this enzyme in the environment increases $CaCO_3$ precipitation rate. Carbonic anhydrase production is ubiquitous among both eukaryotic and prokaryotic organisms (Smith and Ferry 2000). The vital condition for MICP through microbial carbonic anhydrase is the presence of alkalinity, since the process itself has no influence on pH. Therefore, exploiting this MICP pathway in the field of construction materials is limited to cementitious materials. The major advantage of the process is sequestration of CO_2 from atmosphere, which makes the process environmentally friendly.

$$H_2O + CO_2 \xleftrightarrow{Carbonic\ Anhydrase} H_2CO_3 \leftrightarrow HCO_3^- + H^+ \tag{12}$$

$$Ca(OH)_2 + HCO_3^- \leftrightarrow CaCO_3 + OH^- \tag{13}$$

4 MICP Applications Waiting for Further Exploration

4.1 Production of Microbial Bricks

MICP is used to agglomerate and bind loose grains, and the process is called biocementation. Providing the suitable conditions for bacteria and controlling the formation of $CaCO_3$ crystals, significant binding performance could be achieved from $CaCO_3$ crystals. This binder bridges the loose grains, fills their pores, hence increases the overall integrity, and decreases permeability (Kucharski et al. 2006). Initially, biocementation (through ureolysis) was integrated into the field of construction materials as a complementary process to enhance certain properties of construction materials, such as strength and permeability. It was reported that biocementation could improve the compressive strength and water tightness of mortar specimens by filling the pores inside the mortar matrix (Achal et al. 2011a, b). Construction materials such as microbial bricks that solely depend on biocementation have been investigated since 2011 (Dosier 2011). Initial efforts resulted in the production of sandstones having compressive strengths ranging from 6 to 16 MPa (Dosier 2011; Rong et al. 2012). Nowadays, microbial bricks reveal comparable mechanical properties with the regular bricks and commercialization attempts have been made (Dosier 2011).

Following successful implementation, biocementation was also considered as a complementary process for ash brick production. It was reported that compressive strength, water tightness, and freeze-thaw resistance of fly ash bricks and rice husk ash bricks could be improved through biocementation (Dhami et al. 2012). Reported improvements on the compressive strengths of rice husk ash and fly ash bricks were ~30% (from 9.7 MPa to 12.8 MPa) and 26% (from 11.7 MPa to 14.9 MPa), respectively, which resulted in strengths similar to that of commercial red bricks (12.3 MPa). Moreover, the bricks obtained through biocementation were reported to perform better than red bricks in terms of freeze-thaw resistance. Making use of biocementation during the production of ash bricks led to a more compact structure with less air voids and capillaries, and thus, fly ash and rice husk ash biobricks absorbed 35–40% less water compared to conventional red bricks (Dhami et al. 2012). It was reported that biocementation could also work for very fine particles such as cement kiln dust and lime kiln dust (Cuzman et al. 2015a). Recently, through biocementation of such industrial wastes (cement kiln dust and rice husk ash) with sand grains, construction materials such as plasters, bedding mortars, and bricks were also produced (Cuzman et al. 2015a, b).

After achieving significant progress on the production of microbial bricks by means of ureolysis-based MICP, one of the alternative pathways, CO_2 hydrolysis by means of microbial carbonic anhydrase, was used to produce biobricks. The aim of this novel approach was to use steel slag and its CaO content effectively. Steel slag contains tricalcium silicate ($3CaO{\cdot}SiO_2$) and dicalcium silicate ($2CaO{\cdot}SiO_2$) which can react with CO_2 in the presence of water. As a result of this reaction, $CaCO_3$ and calcium silicate hydrate (C–S–H) are formed which are involved in strength development of the steel slag bricks (Wang et al. 2016). However, the strength development is limited with the CO_2 dissolution (carbonation) rate which interferes with achieving strong steel slag bricks in a feasible way. As previously described, microbial carbonic anhydrase enzyme rapidly turns CO_2 into HCO_3^- and facilitates CO_2 dissolution. In a recent study, *Bacillus mucilaginous*, a carbonic anhydrase positive strain, was mixed with steel slag, slake lime (SL:SS—0.3:1), and sand (Wang et al. 2016). It was reported that in the presence of excess amount of CO_2 for 3 h (in carbonation chamber), compressive strength of microbial steel slag bricks was improved by almost 300% compared to abiotic controls (from 6 to 17 MPa). By putting *Bacillus mucilaginous* into play, they decreased the time required for the production of useful steel slag bricks (strength around 17 MPa) from 12 days to 3 h. It was also suggested that required high CO_2 concentrations could be obtained through industrial emissions which would lead to the production of an environmentally friendly and sustainable brick with adequate strength.

4.2 Simultaneous Greenhouse Gas Sequestration and Surface Treatment

MICP was investigated to improve the chemical and physical resistance of the natural stones and cementitious materials (De Muynck et al. 2010). The idea was to decrease porosity and even create a resistant surface layer which acts as a barrier (made of $CaCO_3$) against physical and chemical factors. So far, microorganisms were applied to the surface of the material through immersion, spraying, or brushing techniques. The most used MICP pathway was ureolysis due to its high $CaCO_3$ precipitation rate and independence on external alkalinity. Detailed review studies discussing the surface treatment of construction materials through ureolysis-based MICP are available (Achal and Mukherjee 2015; De Muynck et al. 2010). A more environmentally friendly strategy for MICP-based surface treatments would be the addition of simultaneous greenhouse gas sequestration function. Both aerobic methane oxidation and CO_2 hydrolysis pathways were considered to achieve construction materials with greenhouse gas sequestration function (Ganendra et al. 2014; Qian et al. 2016).

One of the promising approaches was to obtain specific construction materials for livestock barns, because the livestock barns are placed second-most crucial anthropogenic greenhouse gas source due to their methane emission (Conrad 2009). In order to deal with this greenhouse gas source, researchers put the methane-oxidizing bacteria (MOB) in play (Ganendra et al. 2014, 2015a, b). The idea was to use porous masonry as a carrier for MOB and develop alternative construction materials that serve as methane sink (Ganendra et al. 2014). They screened four types of MOB (*Methylomicrobium alcaliphilum*, *Methylomicrobium kenyense*, *Methylosinus trichosporium*, and *Methylocystis parvus*) and three types of porous masonry (autoclaved aerated concrete, two types of limestones, and three types of bricks) in total. Among the combinations, the most promising performance was achieved when *Methylocystis parvus* was immobilized on autoclaved aerated concrete. The methane removal rates of 28.5 ± 3.8 μg CH_4 and 1.7 ± 0.4 μg CH_4 per gram building material per hour were reported at methane availabilities of ~20% (v/v) and ~100 ppmv, respectively (Ganendra et al. 2014). Moreover, it was claimed that by providing $Ca(HCOO)_2$ together with ~20% (v/v) CH_4, one can enhance the performance of *Methylocystis parvus* to the MICP rates similar to the ones obtained by ureolytic bacteria and increase the water tightness of autoclaved aerated concrete by 27% compared to the regular ones (Ganendra et al. 2014). Overall, the MOB-containing autoclaved aerated concrete gave promising results and should be further considered among the environmentally friendly construction materials with improved surface properties.

Another strategy was the use of microbial carbonic anhydrase for the development of construction materials with CO_2 sequestration capability (Qian et al. 2016). Similar to most of the initial attempts, a conservative approach was used to prove the concept. It was revealed that a higher amount of CO_2 could be captured when mortar-containing *Bacillus mucilaginous*, a carbonic anhydrase-producing strain, was applied on the surface of a wallboard. It was claimed that MICP by means of microbial carbonic anhydrase improved the surface properties and the water tightness (Qian et al. 2016). On the one hand, it is crucial to understand the effect of such bacteria on carbonation rate of the structural concrete as high carbonation rates neutralize the protective alkali layer around the steel reinforcement bars and facilitate corrosion (Pourbaix 1973). On the other hand, such coatings may be considered for concrete roads and the walls of industrial chimneys where water/water vapor can be intermittently available and significant amount of CO_2 capture is likely. Since the amount of studies is few in this area, the mechanical properties, economic feasibility of the materials, and the possibilities to enhance CO_2 removal efficiencies are still unknown and should be investigated in the future research.

4.3 Crack Repair in Cementitious Composites

The idea of bacteria incorporation into cementitious materials for the development of microbial self-healing concrete was first mentioned in 2007 (Jonkers 2007). Self-healing concrete can be described as a sustainable concrete that repairs its own defects and prolongs the lifetime of the structures. Significant amount of research was conducted to exploit MICP for the development of self-healing concrete, and the initial results became available in 2011. It was revealed that incorporated bacteria could repair cracks up to 460-μm-crack width in 100 days through aerobic oxidation of calcium lactate (Wiktor and Jonkers 2011). Although the first attempts were based on aerobic oxidation of organic carbons, research on ureolysis-based MICP for self-healing concrete dominated the literature (Wang et al. 2014a, b). Complete sealing of cracks up to 500 μm in 28 days could be mentioned as the prominent outcomes for biomortars developed based on ureolysis-based MICP (Wang et al. 2014b). The ureolytic pathway preference of the researchers was most probably due to the already available literature rich in ureolysis-based MICP and its application for surface treatment. As there are many comprehensive review articles available about ureolysis-based self-healing concrete, it is crucial to elaborate more on the application of alternative pathways in the scope of this chapter. As mentioned previously, the first attempts were made by using aerobic carbon oxidation pathways and the cultures of *Bacillus alkalinitrilicus*, *Bacillus pseudofirmus,* and *Bacillus cohnii* (Wiktor and Jonkers 2011; Jonkers and Schlangen 2009). In subsequent studies, *Bacillus subtilis* strain was also added to the list (Khaliq and Ehsan 2016; Sarkar et al. 2015). It was reported that 56 days wet–dry treatment was enough for cracked (350 μm) mortar specimens containing *Bacillus cohnii* to recover 98% of its original water tightness (Tziviloglou et al. 2016). Sealing of wider cracks in shorter times was also reported. For instance, autonomous sealing of 500-μm-wide crack in 28 days was reported when *Bacillus subtilis* was added into concrete with proper nutrients (Khaliq and Ehsan 2016). Different from the mentioned examples which were cracked after 28 days of curing, crack healing of aged concrete was also tested. For aged concrete (cracking after 56 days) containing *Bacillus pseudofirmus* and Ca lactate (as organic carbon), autonomous sealing of 80–220-μm-wide cracks in 100 days was reported (Stuckrath et al. 2014).

Another MICP pathway for the development of microbial self-healing concrete was based on microbial carbonic anhydrase. So far, *Bacillus mucilaginous* was the only bacteria tested to develop self-healing concrete working based on microbial carbonic anhydrase. Few researches were conducted by using this approach, and they revealed promising results. The first attempt was to heal early age cracks (cracks occurred after 7 days curing). It was reported that concrete-containing *Bacillus mucilaginous* could autonomously seal more than 85% of a 400-μm-wide crack in 10 days (Qian et al. 2015). The rapid sealing was attributed to the positive effect of carbonic anhydrase on CO_2 dissolution. Especially for early age cracks where autogenous healing contribution to the crack closure is significant, an increase in CO_2 dissolution can speed up the process remarkably. Longer sealing times were reported when the cracking age was increased. For similar crack (400 μm wide) occurring after 28 days, the sealing efficiency could reach $\sim$80% only after 28 days of treatment and there was still significant variation ($\pm$20%). The authors attributed the decrease in performance to the loss of bacteria in time, since they were added into the concrete without any protective carriers (Qian et al. 2015). It was revealed in another study that, if the same bacteria were immobilized on protective carriers such as ceramsite, the developed mortar could seal 88% of a 400-μm-wide crack in 28 days (Chen et al. 2016).

One of the vital parameters for the aforementioned processes is the availability of air to provide oxygen for aerobic bacteria and to replenish the CO_2 converted to HCO_3^- by carbonic anhydrase enzyme. It was claimed that for air-dependent processes, continuity of the crack closure toward the deeper parts of the crack is

Fig. 5 Visualization of MICP along the crack depth by splitting the 400 μm crack after 28 days of healing period (unpublished data from Erşan et al. 2016)

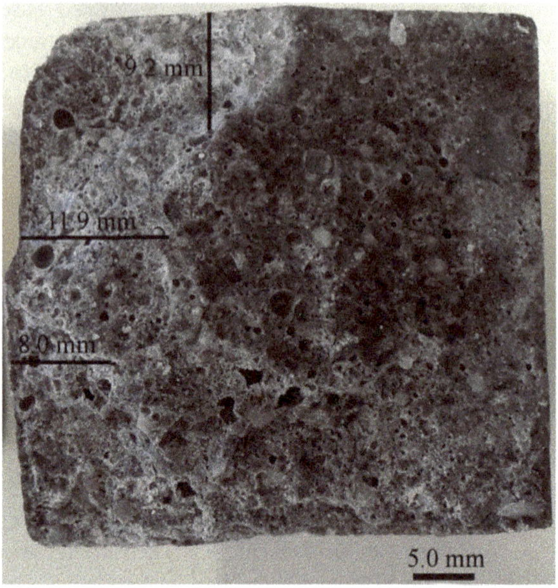

unlikely. For instance, even for early age cracks, 1.1 mm from the surface was reported as the deepest point where MICP could occur (Qian et al. 2015). Although there is no depth data provided for aerobic bacteria, similar depths would not be a surprise due to the anoxic conditions in the deeper zones of the cracks. Therefore, anoxic oxidation of organic carbon was also investigated to develop self-healing concrete. The idea was to promote $CaCO_3$ precipitation in deeper layers of the concrete crack where airborne substances hardly reach. It was revealed that by using concrete admixture, $Ca(NO_3)_2$, as a source for electron acceptor for *Pseudomonas aeruginosa* and *Diaphorobacter nitroreducens*, ~400-μm-wide cracks could be healed in 28 days. The performance of the suggested bacteria in 56 days was enough to close cracks of 500 μm wide (Erşan et al. 2016). Moreover, it was reported that the water tightness of cracked self-healing concrete (450-μm-wide crack) could be improved by ~75% upon 56 days of healing period. The $CaCO_3$ precipitation at deeper zones of the cracks, up to 12 mm from the surface (Fig. 5), was reported when nitrate-reducing bacteria were used (Erşan et al. 2016; Ersan 2016).

Overall, although ureolysis-based MICP was mostly preferred by researchers to develop self-healing concrete, studies revealed that alternative MICP pathways were at least as effective as ureolysis in the development of self-healing concrete. Therefore, considering the fact that ureolysis process leads to some environmental toxic by-products (i.e. ammonia), for aquatic environments in particular, it can be a greener approach to focus on described alternative pathways in further microbial self-healing concrete studies.

4.4 *Improvement of Recycled Concrete Aggregate Properties*

As previously mentioned, by volume, aggregates (fine and coarse) are the most used material in the field of construction. In order to supply the demand in construction industry, aggregate mining pits are continuously built. A sustainable approach aiming to recycle aggregates from construction and demolition wastes was integrated into the construction industry. Unfortunately, recycled concrete aggregates (RCAs) contain part of the hydrated cement paste as well and thus have higher porosity and water absorption (4 to 8 times in coarse, 8 to 12 times in fine aggregates) than the natural aggregates (Qiu et al. 2014). Therefore, their application is limited to road base course, subbase material, production of sidewalk concrete, concrete and asphalt pavements (Cement Concrete and Aggregates Australia 2008). Investigations on quality improvement (through pre-coating with pozzolans, chemical and physical treatments, coating with polyvinyl alcohol) gave promising results; yet, suggested methods are expensive and energy intensive, which detract the approach from sustainability (Qiu et al. 2014). As discussed in previous sections, MICP could improve surface properties of porous structures. Therefore, it was recently considered to improve the quality of RCA and promising results were obtained. Similar to all MICP applications, ureolysis pathway was the first one tested. It was reported that by treating RCA with urea, $CaCl_2$, and *Sporosarcina pasteurii*, approximately 0.35% weight increase could be achieved. Moreover, the water absorption was decreased by 15% (from 8.9 to 7.5%) (Qiu et al. 2014). In a similar study, application of MICP (by means of *Sporosarcina pasteurii*) on old cement mortar decreased the porosity of the aggregates by 32% and decreased the water absorption coefficient by 85% (Pan et al. 2015). It was also claimed that direct tensile strength of asphalt mixture could be improved by 55% when treated RCA was replaced with the untreated ones. In another study, water absorption of RCA was also decreased by 15–20% depending on the grain size. It was claimed that MICP was more effective on finer grains (6/8 mm) and on low-quality RCA (Grabiec et al. 2012). Since the approach is fresh, no information on mechanical properties of concrete-containing MICP-treated RCA is available. Since strength is one of the decisive parameters for structural concrete, further studies should consider investigating these parameters as well. As it was reported that the properties of finer grains could be improved better, it may be worth to try applying MICP on recycled concrete fines (RCF) as well. Recently, RCF was tested to partially replace the silica sand particles in engineered cementitious composites (Li and Yang 2017). They reported compressive strengths of around 50 MPa and strain capacity of at least 0.8% for the developed composites. Improvement of the surface properties of RCF through MICP may lead to better results and pave the way for replacement of silica sand with RCF in cementitious composites.

5 Overlooked Microbial Strategies with Significant Potential

5.1 Microbial Surface Cleaning

Surface properties of construction materials play a crucial role on their durability. Undesired fouling of the building facades, natural stones, and monuments is one of the major problems in the field of construction materials. Cleaning of the material surface from the colonized organisms (lichens, fungi, etc.) requires high precision in order to minimize the material loss. Excessive abrasion, further fouling due to increased surface roughness, discoloration, and reduction in chemical resistance are reported issues commonly appearing on surface cleaning (De Graef et al. 2005). Since microorganisms work at microscale they can provide the sensitivity needed in surface cleaning of construction materials. In the late 1970s, the use of microorganisms for surface cleaning of stones was suggested (Doehne and Price 2010). Later, some researchers investigated biological cleaning of marble stones. Tests conducted by using sulfur-reducing bacteria (*Desulfovibrio desulfuricans*) to remove the black crust from marble surface gave promising results (Cappitelli et al. 2006). Moreover, an enzyme-based cleaner was developed by extracting enzymes from *Trametes versicolor* and applied on marble surfaces to remove the biological staining (Konkol et al. 2009). In order to clean the chemical contaminants (nitrate, sulfate, organic matter) on the surface of artworks, researchers applied engineered biofilms. Biofilms were applied by using inert carrier, and around 90% removal of the chemical substances were reported upon 30 h treatment (Ranalli et al. 2000). Sulfur-oxidizing bacteria from genus *Thiobacillus* were also applied to clean the lichen fouling from concrete surfaces. It was reported that applying *Thiobacillus* on the surface of concrete with moderate and high lichen fouling cleaned the surface as effective as direct application of H_2SO_4 (De Graef et al. 2005). In a recent study, a strictly halophilic fungi *Wallemia sebi* was applied to sandstone surface to remove halite (NaCl) on the surface (Mansour 2017). It was reported that 95% of the NaCl on the sandstone surface could be removed by *Wallemia sebi* without causing significant deterioration. The method is promising for halite removal from the archeological sandstone, monuments, and similar materials. Another approach was to change the properties of concrete so that colonization of certain microorganisms would not be possible. *Paenibacillus polymyxa* was one of the bacteria used for this purpose since they can synthesize an antibiotic agent called fusaricidin (Park et al. 2014). It was reported that the integration of *Paenibacillus polymyxa* into cement paste provided antifungal properties and prevented the colonization of *Aspergillus niger*. Since the uncontrolled growth of the most of these microorganisms can induce biodeterioration of the materials, it is necessary to create strategies that inhibit the further growth of microorganisms after their work is done.

The amount of research on microbial cleaning of material surfaces is limited in number, and it is necessary to revive this subject on the course of improving sustainability in construction. The most crucial point in the application of biological surface cleaning is to develop strategies that always favor the balance between bioremediation and biodeterioration toward bioremediation. Successful progress in this field can create opportunities for replacement of chemical treatments with nature-oriented treatments which are environmentally friendly.

5.2 *Bio-Based Insulation*

One of the construction materials in which the presence of microorganisms is unfavorable is the building insulation materials. They are known to invade the insulation material and cause deterioration at high relative humidity (Sterflinger et al. 2013; Klamer et al. 2004). Moreover, their negative impact on the indoor air quality is a significant concern (Verdier et al. 2014; Pessi et al. 2002). However, a novel approach, using mycelium for the development of insulation boards, has the potential to positively affect the perception of microorganisms in the field of insulation materials. Although the available knowledge about the application is scarce, it is simply the use of mycelium fibers as a binder for lightweight material such as rice husk, corn stalk, wheat grains, straw, dry leaves, and woodchips (Mayoral 2011; Mayoral González and González Diez 2015; Arifin and Yusuf 2013). Depending on the composition, the density of the obtained material varies between 0.2 and 0.3 g cm^{-3} which is adequate for thermal insulation purposes (Arifin and Yusuf 2013). Although the information about mycelium boards' thermal and acoustic insulation properties is missing, the approach seems promising for the development of eco-friendly insulation materials with eight times less CO_2 emission and ten times less energy use compared to the widely used styrofoams (Mayoral 2011).

5.3 *Green Walls with Microorganisms: An Alternative Way to Sequester Carbon Dioxide*

Sustainability in construction materials can be achieved in two ways. The most common approach is to make modifications in the production process and minimize raw material use for production and maintenance. The other approach is to develop and design materials or structures that can also serve for the environment. Green walls and green roofs can be considered in the latter group as they improve the green areas in an urban environment. The most commonly used green wall systems require additional supporters (climbers, steel frames, modular panel, plant cassette, etc.) and growth medium (artificial substrate or potting soil) for the plants (Wong et al. 2010). However, their application had certain problems such as high installation and maintenance costs, low integration with the structure, extra loads on structural units (Manso 2014). One of the reasons for such equipment need was the use of macroscale photosynthetic organisms (plants). Lately, colonization of concrete walls by a photosynthetic single-celled organism, *Chlorella vulgaris*, was suggested to promote the application of green walls (Manso 2014). Researchers developed a cementitious composite made of magnesium phosphate cement (MPC) with a pH between 5.8 and 7 to facilitate the colony growth of *Chlorella vulgaris* (Manso et al. 2016). Results revealed effective algal colonization on MCP concrete, which paves the way for increasing the green space in urban areas. Number of researches investigating the microorganisms-based green walls is limited, and it is among the areas to be explored further. On the one hand, the carbon sequestration efficiency, colonization rate, homogeneity and the stability of the colonized surfaces, and possible insulating effects (thermal and acoustic properties of the colonized material) can be mentioned among the further research topics. On the other hand, strategies to overcome the drawbacks of such material (inadequate strength and undesired pH levels for steel passivation) should also be considered to enable structural use of these materials. Unless the mechanical properties are improved, these green walls are only useful as non-structural construction materials such as exterior cladding (Manso and Aguado 2016; Manso 2014).

5.4 Microbial-Induced Corrosion Inhibition

Steel corrosion in reinforced concrete structures is one of the major reasons of durability issues. Protecting the steel reinforcement bars against corrosive agents can increase the service life of structures. Regular maintenance and the use of chemical corrosion inhibitors can be listed among the measures taken to protect steel rebars. A sustainable alternative to currently used methods is the use of bacteria for the protection of steel rebars. Aforementioned microbial self-healing concrete is one of the ways to avoid regular maintenance of the concrete structures. An overlooked strategy in that approach would be to use bacteria for simultaneous corrosion inhibition of the steel rebars. This was referred in a couple of studies; yet, the amount of available knowledge is still scarce. There were a couple of ideas to inhibit steel corrosion by means of microorganisms. It was claimed that aerobic microorganisms can interfere with steel corrosion since they consume the available oxygen (the available electron acceptor) (Wiktor and Jonkers 2011). Promising results on steel corrosion inhibition (due to consumption of oxygen) were achieved in the early 1990s by using marine isolates *Pseudomonas* sp. S9 and *Serratia marcescens* (Pedersen and Hermansson 1991). However, the aim was not the reinforced concrete and no further study was conducted in concrete environment. In a more relevant study, *Bacillus* sp. CT-5 a ureolytic bacteria that grow on oxygen was tested to inhibit steel corrosion in mortar (Achal et al. 2012). The corrosion current of the steel rebar in microbial mortar was reported as 14.78 mA m^{-2} while in plain mortar it was 60.83 mA m^{-2}, which indicated a remarkable corrosion inhibition. However, the inhibition was attributed to the change in transport properties due to surface consolidation rather than oxygen consumption. The idea of microbial oxygen consumption has been overlooked, and no comprehensive study is available about the impact of microorganisms on steel corrosion in microbial self-healing concrete. A couple of studies may be indirectly correlated with this strategy. For instance, in recent studies, it was shown that due to the microbial activity of the aerobic bacteria, dissolved oxygen concentration decreases near the cracks of microbial self-healing concrete (Tziviloglou et al. 2016; Wang et al. 2015). Yet, the impact of such depletion on steel corrosion remains unclear and requires more attention.

Production of anionic corrosion inhibitors by using bacteria is another alternative to inhibit corrosion. Such approach was reported for inhibition of aluminum and brass corrosion through the production of polyaspartate and γ-polyglutamate by *Bacillus subtilis* and *Bacillus licheniformis*, respectively (Örnek et al. 2002a, b). In a novel strategy, nitrate-reducing bacteria were suggested to simultaneously inhibit steel corrosion and repair concrete cracks. Microbial production of NO_2^-, an intermediate product of the NO_3^- reduction process, was expected to inhibit corrosion during the microbial crack repair process. It was claimed that commercial corrosion inhibitor $Ca(NO_2)_2$ can be replaced with $Ca(NO_3)_2$ and bacteria, while an additional function (self-healing) is provided to concrete (Erşan et al. 2016). An initial attempt has been made, and it was revealed that microbial-produced NO_2^- can indeed be effective on corrosion behavior of the steel. According to the presented results, corrosion potential of the steel was increased from −360 to −140 mV in a week due to microbial production of NO_2^- in the medium. However, a comprehensive research is still needed to determine the effect in more realistic conditions such as testing at higher Cl^- concentrations that can represent marine environment and testing in concrete environment.

There are a couple of other microbial strategies to inhibit steel corrosion which have never been considered in the field of construction materials. For instance, it might be possible to form a protective layer composed of microbial-produced polymeric substances on the steel reinforcement bars (Kip and van Veen 2015). It was claimed that by genetically modifying *Shewanella oneidensis* a protective biofilm and precipitation layer could be obtained on steel samples (Dubiel et al. 2002). In another study, *Pseudomonas flava* was used to create a protective biofilm layer on mild steel (Gunasekaran et al. 2004). It is necessary to consider these previously reported results for developing more environmentally friendly corrosion inhibition strategies applicable to reinforced concrete structures in marine environment.

5.5 Exploiting Microbial-Produced Polymers

The use of microbial-produced polymeric substances in construction materials is common. Microbial biopolymers such as xanthan gum, Welan gum, succinoglycan, protein hydrolysates are added in concrete mainly as set retarder/accelerator, stabilizer, rheology modifier, etc. (Plank 2004). Comprehensive review articles are available about the use of biopolymers in construction materials, and detailed information could be obtained from (Stabnikov et al. 2015; Plank 2004). Different from the use of biopolymers as admixtures, some attempts were made to apply specific bacterial species (*Pseudoalteromonas*, *Paracoccus marcusii*) that can produce biofilm (bacteria and biopolymer) on the surface of the concrete (Lv et al. 2015). In order to promote biofilm formation on the material surface, immersion and spraying techniques were used. It was reported that by soaking mortar specimens in nutrient solutions containing bacteria, a thick biofilm could be achieved and Cl^- ingression to specimens could be reduced by 30% (Lv et al. 2015). However, the composition of the produced biofilm was not characterized; hence, the mechanism behind the barrier effect of the biofilm remained unclear.

Another useful application of microbial biofilm in construction materials was for surface hydrophobization. Currently, water-repellent surface coatings and hydrophobic admixtures are the major components applied to improve water resistance of concrete surfaces and to minimize the ingress of water and corrosive ions. Sustainability concerns in this field also brought microorganisms into the play. The idea was to exploit hydrophobic properties of bacterial secretions to improve water resistance of concrete. Scientists developed a hybrid mortar with water-repellent properties by incorporating bacterial biofilm (produced by *Bacillus subtilis*) into the mortar mix (Grumbein et al. 2016). It was reported that by incorporating 2.5% (w_{wet}/w_{dry}) fresh biofilm into regular mortar mix, water contact angle could be improved from 30° to ~90° and the obtained angle was stable for around 60 min contact time. Addition of biofilm powder (lyophilized) was claimed to be more effective which led to contact angle increase up to 110° (Grumbein et al. 2016). It was assorted that the addition of biofilm into mortar mixture increased the surface roughness homogenously and thus provided a hydrophobic surface. These promising attempts should be further explored, and tests should cover the identification of biofilm composition and the influence of the admixture on mechanical properties. It is known that organic matter has negative influence on the strength development and setting properties of concrete (Erşan et al. 2015) which might be a challenge to overcome.

Microbial polymers can also be used as raw material in the construction industry. Most of the bioplastics in the construction industry are composed of polyhydroxyalkanoate (PHA) which are produced by bacteria as carbon and energy source during fermentation. Produced PHA can be stored inside the cell up to 90% of the cell dry weight (Choi et al. 2016; Jung et al. 2010). These polymers are extracted from bacteria to produce bioplastics and have potential to replace the petro-based polymers.

Fig. 6 Microbial self-healing concrete containing self-protected nitrate-reducing culture (scale bar indicates 1 mm) (Redrafted after Erşan et al. 2015)

6 Challenges and Future Perspectives

The main challenge for implementation of microorganisms and microbial strategies in the field of construction materials is the cost (Plank 2004; Silva et al. 2015; Johnson et al. 2009). So far, all of the studies mentioned above were conducted by using axenic cultures under well-controlled laboratory conditions. Economic concerns arise when the required sterile conditions and specific substrates are considered at industrial scale. In order to overcome the high cost of axenic culture-oriented processes, non-axenic cultures and microbiome engineering should be considered (Johnson et al. 2009; Erşan et al. 2015; Silva et al. 2015). The so-called microbial resource management (MRM) approach enables enrichment of a non-axenic microbial community dedicated to a desired microbial process (Verstraete 2015). In this approach, operational conditions and substrate play the major role on ecological selection of a core community from a diverse biomass by creating a selective pressure. Such approach has already been investigated for PHA production, and it was revealed that bioplastics can be produced from waste streams by using non-axenic cultures which significantly decreases the cost of the process (Johnson et al. 2009; Marang et al. 2016). Upon implementing this approach at industrial scale, microbial-produced PHA can supersede petro-based plastics.

Recently, initial attempts were made to engineer non-axenic communities for the development of microbial self-healing concrete as well. It was shown that by using non-axenic cultures, the cost of microbial self-healing concrete can be decreased by ten times without negatively affecting the crack repair performance (Erşan et al. 2015; Silva et al. 2015). It was reported that even the cracks appearing after 6 months of curing were effectively sealed when microbial mortars were prepared by using non-axenic nitrate-reducing culture (Fig. 6).

7 Conclusions

Overall, microorganisms have proven their potential to create a more sustainable construction material industry. They are useful for the production of construction materials (bricks, bioplastics, insulation boards), for dealing with durability issues (crack repair, corrosion inhibition), for improvement of chemical and physical resistance (hydrophobization, water tightness, wearing resistance, surface cleaning), and for improvement of the air quality (greenhouse gas sequestration, green walls). There are many MICP pathways rather than ureolysis waiting for further exploration and integration into construction materials. Among them, nitrate reduction pathway stands out for the development of microbial self-healing concrete due to its significant $CaCO_3$ precipitation yield, independence on external alkalinity, formation of thicker sealing layers, and possible inhibitory effect on steel corrosion. Furthermore, the development of materials that can sequester GHG deserves more attention to decrease the carbon footprint of the construction materials. It is utmost necessity to integrate all the discussed microbial strategies at industrial scale in order to act in parallel with the EU 2020 strategy about construction materials and to reach UNEP 2020 and 2050 global GHG emission targets. The MRM approach, which prioritizes the engineering of microbial communities rather than engineering of microorganisms, is the most promising way to expedite microorganisms' integration into the field of construction materials at industrial scale.

References

Achal V, Mukherjee A (2015) A review of microbial precipitation for sustainable construction. Constr Build Mater 93:1224–1235. https://doi.org/10.1016/j.conbuildmat.2015.04.051

Achal V, Mukherjee A, Reddy MS (2011a) Microbial concrete: way to enhance the durability of building structures. J Mater Civ Eng 23:730–734. https://doi.org/10.1061/(ASCE)MT.1943-5533.0000159

Achal V, Pan X, Özyurt N (2011b) Improved strength and durability of fly ash-amended concrete by microbial calcite precipitation. Ecol Eng 37:554–559. https://doi.org/10.1016/j.ecoleng.2010.11.009

Achal V, Mukherjee A, Goyal S, Reddy MS (2012) Corrosion prevention of reinforced concrete with microbial calcite precipitation. ACI Mater J 109:157–164

Arifin YH, Yusuf Y (2013) Mycelium fibers as new resource for environmental sustainability. Procedia Eng 53:504–508. https://doi.org/10.1016/j.proeng.2013.02.065

Cappitelli F, Zanardini E, Ranalli G, Mello E, Daffonchio D, Sorlini C (2006) Improved methodology for bioremoval of black crusts on historical stone artworks by use of sulfate-reducing bacteria. Appl Environ Microbiol 72:3733–3737. https://doi.org/10.1128/AEM.72.5.3733-3737.2006

Carbon Dioxide Information Analysis Center (2016) Global Carbon Budget

Cement Concrete & Aggregates Australia (2008) Use of recycled aggregates in construction

Chen H, Qian C, Huang H (2016) Self-healing cementitious materials based on bacteria and nutrients immobilized respectively. Constr Build Mater 126:297–303. https://doi.org/10.1016/j.conbuildmat.2016.09.023

Choi SY, Park SJ, Kim WJ, Yang JE, Lee H, Shin J, Lee SY (2016) One-step fermentative production of poly(lactate-co-glycolate) from carbohydrates in Escherichia coli. Nat Biotechnol 34:1–8. https://doi.org/10.1038/nbt.3485

Conrad R (2009) The global methane cycle: recent advances in understanding the microbial processes involved. Environ Microbiol Rep 1:285–292. https://doi.org/10.1111/j.1758-2229.2009.00038.x

Crow JM (2008) The concrete conundrum. Chem World 62–66 (March)

Cuzman OA, Rescic S, Richter K, Wittig L, Tiano P (2015a) Sporosarcina pasteurii use in extreme alkaline conditions for recycling solid industrial wastes. J Biotechnol 214:49–56. https://doi.org/10.1016/j.jbiotec.2015.09.011

Cuzman OA, Wittig L, Royo Abancens FK, Herrera C, Anastasi NR, Sanchez LA (2015b) Bacterial "Masons" at work with wastes for producing eco-cement. Int J Environ Sci Dev 6:767–774. https://doi.org/10.7763/IJESD.2015.V6.696

De Graef B, De Windt W, Dick J, Verstraete W, De Belie N (2005) Cleaning of concrete fouled by lichens with the aid of Thiobacilli. Mater Struct 38:875–882. https://doi.org/10.1617/14254

De Muynck W, Cox K, De Belie N, Verstraete W (2008) Bacterial carbonate precipitation as an alternative surface treatment for concrete. Constr Build Mater 22:875–885. https://doi.org/10.1016/j.conbuildmat.2006.12.011

De Muynck W, De Belie N, Verstraete W (2010) Microbial carbonate precipitation in construction materials: a review. Ecol Eng 36:118–136. https://doi.org/10.1016/j.ecoleng.2009.02.006

Dhami NK, Reddy MS, Mukherjee A (2012) Improvement in strength properties of ash bricks by bacterial calcite. Ecol Eng 39:31–35. https://doi.org/10.1016/j.ecoleng.2011.11.011

Doehne E, Price CA (2010) Stone conservation: an overview of current research. Getty Conservation Institute, Los Angeles, USA

Dosier GK (2011) Methods for making construction material using enzyme producing bacteria. WO2011137106 A1

Dubiel M, Hsu CH, Chien CC, Mansfeld F, Newman DK (2002) Microbial iron respiration can protect steel from corrosion. Appl Environ Microbiol 68:1440–1445. https://doi.org/10.1128/AEM.68.3.1440-1445.2002

Ehrlich HL, Newman D (2008) Geomicrobiology, 5th edn. CRC Press, New York

Ersan YC (2016) Microbial nitrate reduction induced autonomous self-healing in concrete. Ghent University, Belgium

Erşan YÇ, Silva FB, Boon N, Verstraete W, De Belie N (2015a) Screening of bacteria and concrete compatible protection materials. Constr Build Mater 88:196–203. https://doi.org/10.1016/j.conbuildmat.2015.04.027

Erşan YÇ, Gruyaert E, Louis G, Lors C, De Belie N, Boon N (2015b) Self-protected nitrate reducing culture for intrinsic repair of concrete cracks. Front Microbiol 6:1228. https://doi.org/10.3389/fmicb.2015.01228

Erşan YÇ, Hernandez-Sanabria E, De Belie N, Boon N (2016a) Enhanced crack healing performance of microbial concrete through nitrate reduction. Cem Concr Compos 70:159–170. https://doi.org/10.1016/j.cemconcomp.2016.04.001

Erşan YÇ, Verbruggen H, De Graeve I, Verstraete W, De Belie N, Boon N (2016b) Nitrate reducing $CaCO_3$ precipitating bacteria survive in mortar and inhibit steel corrosion. Cem Concr Res 83:19–30. https://doi.org/10.1016/j.cemconres.2016.01.009

Ettwig KF, Butler MK, Le Paslier D, Pelletier E, Mangenot S, Kuypers MMM, Schreiber F, Dutilh BE, Zedelius J, de Beer D, Gloerich J, Wessels HJCT, van Alen T, Luesken F, Wu ML, van de Pas-Schoonen KT, Op den Camp HJM, Janssen-Megens EM, Francoijs K-J, Stunnenberg H, Weissenbach J, Jetten MSM, Strous M (2010) Nitrite-driven anaerobic methane oxidation by oxygenic bacteria. Nature 464:543–548. https://doi.org/10.1038/nature08883

EU SME Centre (2015) The construction sector in China. Beijing

FMI (2016) US markets construction overview

Ganendra G, De Muynck W, Ho A, Hoefman S, De Vos P, Boeckx P, Boon N (2014a) Atmospheric methane removal by methane-oxidizing bacteria immobilized on porous building materials. Appl Microbiol Biotechnol 98:3791–3800. https://doi.org/10.1007/s00253-013-5403-y

Ganendra G, De Muynck W, Ho A, Arvaniti EC, Hosseinkhani B, Ramos JA, Rahier H, Boon N (2014b) Formate oxidation-driven calcium carbonate precipitation by Methylocystis parvus OBBP. Appl Environ Microbiol 80:4659–4667. https://doi.org/10.1128/AEM.01349-14

Ganendra G, Wang J, Ramos JA, Derluyn H, Rahier H, Cnudde V, Ho A, Boon N (2015a) Biogenic concrete protection driven by the formate oxidation by Methylocystis parvus OBBP. Front Microbiol 6:1–11. https://doi.org/10.3389/fmicb.2015.00786

Ganendra G, Mercado-Garcia D, Hernandez-Sanabria E, Peiren N, De Campeneere S, Ho A, Boon N (2015b) Biofiltration of methane from ruminants gas effluent using autoclaved aerated concrete as the carrier material. Chem Eng J 277:318–323. https://doi.org/10.1016/j.cej.2015.04.128

Grabiec AM, Klama J, Zawal D, Krupa D (2012) Modification of recycled concrete aggregate by calcium carbonate biodeposition. Constr Build Mater 34:145–150. https://doi.org/10.1016/j.conbuildmat.2012.02.027

Grumbein S, Minev D, Tallawi M, Boettcher K, Prade F, Pfeiffer F, Grosse CU, Lieleg O (2016) Hydrophobic properties of biofilm-enriched hybrid mortar. Adv Mater 8138–8143. https://doi.org/10.1002/adma.201602123

Gunasekaran G, Chongdar S, Gaonkar SN, Kumar P (2004) Influence of bacteria on film formation inhibiting corrosion. Corros Sci 46:1953–1967. https://doi.org/10.1016/j.corsci.2003.10.023

Hammes F, Verstraete W (2002) Key roles of pH and calcium metabolism in microbial carbonate precipitation. Rev Environ Sci Biotechnol 1:3–7

Herczeg M, McKinnon D, Milios L, Bakas I, Klaassens E, Svatikova K, Widerberg O (2014) Resource efficiency in the building sector

Horvath A (2004) Construction materials and the environment. Annu Rev Environ Resour 29:181–204

Johnson K, Jiang Y, Kleerebezem R, Muyzer G, Van Loosdrecht MCM (2009) Enrichment of a mixed bacterial culture with a high polyhydroxyalkanoate storage capacity. Biomacromol 10:670–676. https://doi.org/10.1021/bm8013796

Jonkers HM (2007) Self healing concrete: a biological approach. In: van der Zwaag S (ed) self healing materials: an alternative approach to 20 centuries of materials science, pp 195–204

Jonkers HM, Schlangen E (2009) A two component bacteria-based self-healing concrete.In: Concr Repair, Rehabil Retrofit II, pp 215–220

Jung YK, Kim TY, Park SJ, Lee SY (2010) Metabolic engineering of Escherichia coli for the production of polylactic acid and its copolymers. Biotechnol Bioeng 105:161–171. https://doi.org/10.1002/bit.22548

Khaliq W, Ehsan MB (2016) Crack healing in concrete using various bio influenced self-healing techniques. Constr Build Mater 102:349–357. https://doi.org/10.1016/j.conbuildmat.2015.11.006

Kip N, van Veen JA (2015) The dual role of microbes in corrosion. ISME J 9:542–551. https://doi.org/10.1038/ismej.2014.169

Klamer M, Morsing E, Husemoen T (2004) Fungal growth on different insulation materials exposed to different moisture regimes. Int Biodeterior Biodegradation 54:277–282. https://doi.org/10.1016/j.ibiod.2004.03.016

Konkol N, McNamara C, Sembrat J, Rabinowitz M, Mitchell R (2009) Enzymatic decolorization of bacterial pigments from culturally significant marble. J Cult Herit 10:362–366. https://doi.org/10.1016/j.culher.2008.10.006

Kucharski ES, Whiffin V, Mutlaq S (2006) Microbial biocementation. WO2006066326 A1. 1–10

Li J, Yang E-H (2017) Macroscopic and microstructural properties of engineered cementitious composites incorporating recycled concrete fines. Cem Concr Compos 78:33–42. https://doi.org/10.1016/j.cemconcomp.2016.12.013

Liao KH, Chan JKH (2016) What is ecological wisdom and how does it relate to ecological knowledge? Landsc Urban Plan 155:111–113. https://doi.org/10.1016/j.landurbplan.2016.07.006

Lv J, Mao J, Ba H (2015) Influence of marine microorganisms on the permeability and microstructure of mortar. Constr Build Mater 77:33–40. https://doi.org/10.1016/j.conbuildmat.2014.11.072

Manso S (2014) Bioreceptivity optimisation of concrete substratum to stimulate biological colonisation. Universitat Politecnica de Catalunya & Ghent University

Manso S, Aguado A (2016) The use of bio-receptive concrete as a new typology of living wall systems. Matériaux Tech 104:502. https://doi.org/10.1051/mattech/2016028

Manso S, Calvo MA, Aguado A, Belie NDE (2016) Sustainable cements in construction : magnesium phosphate cements to stimulate colonization of building materials. J Sustain Cem Mater 1–10. https://doi.org/10.1080/21650373.2016.1201018

Mansour MMA (2017) Effects of the halophilic fungi Cladosporium sphaerospermum, Wallemia sebi, Aureobasidium pullulans and Aspergillus nidulans on halite formed on sandstone surface. Int Biodeterior Biodegradation 117:289–298. https://doi.org/10.1016/j.ibiod.2017.01.016

Marang L, van Loosdrecht MCM, Kleerebezem R (2016) Combining the enrichment and accumulation step in non-axenic PHA production: cultivation of Plasticicumulans acidivorans at high volume exchange ratios. J Biotechnol 231:260–267. https://doi.org/10.1016/j.jbiotec.2016.06.016

Matthews E, Amann C, Bringezu S, Fischer-Kowalski M, Hüttler W, Kleijn R, Moriguchi Y, Ottke C, Rodenburg E, Rogich D, Schandl H, Schütz H, Van der Voet E, Weisz H (2000) The weight of nations: material outflows from industrial economies. World Resources Institute, Washington DC

Mayoral E (2011) Growing architecture through mycelium and agricultural waste. Int J Constr Environ 1:87–132

Mayoral González E, González Diez I (2015) Bacterial induced cementation processes and mycelium panel growth from agricultural waste. Key Eng Mater 663:42–49. https://doi.org/10.4028/www.scientific.net/KEM.663.42

Nadson G (1903) Microorganisms as geologic agents. I Tr Komisii Isslect Min Vodg

Nadson G (1928) Beitrag zur kenntnis der bakteriogenen kalkablagerung. Arch fur Hydrobiol 19:154–164

NRMCA (2012) Concrete CO_2 Fact Sheet

Örnek D, Jayaraman A, Syrett B, Hsu CH, Mansfeld F, Wood T (2002a) Pitting corrosion inhibition of aluminum 2024 by Bacillus biofilms secreting polyaspartate or??-polyglutamate. Appl Microbiol Biotechnol 58:651–657. https://doi.org/10.1007/s00253-002-0942-7

Örnek D, Wood TK, Hsu CH, Mansfeld F (2002b) Corrosion control using regenerative biofilms (CCURB) on brass in different media. Corros Sci 44:2291–2302. https://doi.org/10.1149/1.1456922

Pan Z-Y, Li G, Hong C-Y, Kuang H-L, Yu Y, Feng F-X, Liu D-M, Du H (2015) Modified recycled concrete aggregates for asphalt mixture using microbial calcite precipitation. RSC Adv 5:34854–34863. https://doi.org/10.1039/C5RA04203H

Park SJ, Park SH, Ghim SY (2014) The effects of Paenibacillus polymyxa E681 on antifungal and crack remediation of cement paste. Curr Microbiol 69:412–416. https://doi.org/10.1007/s00284-014-0604-x

Pedersen A, Hermansson M (1991) Inhibition of metal corrosion by bacteria. Biofouling 3:1–11. https://doi.org/10.1080/08927019109378157

Pessi AM, Suonketo J, Pentti M, Kurkilahti M, Peltola K, Rantio-Lehtimäki A (2002) Microbial growth inside insulated external walls as an indoor air biocontamination source. Appl Environ Microbiol 68:963–967. https://doi.org/10.1128/AEM.68.2.963-967.2002

Plank J (2004) Applications of biopolymers and other biotechnological products in building materials. Appl Microbiol Biotechnol 66:1–9. https://doi.org/10.1007/s00253-004-1714-3

Portland Cement Association (2014) World Cement Consumption

Pourbaix M (1973) Electrochemical equilibria. In: Staehle RW (ed) Lectures on electrochemical corrosion, pp 83–183

Qian C, Chen H, Ren L, Luo M (2015) Self-healing of early age cracks in cement-based materials by mineralization of carbonic anhydrase microorganism. Front Microbiol 6:1–9. https://doi.org/10.3389/fmicb.2015.01225

Qian C, Ren L, Xue B, Cao T (2016) Bio-mineralization on cement-based materials consuming CO2 from atmosphere. Constr Build Mater 106:126–132. https://doi.org/10.1016/j.conbuildmat.2015.10.105

Qiu J, Tng DQS, Yang E-H (2014) Surface treatment of recycled concrete aggregates through microbial carbonate precipitation. Constr Build Mater 57:144–150. https://doi.org/10.1016/j.conbuildmat.2014.01.085

Ranalli G, Matteini M, Tosini I, Zanardini E, Sorlini C (2000) Bioremediation of cultural heritage: removal of sulphates, nitrates and organic substances. In: Ciferri O, Tiano P, Mastromei G (eds) Microbes art. Springer, US, pp 231–245

Rong H, Qian CX, Li LZ (2012) Study on microstructure and properties of sandstone cemented by microbe cement. Constr Build Mater 36:687–694. https://doi.org/10.1016/j.conbuildmat.2012.06.063

Sarkar M, Adak D, Tamang A, Chattopadhyay B, Mandal S (2015) Genetically-enriched microbe-facilitated self-healing concrete–a sustainable material for a new generation of construction technology. RSC Adv 5:105363–105371. https://doi.org/10.1039/C5RA20858K

Silva FB, Boon N, De Belie N, Verstraete W (2015a) Industrial application of biological self-healing concrete: challenges and economical feasibility. J Commer Biotechnol 21:31–38. https://doi.org/10.5912/jcb662

Silva F, De Belie N, Boon N, Verstraete W (2015b) Production of non-axenic ureolytic spores for self-healing concrete applications. Constr Build Mater. https://doi.org/10.1016/j.conbuildmat.2015.05.049

Smith KS, Ferry JG (2000) Prokaryotic carbonic anhydrases. FEMS Microbiol Rev 24:335–366. https://doi.org/10.1016/S0168-6445(00)00030-9

Stabnikov V, Ivanov V, Chu J (2015) Construction biotechnology: a new area of biotechnological research and applications. World J Microbiol Biotechnol 31:1303–1314. https://doi.org/10.1007/s11274-015-1881-7

Sterflinger K, Ettenauer J, Piñar G (2013) Bio-susceptibility of materials and thermal insulation systems used for historical buildings. Energy Procedia 40:499–506. https://doi.org/10.1016/j.egypro.2013.08.057

Stuckrath C, Serpell R, Valenzuela LM, Lopez M (2014) Quantification of chemical and biological calcium carbonate precipitation: performance of self-healing in reinforced mortar containing chemical admixtures. Cem Concr Compos 50:10–15. https://doi.org/10.1016/j.cemconcomp.2014.02.005

The European Cement Association (2011) Activity Report 2011

Tziviloglou E, Van Tittelboom K, Palin D, Wang J, Sierra-Beltran MG, Erşan YÇ, Mors R, Wiktor V, Jonkers HM, Schlangen E, De Belie N (2016) Bio-based self-healing concrete: from research to field application. In: Hager MD, van der Zwaag S, Schubert US (eds) Advances in polymer science. Springer, pp 345–385

Tziviloglou E, Wiktor V, Jonkers HM, Schlangen E (2016b) Bacteria-based self-healing concrete to increase liquid tightness of cracks. Constr Build Mater 122:118–125. https://doi.org/10.1016/j.conbuildmat.2016.06.080

United Nations Environment Programme (2009) Buildings and climate change: summary for decision-makers

Verdier T, Coutand M, Bertron A, Roques C (2014) A review of indoor microbial growth across building materials and sampling and analysis methods. Build Environ 80:136–149. https://doi.org/10.1016/j.buildenv.2014.05.030

Verstraete W (2015) The technological side of the microbiome. Npj Biofilms Microbiomes 1:15001. https://doi.org/10.1038/npjbio

Wang J, Snoeck D, Van Vlierberghe S, Verstraete W, De Belie N (2014a) Application of hydrogel encapsulated carbonate precipitating bacteria for approaching a realistic self-healing in concrete. Constr Build Mater 68:110–119. https://doi.org/10.1016/j.conbuildmat.2014.06.018

Wang J, Soens H, Verstraete W, De Belie N (2014b) Self-healing concrete by use of microencapsulated bacterial spores. Cem Concr Res 56:139–152. https://doi.org/10.1016/j.cemconres.2013.11.009

Wang J, Mignon A, Snoeck D, Wiktor V, Van Vliergerghe S, Boon N, De Belie N (2015) Application of modified-alginate encapsulated carbonate producing bacteria in concrete: a promising strategy for crack self-healing. Front Microbiol 6:1088. https://doi.org/10.3389/fmicb.2015.01088

Wang K, Qian C, Wang R (2016) The properties and mechanism of microbial mineralized steel slag bricks. Constr Build Mater 113:815–823. https://doi.org/10.1016/j.conbuildmat.2016.03.122

Wei S, Jiang Z, Liu H, Zhou D, Sanchez-Silva M (2013) Microbiologically induced deterioration of concrete-a review. Brazilian J Microbiol 44:1001–1007. https://doi.org/10.1590/S1517-83822014005000006

Wiktor V, Jonkers HM (2011) Quantification of crack-healing in novel bacteria-based self-healing concrete. Cem Concr Compos 33:763–770. https://doi.org/10.1016/j.cemconcomp.2011.03.012

Willmott Dixon (2010) The impacts of construction and the built environment. Brief Note 33. https://doi.org/10.3846/1392-3730.2009.15.269-280

Wong NH, Kwang Tan AY, Chen Y, Sekar K, Tan PY, Chan D, Chiang K, Wong NC (2010) Thermal evaluation of vertical greenery systems for building walls. Build Environ 45:663–672. https://doi.org/10.1016/j.buildenv.2009.08.005

Microbially Induced Calcite Precipitation (MICP) for Soil Stabilization

Liang Cheng and Mohamed A. Shahin

Abstract Bio-cementation is a recently developed technique for soil stabilization in geotechnical engineering applications, as it employs microbiological activity that improves the engineering properties of soils. One of the most commonly adopted processes to achieve soil stabilization by biocementation is through microbially induced calcite precipitation (MICP), which is commonly known as "biogrout". This technique utilizes the metabolic pathways of bacteria to form calcite ($CaCO_3$) that binds the soil particles together, leading to increased soil strength and stiffness. Biogrout is environmental-friendly and has the potential to be a better alternate to chemically based grouting materials such as lime or cement. However, there are still many challenges that lay ahead for future research prior to real practical application of this promising technique. In this chapter, some salient chemical and physical factors governing soil treatment by biogrout are described and explained, and possible applications of biogrout in geotechnical engineering are discussed.

Keywords Biocement · Biogrout · Calcite · Soil · Cement

1 Introduction

The historic detrimental impact of weak soils to society is not uncommon. For instance, earthquakes' induced soil liquefaction is frequent in many places around the world, causing billions of dollars of damages, e.g., the Northridge 1994, Kobe 1995, Loma Prieta (San Francisco) 1989, Christchurch 2011, and Japan 2011 (Mote and Dismuke 2011). Unstable slopes of dikes and dunes due to a combination of

L. Cheng (✉)
Biofuels Institute of Jiangsu University, Jiangsu University, Zhenjiang 212013, China
e-mail: ClCheng@ujs.edu.cn

L. Cheng · M. A. Shahin
Department of Civil Engineering, Curtin University, Perth, WA 6845, Australia
e-mail: M.Shahin@curtin.edu.au

V. Achal and A. Mukherjee (eds.), *Ecological Wisdom Inspired Restoration Engineering*, EcoWISE, https://doi.org/10.1007/978-981-13-0149-0_3

seepage and external loading are also common. In addition, failure of marine structures (e.g., pipelines and oil platform) has been always linked to wave-induced instability of weak soil deposits, leading to continuous erosion of beach soil (De Groot and Meijers 1992), which over the next 60 years is estimated to claim one out of four houses within 150 m of the US shoreline (John Heinz III Center for Science Economics and the Environment 2000). All of the above mentioned cases can benefit from soil stabilization (or ground improvement), by enhancing the mechanical and geotechnical properties of in situ soil materials so as to create a solid and strong foundation ground leading to improved soil bearing capacity and reduced excessive settlement.

Currently, more than 40,000 ground improvement projects/year with a cost of $6B USD are carried out worldwide in which chemical soil stabilization by grouts (e.g., cement, epoxy, acrylamide, silicates, and polyurethane) is the most widely employed method. However, chemical grouts require substantial energy for material production and/or installation and also raise the issues of cost, health, and safety and in turn have now been banned in several countries around the world (DeJong et al. 2010). To overcome the abovementioned limitations of chemical soil stabilization, research has been recently aimed at finding alternative methods that can achieve an optimum performance and being environmentally sustainable and economically viable. The current emerging interdisciplinary research at a confluence of microbiology, geotechnical engineering, geochemistry, and crystallography has led to a new frontier knowledge called bio-inspired geotechnical engineering. One of the most commonly adopted processes in this research field is the use of bacteria as catalyst to achieve soil stabilization through microbially induced calcite precipitation (MICP), commonly known as "bio-cementation" or "biogrout". MICP by urea hydrolysis is an innovative and attractive technology for soil stabilization and has undergone a very rapid development over the past ten years or so. The annual publications on this topic of research have grown exponentially in the last few years, and exciting work has been carried out on the optimization of this bio-inspired geotechnical engineering technology at the laboratory scale and commercialization of the product to meet the society needs (Parmar and Singh 2014). The main aim of this chapter is to shed some light on this promising and green technology in the field of soil stabilization, by presenting and discussing some salient chemical and physical governing factors and the most possible applications in geotechnical engineering.

2 Principal of Microbially Induced Carbonate Precipitation (MICP) Process

When the amount of calcium and carbonate ions in solution exceeds the solubility product, precipitation of $CaCO_3$ occurs. It is suggested that biotic action (microbial action) has more contribution in $CaCO_3$ precipitation in most environment on earth

(Castanier et al. 1999), compared to abiotic change (e.g., change in temperature, pressure, or evaporation). Calcium carbonate precipitation is a rather straightforward and spontaneous chemical process controlled mainly by four parameters (Hammes and Verstraete 2002): (1) Ca^{2+} concentration; (2) concentration of dissolved inorganic carbon (DIC); (3) pH of the environment; and (4) presence of nucleation sites. Spontaneous crystal growth on stable nuclei occurs when supersaturated condition is reached, whereby the concentrations of calcium (Ca^{2+}) and carbonate (CO_3^{2-}) ions must exceed the calcite solubility product (K_{sp}) according to Eq. 1, as follows:

$$\text{SI(Supersaturation Index)} = \frac{[Ca^{2+}][CO_3^{2-}]}{K_{sp}} \quad (1)$$

Microbially induced carbonate precipitation (MICP) has been studied over the last several decades. Microorganisms can influence calcium carbonate precipitation by changing all the parameters described above. In MICP process, the role of bacteria or microorganisms in calcium carbonate precipitation is attributed to: producing carbonate (e.g., respiration, hydrolysis, etc.), producing alkalinity (raising the pH of the environment), and acting as nucleation sites in an oversaturated solution (Stocks-Fischer et al. 1999).

In nature, microbially induced carbonate precipitation plays a significant role in metal co-precipitation and cementation. Bacterial cells have been shown to be excellent nucleation sites for growing minerals due to the presence of several negatively charged groups on the cell wall, resulting in positively charged metal ions bound on bacterial surfaces (Douglas and Beveridge 1998; Ehrlich 1998). Many investigations confirmed the precipitation of $CaCO_3$ on the bacterial cell surface (Fujita et al. 2000; Hammes et al. 2003; Warren et al 2001). Several mechanisms have been identified by which bacteria or organisms can induce calcium carbonate precipitation, including photosynthesis by photosynthetic organisms (autotrophic organisms), like algae or cyanobacteria in aquatic environments (McConnaughey and Whelan 1997) (see Table 1). The dissolved CO_2 is consumed by the photosynthetic organisms through an exchange of HCO_3^-/OH^- across the cell membrane, leading to an increase in pH in the local environment around the cells, production of HCO_3^-, and calcium carbonate precipitation in the presence of calcium ions (Hammes and Verstraete 2002). It is estimated that photosynthetic organisms are the major contributors to the production of carbonate rock during almost 70% of earth history (Altermann et al. 2006), and the autotrophic pathway is considered one of the principal sources of biogenic calcium carbonate precipitation (Ehrlich 1996).

Apart from autotrophic photosynthetic organisms, $CaCO_3$ can also be induced by heterotrophic organisms via producing carbonate or bicarbonate and modifying the environment creating such an alkaline environment to favor precipitation. The heterotrophic pathways generally involve nitrogen cycle and sulfur cycle, as explained below:

Table 1 MICP metabolic pathway and the reactions

Metabolic pathway	Reactions	References
Photosynthesis	$2HCO_3^- + Ca^{2+} \rightarrow CH_2O + CaCO_3 + O_2$	Hammes and Verstraete (2002)
Denitrification	$CH_2COO^- + 2.6H^+ + 1.6NO_3^- \rightarrow 2CO_2 + 0.8N_2 + 2.8H_2O$ $CO_2 + 2OH^- + Ca^{2+} \rightarrow H_2O + CaCO_3$	Van Paassen et al. (2010a)
Sulfate reduction	$SO_4^{2-} + 2[CH_2O] + OH^- + Ca^{2+} \rightarrow CaCO_3 + CO_2 + 2H_2O + HS^-$	Castanier et al. (1999)
Urea hydrolysis	$CO(NH_2)_2 + 2H_2O + Ca^{2+} \rightarrow 2NH_4^+ + CaCO_3$	Cheng et al. (2013)

- Nitrogen cycle (denitrification), in which nitrate is used by microorganisms to oxidize organic compounds, increases the pH in the surrounding medium by consuming proton and produces CO_2 from oxidation of organic carbon, which favors carbonate precipitation (van Paassen et al. 2010a) (Table 1).
- Sulfur cycle (sulfate reduction), which is carried out by sulfate-reducing bacteria (SRB) under anoxic conditions, reduces sulfate to sulfide while oxidizing organic carbon to bicarbonate, during which an increase in pH occurs favoring the precipitation of calcium carbonate (Castanier et al. 1999) (Table 1).

MICP through urea hydrolysis pathway is the most interesting and widely investigated mechanism for carbonate precipitation. Many researchers have focused on this approach compared to other pathways, and it was found that urea hydrolysis is the fastest and most effective reaction to produce $CaCO_3$ crystals. The ureolytic bacteria hydrolyze urea to produce ammonia and carbonic acid (Cheng and Cord-Ruwisch 2012). The hydrolysis of ammonia increases the pH of the environment, and the dissociation of carbonic acid generates bicarbonates. Therefore, the urea hydrolysis pathway favors the precipitation of calcium carbonate in the presence of calcium ions in solution (Table 1).

3 Urease-Driven MICP

3.1 Urease Enzyme and Microbial Urease

Urease (urea amidohydrolase; EC 3.5.1.5) is an enzyme that catalyzes the hydrolysis of urea to obtain ammonia and carbon dioxide. Urease from jack bean was the first enzyme crystallized (Summer 1926) and has also been found in bacteria, yeast, and several higher plants. Most bacterial ureases are heteropolymeric enzymes. The metal content analysis for all purified ureases indicated that all ureases contain nickel, which affects the presence of urease activity in ureolytic organisms (Hausinger 1987). Microbial urease activity is widespread in the environment as a result of biological action, including bacteria, yeasts, filamentous bacteria, and algae (Booth and Vishniac 1987; Bekheet and Syrett 1977; Seneca et al. 1962). In general, there are five modes of regulation exist for the synthesis of urease in the microbial system (Mobley et al. 1995) and it is vital to understand the regulation of urease synthesis under different conditions because sufficient urease activity is essential for the success of ground improvement. These modes include:

1. Constitutive expression—where the enzyme expression is constant regardless of the environmental conditions (e.g., *Bacillus pasteurii*, *Sporosarcina urea*, *Morganella morganii*, and the chromosomally encoded ureases of some *E. coli* isolates).
2. Nitrogen regulation—where in the presence of high-quality nitrogen source (NH_3, NH_4^+ or urea), the urease synthesis is repressed. In contrast, the urease

synthesis is activated (re-depressed) under conditions of nitrogen-limiting or nitrogen starvation. The nitrogen regulation of urease expression has been reported to occur in many ureolytic organisms.

3. Induction by urea—where the urease expression is induced by the presence of the substrate urea. Cheng and Cord-Ruwisch (2013) have successfully produced highly urease active bacterial culture from activated sludge using non-sterilized chemostat reactor fed with urea-rich growth medium.
4. Developmental control—where the expression of urease varies relating to the developmental stages. Falkinham and Hoffman (1984) reported that in *Proteus* species, the urease level and urease transcript were high during swarming state, while negligible activity was found in non-swarmer cells.
5. Regulation by pH—where the urease is regulated by the environmental pH through the manner of regulating the rate of urease synthesis (Sissons et al. 1990).

3.2 Metabolism of Urease Positive Bacteria

Urea hydrolysis has been reported to play a novel role in adenosine triphosphate (ATP) generation in several ureolytic bacteria (Mobley and Hausinger 1989). The generation of ATP is governed by the proton motive force (Δp), which includes the transmembrane pH gradient (ΔpH) and charge gradient or membrane potential ($\Delta \psi$). Ureolytic bacteria, most of which are alkaliphiles, utilize urea hydrolysis to generate ATP. Romano et al. (1980) demonstrated that urea added to Ureaplasma cells resulted in a rapid increase in intracellular ATP concentration. Jahns (1996) suggested that in the presence of high concentration of urea (300 mM), *S. pasteurii* utilizes NH_4^+ to drive ATP generation. A schematic model for urease-dependent ATP generation is illustrated in Fig. 1. Apart from the ATP generation, for *P. vulgaris* (soil inhabiting organism), ammonium from the urea hydrolysis can be directly assimilated into biomass via the glutamine synthetase–glutamate synthase (GSGOGAT) pathway or by the action of glutamate dehydrogenase (GDH) (Tyler 1978).

3.3 MICP Induced by Urea Hydrolysis

In solution, dissolved calcium ions are attracted to the negative charged cell wall of ureolytic bacteria. Then, ureolytic bacteria uptake urea and catalyze the hydrolysis of urea to release ammonium, resulting in an increase in pH, and dissolved inorganic carbon (DIC) (i.e., CO_2, HCO_3^-, CO_3^{2-}) in the microenvironment (De Muynck et al. 2010). The escalation of alkalinity helps to facilitate an environment

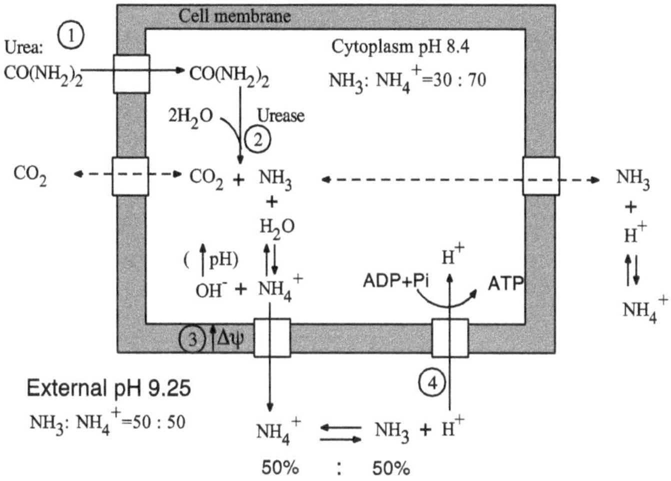

Fig. 1 ATP-generating system coupled with urea hydrolysis process in *S. pasteurii* as suggested by Jahns (1996) and Al-Thawadi (2008). The chemical transport processes related to microbial urea hydrolysis was suggested by Mobley and Hausinger (1989). The reactions are (1) urea diffusion into the bacterial cell according to the concentration gradient; (2) urea is hydrolyzed by urea resulting in an increase in pH of the cytoplasm and a decrease in ΔpH (difference in pH between inside and outside of the cell); (3) efflux of ammonium from the cells according to the concentration gradient, resulting in an increase in the membrane potential (Δψ); and (4) the increased membrane potential drives in protons against the concentration gradient into the cell, resulting in an ATP generation

encouraging the precipitation of calcite in the presence of calcium (Eqs. 2 and 3 in Fig. 2). The reaction is expressed as follows:

$$Ca^{2+} + \mathrm{Cell} \rightarrow \mathrm{Cell} - Ca^{2+} \tag{2}$$

$$CO_3^{2-} + \mathrm{Cell} - Ca^{2+} \rightarrow \mathrm{Cell} - CaCO_3 \tag{3}$$

Figure 3 shows typical MICP-induced $CaCO_3$ crystals at aqueous phase in which spherical and rhombohedral crystals in spherical arrangements are observed through SEM. It seems that the spherical shell dissolved gradually after certain time from being intact, and the thickness of the spherical wall was about 1–2 μm. Al-Thawadi et al. (2012) suggested that the different shapes of $CaCO_3$ crystals may be due to different stages of the sphere development and possible transformation into rhombohedral crystals.

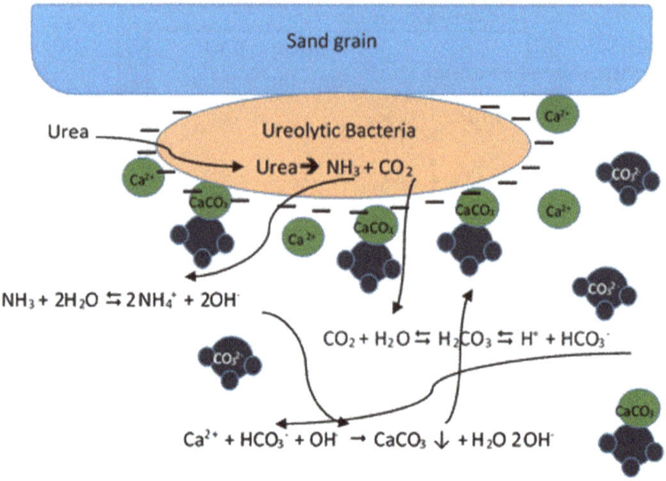

Fig. 2 Overview of bio-mediated calcite precipitation using ureolysis (modified from DeJong et al. 2010)

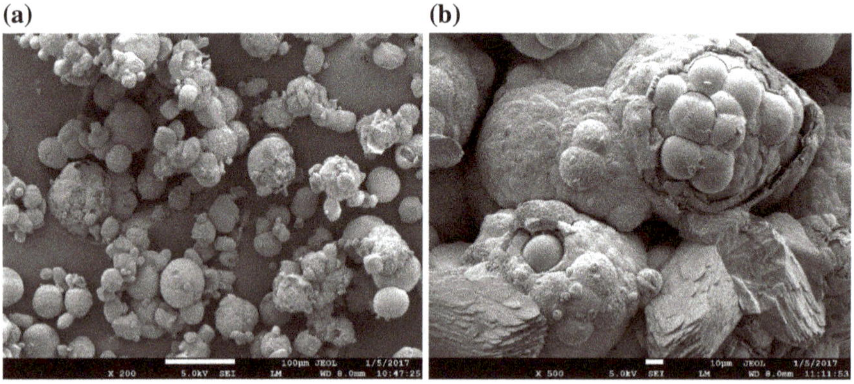

Fig. 3 SEM images: **a** spherical $CaCO_3$ crystals; and **b** decomposed spherical calcite in which small spherical $CaCO_3$ crystals appear

4 Soil Cementation and Stabilization via MICP

4.1 Ex Situ Cultivated Versus In Situ Enriched Ureolytic Bacteria

The general principal of urease-driven MICP for soil cementation and stabilization can be described as follows. Soil bio-cementation is achieved by transporting (injecting, flushing, or percolating) chemical reagents (i.e., urea and $CaCl_2$) and ureolytic bacteria to the locations where strengthening is required. Ureolytic bacteria catalyze the hydrolysis of urea reaction to produce ammonium and carbonate ions. The produced carbonate ions react with calcium ions, often calcium chloride ($CaCl_2$), to form calcite precipitates throughout the soil matrix. The microbially induced calcite crystals bridge the adjacent soil particles, leading to increase in shear strength and stiffness of soils while maintaining high permeability (Cheng et al. 2016; DeJong et al. 2010; De Muynck et al. 2010; van Paassen et al. 2010b).

Highly urease active bacteria are usually cultivated ex situ under sterile conditions to ensure a constant urease activity production. Once the highly urease activity bacteria are cultivated, MICP can be carried out by inducing high concentration of urease active bacteria positioned inside the soil prior to supplying the cementation solution. Submersed sequential flushing of bacterial and cementation solution (van Paassen et al. 2010b; Whiffin et al. 2007; DeJong et al. 2006) or surface percolation of cultivated bacteria and cementation solution (Cheng and Cord-Ruwisch 2012, 2014) have been developed in the literature in accordance with different applications.

The ex situ cultivated bacteria usually face a number of challenges when placed inside the soil pores, such as the reduction in population from predation and competition, stress from abiotic factors (such as pH, osmotic pressure, temperature, and availability of suitable nutrients), and bacteria capsulation by calcium carbonate crystal formation (Cheng et al. 2017; Evans et al. 1993). Usually, the urease activity decreases gradually and is difficult to revert, which necessitates repeated injection of large quantities of ureolytic bacteria that are normally cultivated in an ex situ reactor under sterilized conditions.

The reinjection of bacteria becomes incrementally difficult due to the progressing calcite buildup with the flush of each cementation solution, leading to accumulation of bacteria around the injection end. This can eventually result in a surface or injection point blockage (Whiffin et al. 2007; Cheng and Cord-Ruwisch 2014). In situ enrichment of urease active bacteria can avoid the repeated injection of large quantities of ureolytic bacteria. Cunningham et al. (2009) tested the in situ establishment of pure *Sporosarcina pasteurii* biofilm by introducing inoculum and growth medium together into sand columns and examined the recovery of biofilm by introducing additional growth medium. Cheng et al. (2017) applied bio-cementation in situ by combining the surface percolation of nutrients and cementation solution (Urea/$CaCl_2$) with in situ cultivation of indigenous soil urease positive microorganisms under non-sterile conditions. The loss of ureolytic activity during the bio-cementation process could be also recovered by providing more growth medium under selective enrichment conditions, enabling the in situ enriched ureolytic microorganisms to increase in numbers and urease activity in such a way that continued cementation is possible. Burbank et al. (2011) developed an approach for in situ cultivation of indigenous soil ureolytic bacteria prior to bio-cementation by adding an organic carbon in the form of molasses and urea to soil at neutral pH. Using this method, an increase in soil strength of up to two times was accomplished at a depth of around 1.2 m after about 25 days.

4.2 Pure Strain Versus Enrichment

As suggested by DeJong et al. (2006), a suitable urease active bacteria for soil stabilization must be able to (1) meet the biosafety regulation (non-pathogenic strain); (2) produce highly active urease activity in the presence of salt; and (3) penetrate into deep location of soil and adhere on sand particle surface avoiding being washed away. These rules guided several researchers to screen and enrich the most suitable bacteria. Al-Thawadi and Cord-Ruwisch (2012) tested 13 isolates obtained from the enrichment of local soils. Three isolates were most closely related to *Bacillus* species, within which the bacterium (*Bacillus sp.* MCP 11) producing the highest urease activity in the presence of cementation solution (urea and calcium chloride, up to 1 M) was used to produce biocement. Although another type of bacterium (MCP 4) has similar urease activity, the flocculation behavior of this strain caused uneven cementation of sand column, whereby the clogging was observed around the injection end. This is presumably due to the uneven distribution of the bacterial cells because the bacteria tend to flocculate around the injection end and clog the sand (Al-Thawadi 2008).

In most bio-cementation research, the gram-positive bacterial species *Sporosarcina pasteurii* (former *Bacillus pasteurii*), especially the strain *S. pasteurii ATCC11859* (*DSMZ33, LMG 7130*) (Achal et al. 2009; Van Paassen et al. 2010b; Tobler et al. 2014; Whiffin et al. 2007) was used because of its high urease activity often larger than 1.1 mol-urea/L/h. Other than *Sporosarcina pasteurii*, many other bacterial species were also tested for MICP, such as *B. cereus* (Castanier et al. 2000); *B. megaterium* (Bang et al. 2001), *B. subtilis* (Reddy et al. 2010), *Bacillus pumilus* (Daskalakis et al. 2015), *B. sphaericus* (De Muynck et al. 2008), and *Bacillus sp.* (DSM 23526) (Al-Thawadi and Cord-Ruwisch 2012). The strain *Bacillus sp.* (DSM 23526) (Al-Thawadi and Cord-Ruwisch 2012) was isolated selectively from sludge and soil samples in Western Australia.

Although the pure strain provides relatively constant urease activity of each bacterial culture, enabling reproducible bio-cementation process, the cultivation of urease active bacteria by traditional biotechnology production is very costly which renders an economical large-scale application of bio-cementation. Cheng and Cord-Ruwisch (2013) developed a selective enrichment technique that uses aerobic chemostat inoculated with an activated sludge for continued selective cultivation of highly active ureolytic bacteria culture (i.e., 60 U/mL) by applying high pH and urea/ammonia concentrations as the key selective conditions. It was shown that such highly urease active enrichment could successfully cement sand column up to about 2000 kPa (unpublished result). The SEM of bio-cemented sand using the enrichment revealed substantial crystals precipitates within the sand matrix, indicating a successful bio-cementation, as shown in Fig. 4. Although relatively stable urease activity of the enrichment was obtained constantly in laboratory scale bioreactor, for large-scale production of the enrichment of bacterial culture, the urease activity must be examined before any MICP application.

(a) (b)

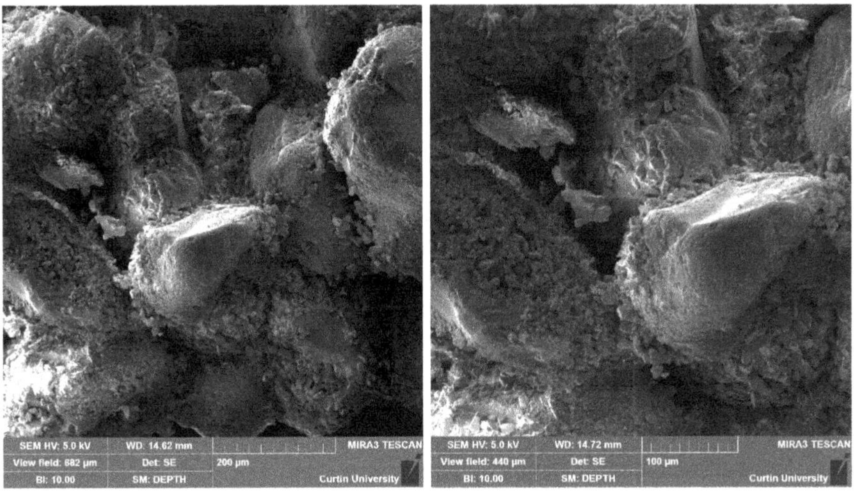

Fig. 4 SEM images of bio-cemented sand using highly urease active enrichment (60 U/mL) as urea hydrolysis catalyst **a** overview of treated sample; **b** calcite bonding

4.3 Injection Versus Surface Percolation Versus In Situ Premixing

Ground improvement by MICP technology has been demonstrated to be a carbon neutral technology. The exploitations of bacterial capacity to produce carbonate in situ in the presence of calcium ions have been tested in waterlogged soils (DeJong et al. 2006; Whiffin et al. 2007; van Paassen et al. 2010b), using heavy machinery and hydraulic injection of the cementation solution. The injection method is the most commonly preferred MICP treatment approach as the operational parameters, such as flow rate, flow path, hydraulic pressure can be well controlled during the treatment. The injection method also allows the bio-cementation treatment to be conducted from both the vertical and horizontal directions. A method of two-phase injection of bacterial suspension and cementation solution for bacterial immobilization has been developed by Whiffin et al. (2007), which was applied to stabilization of a 5 m vertical sand column. Van Paassen et al. (2010b) applied the horizontal injection method for treatment of a large-scale experiment of 100 m^3 sand, which indicated varied strength of products from loosely cemented sand to moderately strong rock with unconfined compressive strengths of 0.7–12 MPa. This large-scale test associated with other laboratory experiments indicated the major drawback of the injection method, including the uneven distribution of bacteria and $CaCO_3$ precipitation. This usually results in a non-uniform cementation of MICP treated soils in which clogging around the injection end and insufficient depth of cementation are common. Cheng and Cord-Ruwisch (2014) described that the immobilization of high bacterial numbers at the injection end contributed to the clogging. To improve the transportation of bacteria within soil and enhance the bacteria delivery to reach long distance, Harkes et al. (2010) suggested that if the bacterial cells attach too quickly to sand causing a consequent accumulation close to the inject point, a low salinity flush can be used to remobilize the bacterial cells and enhance homogeneity of the bacterial distribution.

Bio-cementation can also be performed using the surface percolation of reagents by simply spraying or trickling bacterial suspension and cementation solution alternately onto the soil surface followed by the cementation solution penetration into the soil driven by gravity. The surface percolation technique can be applied in the geoengineering practice for stabilization of unsaturated soil such as surface dry sand (Cheng and Cord-Ruwisch 2012, 2014). The surface percolation method does not require heavy machinery due to the free draining of water movement. However, this method is not suitable for fine-grained soils (e.g., silt or clay) due to the low permeability and infiltration rate. The low infiltration rate of bacterial suspension and cementation solution results in limited cementation depth, which is likely due to the reaction of the cementation solution during the slow infiltration leading to less reagents moving to deeper areas. Whiffin et al. (2007) suggested that fast flow rate would move the cementation reagents further into the treated soil column. However, high flow rate can only be obtained in highly permeable soil (e.g., coarse sand) by the surface percolation technique. The surface percolation method also encounters the problem of uncontrollable flow of reagents. At free-draining environment, the liquid infiltration driven by gravity force has preferential flow paths, where the flow resistance is lower than the other areas (Cheng and Cord-Ruwisch 2014). Therefore, soils continuously receive reagents along these preferential flow paths, leading to reasonably higher content of $CaCO_3$ and strength than in the other areas. Cheng and Cord-Ruwisch (2014) found that a reasonably homogeneous cementation of a silica sand layer up to 200 mm deep was achieved, and heterogeneous cementation was obtained at deep locations. Gomez et al. (2015) studied a field-scale surficial application of MICP and achieved a stiff crust of about 2.5 cm thick. Therefore, the surface percolation technique may be more suitable for surface ground improvement of dust control and erosion resistance (Gomez et al. 2015).

According to the literature, it is difficult to achieve MICP-induced cementation that is highly reproducible and homogeneous via the above-mentioned commonly used injection or surface percolation methods. This is because when the bacteria travel through the pore space of soils, they are likely to be filtered through the soil grains with long linear reduction of microbe concentration along the injection path (Ginn et al. 2001). To overcome the shortage of non-uniform bacteria distribution, the bacteria suspension can be premixed with soil mechanically to reach a desired homogeneity. Using the premixing method, Yasuhara et al. (2012) obtained bio-cemented sand samples of UCS values ranging from 400 kPa to 1.6 MPa. Zhao et al. (2014) found that when using the premixing method almost 83% of the $CaCO_3$ was homogeneously distributed throughout the treated sand columns. Cheng and Shahin (2016) proposed a novel premixing approach using urease active bioslurry instead of bacterial suspension. Compared with the conventional MICP method, one of the major advantages of using the proposed bioslurry approach is the homogeneous distribution of the bacteria/urease activity. The homogeneous distribution of urease activity was due to the uniform distribution of bioslurry after mixing with soil. Although the premixing method solved the homogeneity problem, it remains unfavorable because the mechanical mixing would cause disturbance of the local soil. This is critical because soil disturbance may lead to a pseudo stress development in the soil sample as a result of the vigorous mixing between the soil and reagent solution. Also, the unmeasured stresses applied during mixing of soil samples complicate the soil stress history and make it difficult to discern during mechanical testing (Mujah et al. 2017).

4.4 MICP Improvement of Engineering Properties of Soil

As mentioned earlier, MICP improves the mechanical and geotechnical engineering properties of soils such as shear strength and stiffness. This is attributed to the produced $CaCO_3$ crystals, which are precipitated at the soil particles contacts forming bridges that link the adjacent soil particles together, leading to increased soil strength and stiffness. This process forms cemented soil that is very similar to most rocks (i.e., limestone) because MICP simulates the natural biological digenesis from sand to sandstone within a short time instead of million years.

Shear strength of soils can be defined as the magnitude of shear stress that a soil can sustain, which is mainly attributed to the soil friction due to interlocking of particles as well as cohesive bonding at the particle contacts (Mujah et al. 2017). Duraisamy and Airey (2012) found that the shear strength of bio-cemented soil was mainly enhanced due to the increase in soil cohesion resulted from the increase in the $CaCO_3$ content, while the friction angle was not significantly changed. Montoya and DeJong (2015) investigated the stress–strain behavior of bio-cemented sand and found that as the cementation increased the peak shear stress ratio increased, and the peak ratio was reached at lower axial strains when subjected to undrained loading. The critical state stress ratio was not significantly affected by the cementation. Montoya and DeJong (2015) also found as the cementation increased the stress–strain behavior transitioned from strain hardening to strain softening. Chou et al. (2011) reported that MICP treatment largely increased the soil friction angle, while a minor increase in soil cohesion was found. Cheng et al. (2013) also investigated the cohesion and friction angle of bio-cemented soil treated under different degrees of saturation and found that the precipitated crystals contributed more to improving the soil cohesion rather than friction angle.

In terms of soil stiffness, which is the ratio of stress over strain (i.e., elastic modulus E), Montoya and DeJong (2015) studied the stress–strain behavior of MICP treated sand and found that the soil stiffness was dramatically improved with the increase in $CaCO_3$ content. Lee et al. (2013) performed MICP on residual soil and found that the stiffness behavior of biocement treated residual soil is similar to that of bio-cemented natural sand. By comparing the elastic modulus of bio-cemented sand with other types of geomaterials such as concrete, gravel, and soft rock, Cheng et al. (2013) found that the bio-cemented sand is the most flexible (ductile) among the materials compared.

The unique mechanism of MICP cementation can facilitate retention of sufficient soil permeability after treatment much better than other cementitious materials such as ordinary Portland cement (OPC). Cheng et al. (2013) found that soil samples treated by bio-cementation had higher strength in the range of lower $CaCO_3$ content (<0.1 g/g sand) compared to those treated using OPC, while the permeability of bio-cemented soil samples was significantly higher than that of the OPC treated samples. The significant loss of permeability in the OPC treated samples is due to the occupation of the pore space by the water-insoluble hydrates formed from the cement hydration reaction with the pore water. In contrast, the loss of permeability in bio-cemented soil samples is due to the pore spaces becoming occupied by the calcite crystals, which only causes a smaller volume change compared to the cement hydrates. Cheng et al. (2013) also suggested that it was preferable to conduct the MICP process under lower saturation conditions compared to full saturation, as it enables improved mechanical behavior and maintains relatively high residual permeability.

MICP technique can also be used for soil clogging to reduce the hydraulic conductivity of porous soil media. Ivanov and Chu (2008) were the first to introduce the concept of MICP bio-clogging, by filling the pores of soil with MICP-induced $CaCO_3$ crystals, resulting in a significant reduction in permeability of treated soil from 5×10^{-5} m/s to 1.4×10^{-7} m/s. However, to achieve a successful MICP bio-clogging, Chu et al. (2014) suggested that an amount of $CaCO_3$ precipitates >1.3% w/w is required. Bio-clogging by MICP can be applied to solve the major problem of excessive seepage from aquaculture pond or reservoir in areas of highly permeable soils. In such an application, bio-clogging by MICP can serve as a pore-filling material to cut off seepage by sealing or constructing water ponds.

4.5 *Effective Crystals and Efficiency of Cementation*

The significance of the $CaCO_3$ crystal precipitation pattern on engineering properties of MICP treated soils was highlighted by many researchers (e.g., Al Qabany and Soga 2013; Okwadha and Li 2010; Cheng et al. 2013, 2016). It has been suggested that the shear strength of MICP treated soil may not be directly proportional to the amount of calcite produced. Only those crystals filling the gaps between the sand grains and forming effective bridges contribute to the pathway of the load transfer between soil particles and thus improve the soil stiffness and strength. Cheng et al. (2013) found that bridging crystals were formed by restricting the pore water at the contact points between the sand grains under low degree of saturation conditions, leading to higher efficiency in terms of strength improvement compared with crystals randomly precipitated. Al Qabany and Soga (2013) demonstrated that higher soil strength was obtained using low concentration of cementation solution, which could be attributed to the larger amount of precipitation at the particle contacts. Cheng et al. (2016) claimed that the efficacy (i.e., strength per mass of calcite) of larger calcite crystals precipitated at the gaps between the sand grains is higher than that obtained from smaller crystals randomly precipitated throughout the sand matrix.

The precipitation of $CaCO_3$ in MICP is a very complex process because of the involvement of bacteria acting as nucleation site and carbonate ion producer. Cheng et al. (2016) studied the effect of urease enzyme concentrations on the efficacy of bio-cementation. The results indicated that the $CaCO_3$ produced at high urease activity formed small crystals (2−5 μm) and thin coating layers ($CaCO_3$ envelope) made of agglomerated those small crystals. However, the crystals produced at low urease activity were agglomerated and formed large clusters (approximately 20 −50 μm), which filled in the gaps between the sand grains (see Fig. 5).

The effect of temperature on the pattern of MICP-induced crystals was also examined by Cheng et al. (2016). The crystals microstructure analysis indicated that MICP treatment at 50 °C resulted in small individual crystal (2−5 μm) spatially distributing over the entire sand grain surface as a coating-like layer, leading to large gaps between the sand grains and ineffective bonding behavior. At the ambient temperature, the average size of the produced crystals increased by 10 times (20–50 μm) compared to those formed at 50 °C. These large crystals were found to precipitate on the grain surface, covering the contact areas of the sand grains. Although some preliminary conclusions have been drawn from previous studies, further investigation on this complex MICP process is worth carrying on in the near future. The $CaCO_3$ crystallographic patterns (i.e., size, shape, and distribution), which are attributed to the competition between the crystal growth and crystal nucleation, play a significant role in producing different strength response and determining the final engineering properties of MICP treated soils (Al Qabany and Soga 2013).

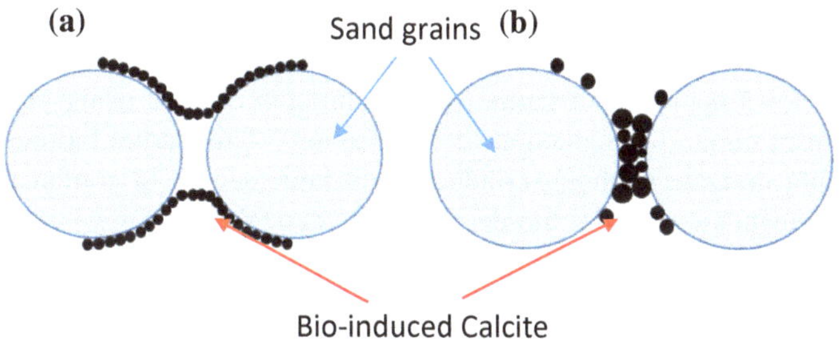

Fig. 5 Schematic diagrams of calcite bonding formed at different concentration of urease enzyme/ bacteria: **a** high concentration and **b** low concentration

4.6 Durability of MICP Treated Soils

The durability of MICP treated soils has not been extensively studied in the literature. In MICP treated soils, it is expected that similar to limestone, the $CaCO_3$ crystals that bond the soil particles together can be eroded by the acid solution, resulting in destruction of soil bonding and severe deterioration of the soil mechanical properties (Cheng et al. 2013). Soils at high latitude or elevation are usually subjected to freeze-thaw (FT) cycles during the winter season. Under this environmental condition, the building foundations are prone to suffer significant structural damages. This is because FT cycles induce uneven stresses within the soil, resulting in a decrease in soil stability. Due to the significant improvement in shear strength, it can be confirmed that MICP can be used as a viable solution for soil improvement against FT cycles. The high durability of MICP treated soils against FT cycles is attributed to the sufficient contact points in the soil matrix and sufficient soil permeability (Cheng et al. 2016). This characteristic enhances the efficacy of MICP cementing agents in terms of bridging the particle-to-particle contacts and in the meanwhile allows a rapid water mass transfer in the soil matrix.

4.7 Approaches Other Than Urease-Driven MICP

Apart from the urease-driven MICP process for bio-cementation, bio-denitrification is also popular as it can increase the pH and induce precipitation of $CaCO_3$ (Hamdan et al. 2011). This process involves denitrifying bacteria, which reduces the nitrate ions to nitrogen gas using calcium salts of fatty acids (calcium acetate) as electron donor and carbon source. Although this process leads to calcium carbonate precipitation, it still requires further optimization for practical application purposes as the rate is rather slow compared to the urea hydrolysis process (van Paassen et al. 2010a). Furthermore, nitrogen gas a product caused due to the reduction of nitrate is known to cause dilation problem to the surrounding soil. However, He et al. (2013) proposed a novel approach of partial bio-desaturation of sand using nitrogen gas produced from nitrate bio-reduction (denitrification). This partial bio-desaturation method could mitigate earthquake-induced soil liquefaction. In combination with MICP bio-cementation, partial bio-desaturation via denitrification was proposed to be a new method of liquefaction mitigation alternative to colloidal silica grouting, bentonite suspension grouting, and air injection. Pham et al. (2016) tested a combined process of biological denitrification and MICP and found that more efficient conversion of nitrate/nitrite was achieved compared to denitrification without MICP precipitation.

Microalgae are also another possible microorganisms that can be used in bio-cementation via photosynthetic metabolism. It was found by Heath et al. (1995) that deposition of calcium minerals occurred as a result of photosynthesis, which caused an increase in pH of microenvironment to approximately pH 9.0. Kamennaya et al. (2012) suggested that in cyanobacteria, calcification may be associated with the carbon dioxide-(CO_2) concentrating mechanism (CCM), which allows the cells to raise the concentration of CO_2 at the site of the carboxylating enzyme ribulose 1, 5-bisphosphate carboxylase/oxygenase (Rubisco) up to 1000-fold over that in the surrounding medium. Dittrich et al. (2004) found that *Synechococcus cyanobacteria*, the eukaryotic *Mychonastes sp.*, and *Chlorella sp.* could induce the precipitation of $CaCO_3$.

5 MICP for Civil Engineering Applications Other Than Soil Improvement

Other than soil improvement, MICP can also be used in many other civil engineering applications, as follows:

Dust and wind erosion control: MICP-induced soil surface crust can diminish water infiltration rate and enhance water and wind erosion resistance (Maleki et al. 2016). Due to the formation of crystal bonding between the soil particles, bio-cementation can also be used to temporarily reduce the dust on unpaved roadways and other surfaces.

Bio-clogging for fractured rock permeability reduction: MICP offers an attractive alternative to traditional grouting methods for creating barriers to groundwater flow. Cuthbert et al. (2013) presented the first field trail applying MICP to reduce the fractured rock permeability in the subsurface. The results showed that a significant reduction in transmissivity of a single fracture over an area of several m^2 was achieved as a result of large quantities of in situ calcite precipitation.

Internal erosion control: MICP treatment can facilitate the reduction of internal erosion and volumetric contraction of sand-clay mixtures. Jiang et al. (2017) found that carbonate precipitation increases the erosion resistance of sand–clay mixtures by absorbing/coating fine particles directly and bridging the contacts of coarse particles.

Corrosion protection by surface bio-coating: MICP can induce a protective layer of calcium carbonate on concrete surface in a much shorter period, compared to that needed for a thin $CaCO_3$ covering layer formed naturally on the surface of cement-based materials (long-term exposure to air). The MICP-induced $CaCO_3$ coating layer on the surface of cement-based materials can be used to repair cracks in concrete, protect concrete from corrosion, or change the surface properties of recycled concrete aggregates (Ramachandran et al. 2001; De Muynck et al. 2008; Kim and Lee 2015; Shirakawa et al. 2011; van Tittelboom et al. 2010). It is noteworthy that only the MICP produced calcium carbonate crystals can adhered to the surface but chemically produced $CaCO_3$ cannot (Stabnikov et al. 2011).

Bio-self-healing of concrete: Bio-self-healing of concrete is concrete mixed with calcite-precipitating bacteria that can be awakened and thrived to produce $CaCO_3$ crystals when cracks occur. The microbially induced $CaCO_3$ crystals are capable of filling deep micro-cracks as well as restricting crack development (Jonkers et al. 2010; Wang et al. 2012). MICP-based bio-self-healing approach prevails to other treatment methods due to the efficient bonding capacity, compatibility with concrete composition, and suitability. As suggested by Seifan et al. (2016), bio-self-healing treatment provides more safer, sustainable, long-standing, and economical construction materials.

6 Conclusions

Bio-inspired technology has been undergoing rapid development over the last decade for civil engineering applications, from laboratory tests to several successful field trials. This technology is interdisciplinary and integrates several research disciplines including microbiology, geochemistry, material science, and geotechnical engineering. Many researchers have focused on the urease-driven microbially induced calcite precipitation (MICP), one of the most commonly used bio-inspired processes, for soil improvement and stabilization. However, the issue of the urea hydrolysis by-product of ammonia and energy required for the chemical reagents' production (i.e., bacteria, enzyme, urea, and calcium chloride) still serves as the main obstacle for large-scale application of this technique in civil engineering. The fundamental mechanism of an effective calcite ($CaCO_3$) precipitation and its related physical–environmental–biological conditions are still not well established. Thus, more research focusing on the optimizing of the MICP process at both the micro- and macro-levels is vital before the direct application of this technology in the field.

Compared to MICP approach, many other biogeochemical processes have received little attention and not yet been fully discovered. For example, partial desaturation by nitrogen gas through biological denitrification has shown promising result of mitigation soil liquefaction. However, the retention of nitrogen gas bubbles within soil matrix is still challenging. The long-term stability of the gas bubbles in sand during upward or horizontal flows of groundwater is the most important for practical applications of denitrification for sand clogging. Bio-clogging via in situ biopolymer generation shows potential in reduction in hydraulic conductivity and formation of hydraulic barriers. However, further investigation and development are required, such as effect of biodegradability of biopolymer on long-term engineering performance.

It is noteworthy that despite the relative success of bio-inspired technologies including MICP in soil stabilization and other geotechnical engineering applications, they are not expected to replace all conventional ground improvement techniques (DeJong et al. 2013). However, it should also be noted that these new and green interdisciplinary technologies offer a sustainable solution for a better tomorrow. Regardless of the current challenges, bio-inspired technologies including MICP have an unexploited potential for relieving some of the present concerns in addressing broad societal priorities, such as climate change, underground space, infrastructure, sustainability, energy, and economic stability (DeJong et al. 2013).

References

Achal V, Mukherjee A, Basu PC, Reddy MS (2009) Strain improvement of *Sporosarcina pasteurii* for enhanced urease and calcite production. J Ind Microbiol Biotechnol 36:981–988

Al Qabany A, Soga K (2013) Effect of chemical treatment used in micp on engineering properties of cemented soils. Géotechnique 63(4):331–339

Altermann W, Kazmierczak J, Oren A, Wright DT (2006) Cyanobacterial calcification and its rock-building potential during 3.5 billion years of earth history. Geobiology 4:147–166

Al-Thawadi S (2008) High strength in situ biocementation of soil by calcite precipitating locally isolated ureolytic bacteria. Ph.D. thesis. Mudroch University, Perth, Western Australia

Al-Thawadi S, Cord-Ruwisch R (2012) Calcium carbonate crystals formation by ureolytic bacterial isolated from Australian soli and sludge. J Adv Sci Eng Res (JASER) 2:12–26

Al-Thawadi S, Cord-Ruwisch R, Bouodina M (2012) Consolidation of sand particles by nanoparticles of calcite after concentrating ureolytic bacteria in situ. Int J Green Nanotechnol 4:28–36

Bang SS, Galinat JK, Ramakrishnan V (2001) Calcite precipitation induced by polyurethane-immobilized *Bacillus pasteurii*. Enzyme Microbial Technol 28:404–409

Bekheet IA, Syrett PJ (1977) Urea-degrading enzymes in algae. Brit Phycol J 12:137–143

Booth JL, Vishniac HS (1987) Urease testing and yeast taxonomy. Can J Microbiol 33(5):396–404

Burbank MB, Weaver TJ, Green TL, Williams BC, Crawford RL (2011) Precipitation of calcite by indigenous microorganisms to strengthen liquefiable soils. Geomicrobiol J 28:301–312

Castanier S, Le Metayer-Levrel G, Perthuisot JP (1999) Ca-carbonates precipitation and limestone genesis—the microbialgeologist point of view. Sediment Geol 126:9–23

Castanier S, Le Métayer-Levrel G, Orial G, Loubière JF, Perthuisot JP (2000) Bacterial carbonatogenesis and applications to preservation and restoration of historic property. In: Ciferri O, Tiano P, Mastromei G (eds) Of microbes and art. The role of microbial communities in the degradation and protection of cultural heritage. Kluwer Academic/Plenum Publisher, New York, pp 203–218

Cheng L, Cord-Ruwisch R (2012) In situ soil cementation with ureolytic bacteria by surface percolation. Ecol Eng 42:64–72

Cheng L, Cord-Ruwisch R (2013) Selective enrichment and production of highly urease active bacteria by non-sterile (open) chemostat culture. J Ind Microbiol Biotechnol 40:1095–1104

Cheng L, Cord-Ruwisch R (2014) Upscaling effects of soil improvement by microbially induced calcite precipitation by surface percolation. Geomicrobiol J 31:396–406

Cheng L, Shahin MA (2016) Urease active bio-slurry: a novel soil improvement approach based on microbially induced carbonate precipitation. Can Geotech J 53:1–10

Cheng L, Cord-Ruwisch R, Shahin MA (2013) Cementation of sand soil by microbially induced calcite precipitation at various degrees of saturation. Can Geotech J 50:1–10

Cheng L, Shahin MA, Mujah D (2016) Influence of key environmental conditions on microbially induced cementation for soil stabilisation. J Geotech Geoenviron Eng. https://doi.org/10.1061/(ASCE)GT.1943-5606.0001586

Cheng L, Shahin MA, Cord-Ruwisch R (2017) Surface percolation for soil improvement by bio-cementation utilising in situ enriched indigenous aerobic and anaerobic ureolytic soil microorganisms. Geomicrobiol J 34(6):546–556

Chou CW, Seagren EA, Aydilek AH, Lai M (2011) Biocalcification of sand through ureolysis. J Geotech Geoenviron Eng 137(12):1179–1189

Chu J, Ivanov V, Naeimi M, Stabnikov V, Liu HL (2014) Optimization of calcium-based bioclogging and biocementation of sand. Acta Geotech 9:277–285

Cunningham AB, Gerlach R, Spangler L, Mitchell AC, Parks S, Phillips A (2009) Reducing the risk of well bore leakage of co_2 using engineered biomineralization barriers. Energy Procedia 4:5178–5185

Cuthbert MO, McMillan LA, Handley-SidhuS Riley MS, Tobler DJ, Phoenix VR (2013) A field and modeling study of fractured rock permeability reduction using microbially induced calcite precipitation. Environ Sci Technol 47:13637–13643

Daskalakis MI, Rigas F, Bakolas A, Magoulas A, Kotoulas G, Katsikis I, Karageorgis AP, Mavridou A (2015) Vaterite bio-precipitation induced by *Bacillus pumilus* isolated from a solutional cave in Paiania, Athens, Greece. Int Biodeterior Biodegrad 99:73–84

De Groot MB, Meijers P (1992) Liquefaction of trench fill around a pipeline in the seabed. BOSS 92: behaviour of offshore structures. BPP Technical Services, London, pp 1333–1344

De Muynck W, Cox K, Verstraete W, De Belie N (2008) Bacterial carbonate precipitation as an alternative surface treatment for concrete. Constr Build Mater 22:875–885

De Muynck W, De Belie N, Verstraete W (2010) Microbial carbonate precipitation in construction materials: a review. Ecol Eng 36:118–136

DeJong J, Fritzges M, Nusstein K (2006) Microbially induced cementation to control sand response to undrained shear. J Geotech Geoenviron Eng 32:1381–1392

DeJong JT, Mortensen BM, Martinez BC, Nelson DC (2010) Bio-mediated soil improvement. Ecol Eng 36:197–210

DeJong JT, Soga K, Kavazanjian E, Burns S, van Paassen LA, Al Qabany A et al (2013) Biogeochemical processes and geotechnical applications: progress, opportunities and challenges. Geotechnique 63:287–301

Dittrich M, Kurz P, Wehrli B (2004) The role of autotrophic picocyanobacteria in calcite precipitation in an oligotrophic lake. Geomicrobiol J 21:45–53

Douglas S, Beveridge TJ (1998) Mineral formation by bacteria in natural microbial communities. FEMS Microbiol Ecol 26(2):79–88

Duraisamy Y, Airey DW (2012) Strength and stiffness of bio-cemented liquefiable sand soil. In: Proceedings of international conference on ground improvement and ground control, Singapore, pp 1233–1239

Ehrlich HL (1996) How microbes influence mineral growth and dissolution. Chem Geol 132(1–4):5–9

Ehrlich HL (1998) Geomicrobiology: its significance for geology. Earth-Sci Rev 45:45–60

Evans J, Wallace C, Dobrowolski N (1993) Interaction of soil type and temperature on the survival of *Rhizobium leguminosarum bv.viciae*. Soil Biol Biochem 25:1153–1160

Falkinham JO III, Hoffman PS (1984) Unique developmental characteristics of the swarm and short cells of *Proteus vulgaris* and *Proteus mirabilis*. J Bacteriol 158(3):1037–1040

Fujita Y, Ferris FG, Lawson RD, Colwell FS, Smith RW (2000) Subscribed content calcium carbonate precipitation by ureolytic subsurface bacteria. Geomicrobiol J 17:305–318

Ginn TR, Murphy EM, Chilakapati A, Seeboonruang U (2001) Stochastic-convective transport with nonlinear reaction and mixing: Application to intermediate-scale experiments in aerobic biodegradation in saturated porous media. J Contam Hydrol 48:121–149

Gomez MG, Martinez BC, DeJong JT, Hunt CE, deVlaming LA, Major DW, Dworatzek SM (2015) Field-scale bio-cementation tests to improve sands. Ground Improv 168(GI3):206–216

H. John Heinz III Centre for Science Economics and the Environment (2000) Evaluation of erosion hazards—contract EMW-97-CO-0375. Federal Emergency Management Agency (FEMA), USA

Hamdan N, Kavazanjian E, Rittman BE, Karatas I (2011) Carbonate mineral precipitation for soil improvement through microbial denitrification. In: Han J, Alzamora DA (eds) Geo-frontiers 2011, Dallas, Texas, pp 3925–3934

Hammes F, Verstraete W (2002) Key roles of pH and calcium metabolism in microbial carbonate precipitation. Rev Environ Sci Biotechnol 1:3–7

Hammes F, Seka A, De Knijf S, Verstraete W (2003) A novel approach to calcium removal from calcium-rich industrial wastewater. Water Res 37:699–704

Harkes MP, van Paassen LA, Booster JL, Whiffin VS, van Loosdrecht MCM (2010) Fixation and distribution of bacterial activity in sand to induce carbonate precipitation for ground reinforcement. Ecol Eng 36:112–117

Hausinger RP (1987) Nickel utilization by microorganisms. Microbiol Rev 51:22–42

He J, Chu J, Ivanov V (2013) Mitigation of liquefaction of saturated sand using biogas. Geotechnique 63(4):267–275

Heath CR, Leadbeater BSC, Callow ME (1995) Effects of inhibitors on calcium carbonate deposition mediated by freshwater algae. J Appl Phycol 7:367–380

Ivanov V, Chu J (2008) Applications of microorganisms to geotechnical engineering for bioclogging and biocementation of soil in situ. Rev Environ Sci Biotechnol 7:139–153

Jahns T (1996) Ammonium/urea-dependent generation of a proton electrochemical potential and synthesis of ATP in *Bacillus pasteurii*. J Bacteriol 178:403–409

Jiang NJ, Soga K, Kuo M (2017) Microbially induced carbonate precipitation (MICP) for seepage-induced internal erosion control in sand-clay mixtures. J Geotech Geoenviron Eng 143 (3):04016100

Jonkers HM, Thijssen A, Muyzer G, Copuroglu O, Schlangen E (2010) Application of bacteria as self-healing agent for the development of sustainable concrete. Ecol Eng 36:230–235

Kamennaya NA, Ajo-Franklin CM, Northen T, Jansson C (2012) Cyanobacteria as biocatalysts for carbonate mineralization. Mineral 2:338–364

Kim HK, Lee HK (2015) A case study: bacterial surface treatment of normal and lightweight concrete. In: Pacheco-Torgal F, Labrincha JA, Diamanti MV, Yu CP, Lee HK (eds) Biotechnologies and biomimetics for civil engineering. Springer, pp 359–372

Lee ML, Ng WS, Tanaka Y (2013) Stress-deformation and compressibility responses of bio-mediated residual soils. Ecol Eng 60:142–149

Maleki M, Ebrahimi S, Asadzadeh F, Tabrizi ME (2016) Performance of microbial-induced carbonate precipitation on wind erosion control of sandy soil. Int J Environ Sci Technol 13 (3):937–944

McConnaughey TA, Whelan JF (1997) Calcification generates protons for nutrient and bicarbonate uptake. Earth-Sci Rev 42:95–117

Mobley HLT, Hausinger RP (1989) Microbial ureases: significance, regulation and molecular characterisation. Microbiol Rev 53(1):85–108

Mobley HLT, Island MD, Hausinger RP (1995) Molecular biology of microbial ureases. Microbiol Rev 59(3):451–480

Montoya BM, DeJong JT (2015) Stress-strain behavior of sands cemented by microbially induced calcite precipitation. J Geotech Geoenviron Eng 141:04015019

Mote IT, Dismuke NJ (2011) Screeing-level liquefaction hazard maps for Australia. In: Australian earthquake engineering society 2011 conference, 18–20 Nov

Mujah D, Shahin MA, Cheng L (2017) State-of-the-art review of bio-cementation by microbially induced calcite precipitation (MICP) for soil stabilization. Geomicrobiol J 34(6):524–537

Okwadha GD, Li J (2010) Optimum conditions for microbial carbonate precipitation. Chemosphere 81(9):1143–1148

Parmar N, Singh A (2014) Geobiotechnology. In: Parmar N, Singh A (eds) Geomicrobiology and biogeochemistry. Springer, Germany, pp 1–15

Pham VP, Nakano A, van der Star WRL, Heimovaara TJ, van Paassen LA (2016) Applying MICP by denitrification in soils: a process analysis. Environ Geotech. https://doi.org/10.1680/jenge.15.00078

Ramachandran SK, Ramakrishnan V, Bang SS (2001) Remediation of concrete using microorganisms. ACI Mater J 98:3–9

Reddy S, Rao M, Aparna P, Sasikala C (2010) Performance of standard grade bacterial (*Bacillus subtilis*) concrete. Asian J Civil Eng (Build Housing) 11:43–55

Romano N, Tolone G, Ajello F, La Licata R (1980) Adenosine 5'-triphosphate synthesis induced by urea hydrolysis in ureaplasma urealyticum. J Bacteriol 144:830–832

Seifan M, Samani AK, Berenjian A (2016) Bioconcrete: next generation of self-healing concrete. Appl Microbiol Biotechnol 100:2591–2602

Seneca H, Peer P, Nally R (1962) Microbial Urease. Nature 193:1106–1107

Shirakawa MA, Cincotto MA, Atencio D, Gaylarde CC, John VM (2011) Effect of culture medium on biocalcification by *Pseudomonas putida*, *Lysinibacillus sphaericus* and *Bacillus subtilis*. Braz J Microbiol 42:499–507

Sissons CH, Perinpanaygan ER, Hancock EM, Cutress TW (1990) pH regulation of urease level in *Streptococus salivarius*. J Dent Res 69:1131–1137
Stabnikov V, Naeimi M, Ivanov V, Chu J (2011) Formation of water-impermeable crust on sand surface using biocement. Cem Concr Res 41:1143–1149
Stocks-Fischer S, Galinat JK, Bang SS (1999) Microbiological precipitation of $CaCO_3$. Soil Biol Biochem 31:1563–1571
Summer JB (1926) The isolation and crystallization of the enzyme urease. J Biol Chem 69:435–441
Tobler DJ, Cuthbert MO, Phoenix VR (2014) Transport of *Sporosarcina pasteurii* in sandstone and its significance for subsurface engineering technologies. Appl Geochem 42:38–44
Tyler B (1978) Regulation of the assimilation of nitrogen compounds. Annu Rev Biochem 47:1127–1162
van Paassen LA, Daza CM, Staal M, Sorokin DY, van der Zon WH, van Loosdrecht MCM (2010a) Potential soil reinforcement by biological denitrification. Ecol Eng 36(2):168–175
van Paassen L, Ghose R, van der Linden T, van der Star W, van Loosdrecht M (2010b) Quantifying biomediated ground improvement by ureolysis: large-scale biogrout experiment. J Geotech Geoenviron Eng 136:1721–1728
van Tittelboom K, De Belie N, De Muynck W, Verstraete W (2010) Use of bacteria to repair cracks in concrete. Cem Concr Res 40:157–166
Wang JY, De Belie N, Verstraete W (2012) Diatomaceous earth as a protective vehicle for bacteria applied for self-healing concrete. J Ind Microbiol Biotechnol 39:567–577
Warren LA, Maurice PA, Parmar N, Ferris FG (2001) Microbially mediated calcium carbonate precipitation: implications for interpreting calcite precipitation and for solid-phase capture of inorganic contaminants. Geomicrobiol J 18:93–115
Whiffin VS, van Paassen LA, Harkes MP (2007) Microbial carbonate precipitation as a soil improvement technique. Geomicrobiol J 24:417–423
Yasuhara H, Neupane D, Hayashi K, Okamura M (2012) Experiments and predictions of physical properties of sand cemented by enzymatically-induced carbonate precipitation. Soils Found 52 (3):539–549
Zhao Q, Li L, Li C, Li MD, Amini F, Zhang HZ (2014) Factors affecting improvement of engineering properties of micp-treated soil catalyzed by bacteria and urease. J Mater Civ Eng 26(12):04014094

Utilization of Microbially Induced Calcite Precipitation for Sand Solidification Using *Pararhodobacter* sp.

G. G. N. N. Amarakoon and Satoru Kawasaki

Abstract Biomineralization is an environment-friendly technology to improve engineering properties of soil. One of common bio-mineralization processes is microbially induced calcite precipitation (MICP). In this chapter, sand solidification tests were conducted using *Pararhodobacter* sp., which is a local ureolytic bacteria isolated from the sand near beachrock in Okinawa, Japan. The goal of this paper is to perform solidification of the specimen having an estimated unconfined compressive strength (UCS) of more than several MPa for soil improvement and to investigate the influence of various factors on engineering properties (i.e., curing temperature, injection interval of cementation solution, Ca^{2+} concentration, curing time, bacterial population, re-injection of bacteria and particle size of sand) of treated soils catalyzed by ureolytic bacteria. The result of estimated UCS value showed that all the studied factors have an obvious effect on the MICP-treated sand. More than 3 MPa of the estimated UCS value was obtained from the solidified samples, and also, it was obtained more than 10 MPa of estimated UCS value for the testing cases of changing concentration of cementation media and re-injection of the bacterial solution after 7 days of curing period. UCS, SEM-EDX, X-ray CT, $CaCO_3$ content of the sample and color measurement tests were conducted. Completely solidified samples were obtained by changing different testing conditions. The results indicate that the average estimated UCS value varied from 3.1 to

Part of this book chapter has been published as: G. G. N. N. Amarakoon and S. Kawasaki (2018) Factors affecting sand solidification using MICP with *Pararhodobacter* sp. Materials Transactions 59 (1) 72–81. The Mining and Materials Processing Institute of Japan granted permission for the same to use in this book.

G. G. N. N. Amarakoon (✉)
Graduate School of Engineering, Hokkaido University, Kita 13, Nishi 8, Kita-ku, Sapporo, Hokkaido 060-8628, Japan
e-mail: nishanirmi@gmail.com

S. Kawasaki
Faculty of Engineering, Hokkaido University, Kita 13, Nishi 8, Kita-ku, Sapporo, Hokkaido 060-8628, Japan

V. Achal and A. Mukherjee (eds.), *Ecological Wisdom Inspired Restoration Engineering*, EcoWISE, https://doi.org/10.1007/978-981-13-0149-0_4

4.4 MPa. Overall, the results of this study will contribute to the application of a new technique for sand improvement and bio-stimulation.

Keywords Ureolytic bacteria · Microbially induced calcite precipitation
Unconfined compressive strength · Scanning electron microscopy
Energy-dispersive X-ray spectrometry

1 Introduction

Present soil improvement applications comprise soil replacement, preloading for consolidation, chemical admixture, and grouting stabilization. These techniques are time-consuming, expensive, and environmentally harmful (DeJong et al. 2010). In addition, coastal erosion is a significant problem throughout the world.

Breakwater construction is used for preventing coastal erosion. Production of cement, which is a major construction material for breakwater construction, is energy-consuming and environmentally unfriendly. During the process of cement production, it releases a large amount of CO_2. In addition, the process is time-consuming and also cement involves heating ingredient up to a temperature of about 1500 °C. Therefore, additional studies to discover alternative techniques for soil improvement are vital for achieving optimum performance, economic viability, and environmental sustainability.

Biomineralization is a promising and environmentally innocuous technology to improve soil engineering properties. It naturally happens and is induced by non-pathogenic organisms that are native to the soil environment (DeJong et al. 2006). It consists of eco-friendly cement material and gives the advantages of less energy to produce and less release of CO_2 (Khanafari et al. 2011). Therefore, these advantages are vital for achieving ecological balance and environmental sustainability. One common biomineralization process is microbially induced calcite precipitation (MICP), which can bind sand grains together and improve the engineering properties of sand. MICP is an effective method for the removal of CO_2 from the environment (Ferris et al. 1994; Mitchell et al. 2010). In this method, CO_2 is converted into carbonate minerals that can form different crystals such as calcite, aragonite, dolomite, and magnesite. This method is safer and more eco-friendly than conventional methods of sequestering CO_2 from the atmosphere.

Improvement of soil mechanical properties by MICP is currently of particular interest to engineers and microbiologists and has been demonstrated by several researchers at varying scales (DeJong et al. 2006; Whiffin et al. 2007; Van Paassen et al. 2010). The technique can alter soil characteristics to increase shear strength and stiffness while maintaining adequate permeability (Burbank et al. 2011). The technique involves introducing aerobically cultivated bacteria with highly active urease enzyme into soil, harnessing the urease enzyme to catalyze the hydrolysis of urea to produce ammonium and carbonate ions. The chemical reaction involved in this process is shown as follows (Eq. (1)):

$$CO(NH_2)_2 + 2H_2O \rightarrow 2NH_4^+ + CO_3^{2-} \quad (1)$$

In the presence of an introduced calcium source, often calcium chloride ($CaCl_2$), the calcium carbonate ($CaCO_3$, calcite) forms throughout the soil matrix based on the following chemical reaction (Eq. (2)):

$$Ca^{2+} + CO_3^{2-} \rightarrow CaCO_3 \downarrow \quad (2)$$

The produced microbially induced $CaCO_3$ precipitates bridge adjacent soil particles by cementing the soil grains together to form cemented sand illustrative of calcareous rock (DeJong et al. 2006).

The engineering properties of MICP-treated soil may vary because MICP is a complex biochemical process, which can be affected by many factors. The MICP contains two key steps, as above equations. The urea hydrolysis is mainly dependent on the concentration of ureolytic bacteria and the available substrate (e.g., urea), whereas calcite precipitation relates to available Ca^{2+} (Mortensen et al. 2011). In accordance with the growth of nutrient concentration and incubation time, the $CaCO_3$ content increases. The particle size also has an effect on MICP-bonded soil. The efficiency of MICP is related to the permeability of the soil being sufficient to allow chemicals to flow to the bacteria, and also, the cement effect of $CaCO_3$ precipitation away particles (Mitchell and Santamarina 2005; Rebata-Landa 2007). Rebata-Landa (2007) showed a relation between grain size and $CaCO_3$ content; maximum carbonate deposition observed on grains was approximately 100 μm in size. Qabany et al. (2012) also found well-graded and coarser sands had a higher rate of precipitation than finer and poorly graded soils.

In this study, solidification tests (small size: syringe test and middle size: model test) were conducted on silica sand using the ureolytic bacteria; *Pararhodobacter* sp. that was isolated from the soil near beachrock in Sumuide, Nago, Okinawa, Japan (Fig. 1). The goal of this study was to solidify a specimen having an

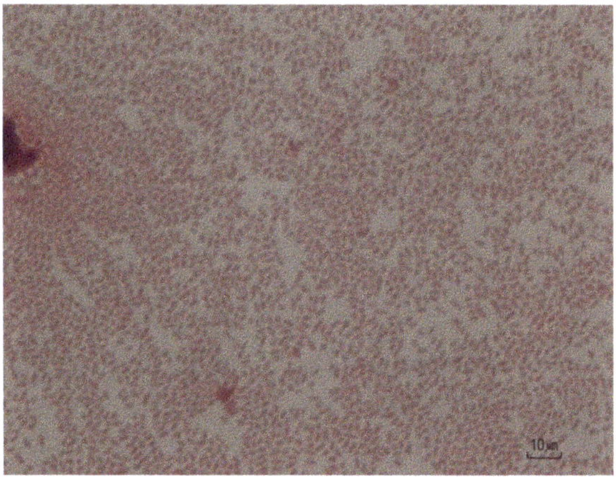

Fig. 1 Grain stain of *Pararhodobactor* sp.

estimated unconfined compressive strength (UCS) of more than several MPa to improve soil properties and investigate the influence of various factors on the engineering properties of treated soil catalyzed by ureolytic bacteria.

2 Syringe Solidification Test Using MICP Method

2.1 Methodology

First, 100 mL NH_4–YE medium solution ((growth media; ATCC 1376), which contained the deionized water consisting of 0.13 M tris buffer (pH = 9.0), 10 g $(NH_4)_2SO_4$, and 20 g yeast extract), was mixed with 0.1 g of *Pararhodobacter* sp. (Fig. 1) and then incubated at 30 °C with gentle shaking at 160 rpm for 72 h. Next, 45 g of silica sand (e.g., Mikawa sand, 97.77% quartz, Table 1) was dried at 110 ° C for 2 days and then placed in a 35 mL syringe (diameter, ϕ = 2.5 cm). Subsequently, 16 mL (more than the estimated 14 mL initial pore volume in the sample) of the bacterial culture and 20 mL of the cementation solution (Table 2) were sequentially added to the syringe and drained, leaving about 2 mL of solution above the surface of the sand to maintain the wet conditions. After curing, the cementation solution was added and drained at fixed intervals. The Ca^{2+} concentration and pH of the drainage were also measured to determine temporal variations of these parameters in the samples. Each test condition is described in Sect. 2.2.

After 14 days of curing, the needle penetration inclination (N_P) of each syringe sample (ϕ = 2.5 cm, height, h = 7 cm) was measured using a needle penetration device (SH-70, Maruto Testing Machine Company, Tokyo, Japan), and the UCS was estimated from the N_P according to the following regression equation.

Table 1 Physical properties of sand types

Sand type	Mizunami sand	Mikawa sand	Toyoura sand
Soil particle density (ρ_s) (g/cm^3)	2.67	2.66	2.64
Minimum density (ρ_{min}) (g/cm^3)	1.348	1.256	1.335
Maximum density (ρ_{max}) (g/cm^3)	1.491	1.476	1.645
Mean diameter (D_{50}) (μm)	1200	600	200

Table 2 Chemical compositions for cementation media

Chemical	Chemical concentration (g/L)—0.5 M Ca
Nutrient broth (g)	3
NH_4Cl (g)	10
$NaHCO_3$ (g)	2.12
$(NH_2)_2CO$ (g)	30.03
$CaCl_2$ (g)	55.49

$$\log(y) = 0.978\log(x) + 2.621 \quad (3)$$

where the correlation coefficient is 0.941, x is N_P (N/mm), and y is UCS (MPa).

2.2 Test Conditions

In the syringe test, 15 test cases were conducted, as shown in Table 3. The test cases were summarized according to the investigation purpose. In case 1–3, the effect of a bacterial population was investigated. In cases 4 and 5, the effect of re-injection of bacteria was identified. In cases 3, 6, and 7, the effect of curing time was investigated. Cases 3, 8, and 9 were also used to investigate the effect of temperature. Moreover, cases 10 and 11 were used to identify the effect of the injection interval of the cementation media; to investigate the effect of cementation media concentration, cases 10, 12, and 13 were used. Cases 3, 14, and 15 were used to investigate the effect of particle size on the solidification.

2.3 Results

2.3.1 Bacterial Population and Re-injection of Bacteria

Figure 2a shows that the estimated UCS of the sample increases with an increase in bacterial population. This finding indicates that bacteria play a key role in MICP, i.e., (1) producing an enzyme to hydrolyze urea and (2) acting as nucleation sites for the formation of $CaCO_3$ crystals (mainly calcite; Fujita et al. 2000). More bacteria in the solution will increase the concentration of enzymes and provide more nucleation sites for the MICP.

Figure 2b shows that the estimated UCS value is larger when the bacteria solution is re-injected after 7 days than when it is not. The reason was that the pH value could be controlled when bacteria were re-injected. The optimal growth for *Pararhadobacter* sp. is achieved at 30–40 °C and pH 7.0–8.5 (Foesel et al. 2011). Moreover, a low concentration of Ca^{2+} concentration also was maintained, which means the calcite precipitation was increased when bacteria were re-injected.

Table 3 Experimental conditions for syringe solidification test

Case no.	Temp. (° C)	Population of bacteria (g)	Injection	Re-injection of bacteria (after 7 days)	Curing days	Concentration of cementation media	Sand material with particle size (mm)
1	30	0.1 with centrifuge	1		14	0.5 M	Mikawa-0.6
2		0.3 with centrifuge					
3		1.0 with centrifuge					
4		0.1 with centrifuge		x			
5		1.0 with centrifuge		x			
6					7		
7					21		
8	25				14		
9	35						
10	30						
11			2				
12			1			0.3 M	
13						0.7 M	
14						0.5 M	Mizunami-1.2
15							Toyoura-0.2

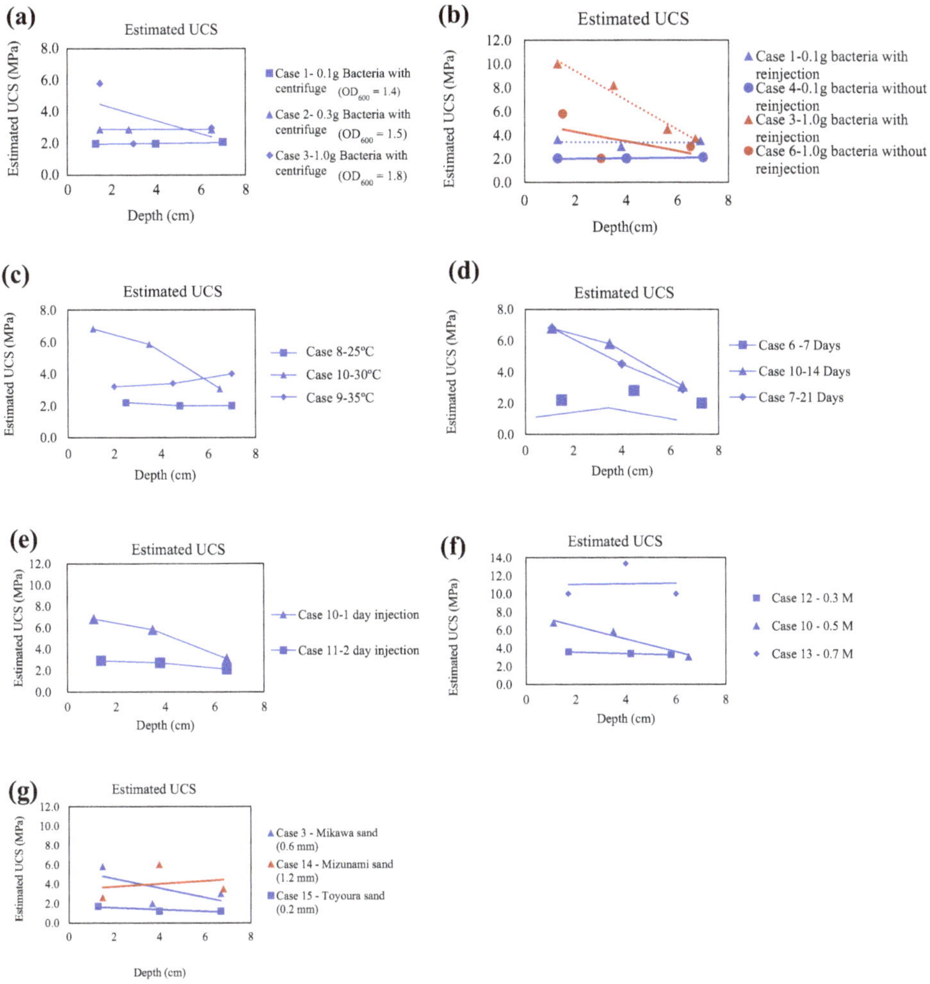

Fig. 2 Results of MICP-treated sample catalyzed by *Pararhodobacter* sp. under different conditions. Effect of **a** bacterial population, **b** re-injection of bacteria, **c** curing temperature, **d** curing time, **e** injection interval, **f** concentration of cementation media, and **g** particle size of sand

2.3.2 Curing Temperature and Curing Time

Figure 2c shows the effect of temperature on estimated UCS. In this investigation, three test cases were used: 25 °C, 30 °C, and 35 °C, and the average estimated UCS values were 2.1 MPa, 5.2 MPa, and 3.5 MPa, respectively. Thus, it is clear that the estimated UCS value was higher at 30 °C than at 25 °C or 35 °C.

Figure 2d shows the results of experiments conducted for the following curing periods: 7 days, 14 days, and 21 days. The average estimated UCS of the sample after a curing period of 7 days (2.3 MPa) was less than average estimated UCS values after 14 and 21 days (average estimated UCS 5.2 MPa and 4.7 MPa, respectively).

2.3.3 Injection Interval and Concentration of Cementation Media

For the two test cases, cementation media was added to the syringe every day or after every 2 days (case 10 and case 11, respectively). In this study, the strength of the sample prepared by the daily addition of cementation media (average estimated UCS 5.2 MPa) was larger than that of the sample prepared by adding cementation media at 2-day intervals (2.6 MPa) for 14 days (Fig. 2e). This was because the Ca^{2+} concentration of the cementation media ($CaCl_2$) was larger in case 10 than in case 11. However, with the depth of the syringe sample (top to bottom), the strength reduced.

Figure 2f shows the results using different concentrations of cementation media for the solidification of syringe samples. The strength of the sample increased with an increase in the concentration of the cementation media solution. Here, 0.3, 0.5, and 0.7 M concentrations were used. If the concentration of the cementation media is increased, the rate of $CaCO_3$ precipitation increases, as does the estimated UCS.

2.3.4 Particle Size

From the results (Fig. 2g), the average estimated UCS value (4 MPa) was larger in the Mizunami sand with 1.2 mm mean diameter than in the Mikawa sand sample, which has a mean diameter 0.6 mm (3.6 MPa), and the Toyoura sand sample with a mean diameter 0.2 mm (1.4 MPa). The particle size of Toyoura sand is smaller than that of the other two sand samples, which may decrease the bacterial and hydrolysis processes. As a result, $CaCO_3$ precipitation will decrease and the UCS value will be lower in the Toyoura sand sample than in the Mikawa or Mizunami sand samples. Therefore, the particle size of the sample mainly affects the solidification of the sample, and the applicable range of sand particle size is 0.6–1.2 mm (mean diameter) according to our study.

2.4 Discussion

2.4.1 Suggested Formula for Prediction of Estimated UCS

Multiple regression analysis was conducted in this study to analyze the relative importance of each test condition to the estimated UCS and to determine experimentally a formula that can predict estimated UCS as a useful reference for future cementation tests and field tests.

Table 4 Results of the multiple regression analysis of data from the syringe solidification test

	Coefficients	Standard error	t stat	P value
Intercept	2.259	1.198	1.8857	0.0887
Bacterial population (g) $\boldsymbol{B_p}$	3.370	2.519	1.3381	0.2105
Concentration of cementation media (M) $\boldsymbol{C_{ca}}$	9.749	3.164	3.0807	0.0116
Curing time (Days) $\boldsymbol{D}$	0.231	0.109	2.1251	0.0595
Injection interval (Days) $\boldsymbol{I_i}$	1.165	1.184	0.9839	0.3484
Particle size (mm) $\boldsymbol{M}$	1.175	1.888	0.6224	0.5476
Temperature (°C) $\boldsymbol{T}$	0.061	0.056	1.0873	0.3024

In this study, multiple regression analysis was conducted using the results of the syringe solidification. In this analysis, the syringe test conditions were set as explanatory variables and the measured UCS was the objective variable. The following relational expression (Eq. (4)) and Table 4 were generated by this analysis.

Equation for estimated UCS value is as follows:

$$q_{eu} = 3.37B_p + 9.75C_{ca} + 0.23D + 1.16I_i + 1.17M + 0.06T + 2.26 \quad (4)$$

where q_{eu} = Estimated UCS (MPa)
B_p = Bacterial population (g)
C_{ca} = Concentration of cementation media (M)
D = Curing time (Days)
I_i = Injection interval (Days)
M = Particle size (mm)
T = Temperature (°C)

In Table 4, the partial regression coefficient indicated the coefficient of each multiple regression equations, which is set so that the theoretical value is close to the measured value. Additionally, the standard error was determined as follows: Only one set of all intended experiments was conducted, but it was assumed that several sets were conducted. The frequency distributions of the partial regression coefficients and the constant term were then obtained by multiple regression analyses against each set of all intended experiments. The standard deviation of the normally distributed histogram obtained by calculating the frequency distributions is the standard error. The t value was obtained by dividing the partial regression coefficient by the standard error. From the t values, the degree of importance of each explanatory variable to the objective variable can be judged. The P value was twice as much as the upper probability of the t value on the t distribution. In this study, a $P \leq 0.01$ and $0.01 < P \leq 0.05$ was considered to indicate that both explanatory variables are important, because the significance levels were 1% and 5%, respectively, while a $P > 0.05$ indicated that the explanatory variable was not important.

Table 5 Results of the multiple regression re-analysis of data from the syringe solidification test

	Coefficients	Standard error	t stat	P value
Intercept	−0.09	1.38	−0.0617	0.9547
Concentration of cementation media (M) $\boldsymbol{C_{ca}}$	13.99	2.81	4.9696	0.0157
Curing time (Days) $\boldsymbol{D}$	0.37	0.10	3.9071	0.0298

As shown in Table 4, the concentration of cementation media, curing time, and injection interval were significant. Based on each t value, the concentration of cementation media had the highest degree of importance followed by curing time, injection interval, bacterial population, curing temperature, and particle size.

To suggest a more reliable formula for prediction of the UCS than Eq. (4), the more important explanatory variables against the objective variable were selected. Because the P values of the concentration of cementation media, curing time, injection interval was less than 0.05. These conditions were selected as explanatory variables, and the UCS of specimens generated at a curing temperature of 30 °C, 1.0 g of bacterial population and 0.6 mm particle size diameter (Mikawa sand) were used as objective variables. The results of this multiple regression re-analysis are shown in Eq. (5) and Table 5.

$$q_{eu} = 13.99C_{ca} + 0.37D - 0.09 \quad (5)$$

The P values of three explanatory variables (Table 5) were lower than 0.05. Therefore, Eq. (5) can be considered a reliable formula for prediction of UCS. However, it should be noted that this formula is only reliable for samples generated using the curing temperature, bacterial population, and particle size of the sand material described above. In addition, the conditions of the explanatory variables can change within the range of syringe tests. Hence, further cementation tests need to be performed to develop a reliable formula for prediction of UCS under more varied conditions. Nevertheless, Eq. (5) will be useful for further cementation tests and field tests.

3 Model Test for Sand Solidification Using MICP Method

3.1 Methodology

In this study, three test cases were carried out. All test cases were conducted at room temperature (25 °C), and cementation media was added daily. A plastic container (length: 20 cm × width: 12.5 cm × height: 14 cm) with a permeable plastic plate at the bottom was used according to Fig. 3, and the methodology was

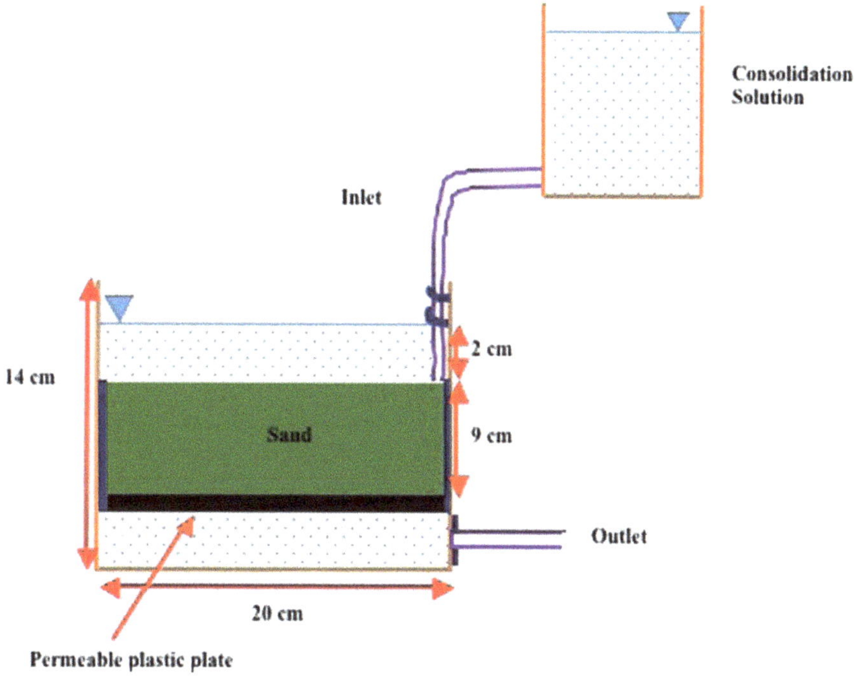

Fig. 3 Concept of the laboratory model experiment for sand solidification using ureolytic bacteria

the same as that for the syringe test. Here, 3320 g of Mikawa sand or 3355 g of Mizunami and 1500 mL of cementation media were used. After solidification was achieved, the sample was opened and cylindrical samples with the diameter of 3 cm and a height of 6 cm were cored. The following tests were then conducted on the cored samples: the UCS test, the needle penetration test, X-ray deflection (XRD), scanning electron microscopy (SEM) (Super Scan SS-550, Shimadzu Corporation, Kyoto, Japan), energy-dispersive X-ray spectroscopy (EDX) (SEDX-500, Shimadzu Corporation), and X-ray computed tomography (X-CT). $CaCO_3$ content was also determined. In addition, the sample color was measured by colorimeter to identify the effect of color on strength of the sample.

3.2 *Test Conditions*

In the model test study, three test cases were carried out by changing curing time and the particle size of the sand material (Table 6).

Table 6 Experiment conditions for model test

Test case no.	Temperature (°C)	Injection interval 1 Day	Curing time	Population of bacteria (g)	Adding bacteria (after 7 days)	Height of sample (cm)	Sand material (particle size mm)
1	25	x	14	13		9	Mikawa (0.6)
2			21		x		
3			14				Mizunami (1.2)

3.3 *Results*

3.3.1 pH and Ca^{2+} Concentration

For test cases 1 and 2 (Table 6), pH decreased with time. However, after 7 days of curing, the pH of test case 2 increased and maintained a value of 7 due to re-injection of bacteria (Fig. 4a). It was very important to maintain the pH at 7 because the optimal pH for *Pararhodobacter* sp. is 7–8. If the pH value is maintained between 7 and 8, the bacterial activity remains high and increases the rate of hydrolysis process, and this gives finally the high $CaCO_3$ precipitation. The results of the Ca^{2+} concentration of the outlet solution showed the evidence for high precipitation of $CaCO_3$. Ca^{2+} concentration increased with time when no bacterial solution was added after 7 days in test case 1 (Fig. 4b). However, in test case 2, the concentration of Ca^{2+} in the outlet solution was small during the curing period.

3.3.2 Color Observation with the Time

Using a colorimeter, color measurements were obtained at four phases of the sample box, as shown in Fig. 5. The measurements were taken 2 cm each from the top and bottom levels of the sample sand layer. Four measurements were taken at each point, and the average value was obtained. With time, mostly at each point, the color (ΔL*) increased (Fig. 6).

In addition, for test case 3 (Mizunami sand sample), color was measured by colorimeter at each point where the NPT were conducted. Figure 7 shows the results of estimated UCS and color *L** of the sample. The color (*L**) of the sample is increased with an increase of estimated UCS. Therefore, there was a close relationship between color *L** and estimated UCS value.

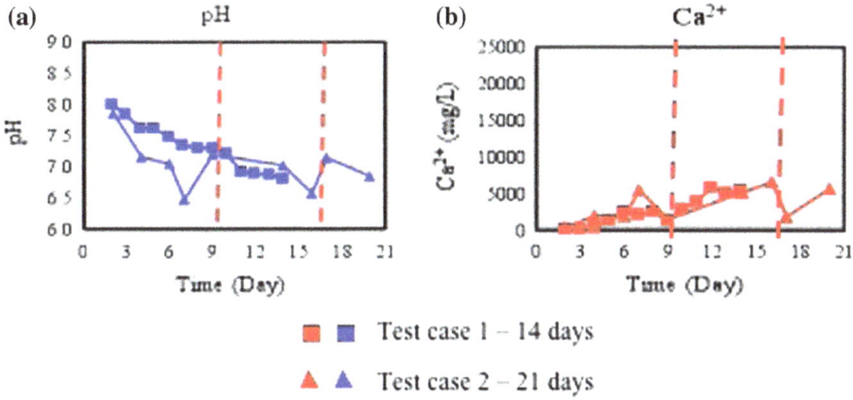

Fig. 4 **a** Changing pH value with the time and **b** Ca^{2+} concentration of the outlet solution of the test case 1 and test case 2

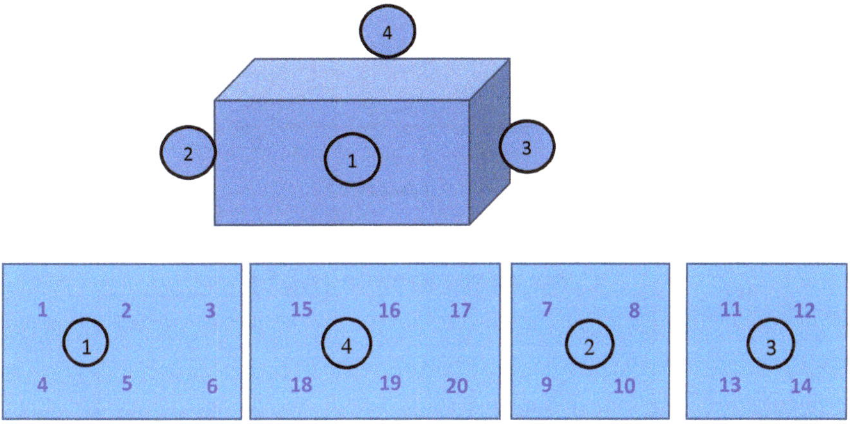

Fig. 5 Measurement point of the color using a colorimeter

3.3.3 Estimated UCS Results

The solidified samples (Fig. 8) were cored vertically and horizontally. The cored samples were cylindrical shaped with the diameter of 3 cm and the height of 6 cm. The strength decreased with the depth in vertically cored samples, as shown in Fig. 9a and b. Moreover, the estimated UCS value of the samples varied from 10 to 3 MPa. In addition, the strength of horizontally cored samples was nearly same as that of the vertically cored samples or the differences in strength were small (Fig. 9c).

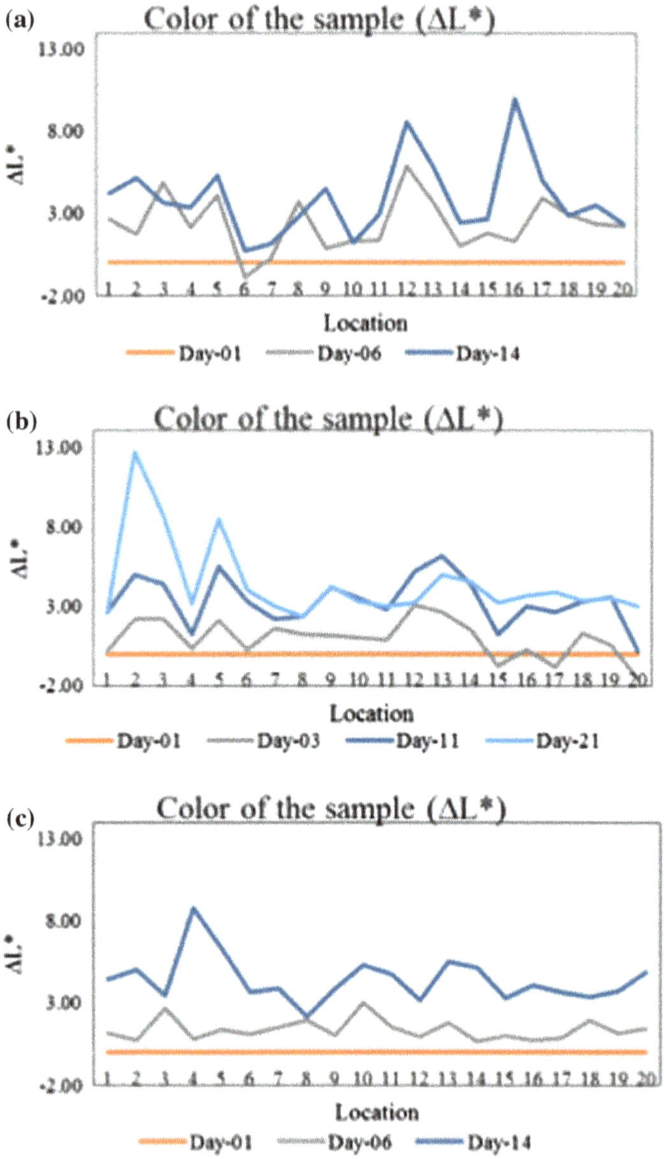

Fig. 6 Results of color of the 20 points of the samples; **a** test case 1, **b** test case 2, and **c** test case 3

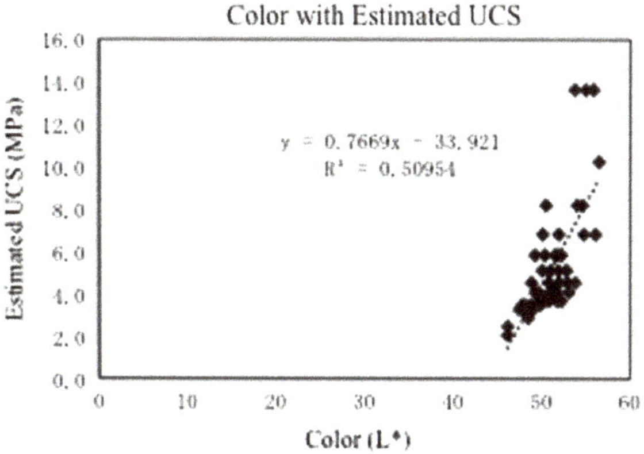

Fig. 7 Relationship between estimated UCS and color of the Mizunami sand sample

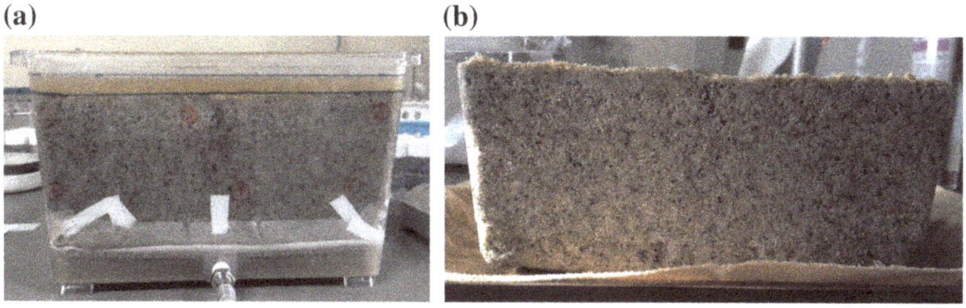

Fig. 8 Lab model sample photos (**a**) before open the sample after 14 days, and (**b**) after open the solidified sample

3.3.4 SEM Images and EDX Results

The SEM images (Fig. 10) show that the $CaCO_3$ crystals were mainly irregular, similar to observations made by Qabany et al. (2012). However, the crystal size (approximately 10–100 μm), we observed, was larger than the size indicated by Qabany et al. (2012); this might be attributed to the different sample preparation method we used. Most of crystal particles were dispersed and uniformly distributed in the gaps between sand particles, and sometimes, there were overlapping structures on the crystal surface.

The results of EDX analysis (Fig. 11) demonstrated that the dominant minerals were SiO_2 and $CaCO_3$, and Ca, O, and C were the main elements in the mineral precipitations. Studies by Van Paassen et al. (2009) and Qabany et al. (2012) confirm that the crystals observed were actually $CaCO_3$ crystals precipitated in the silica sand.

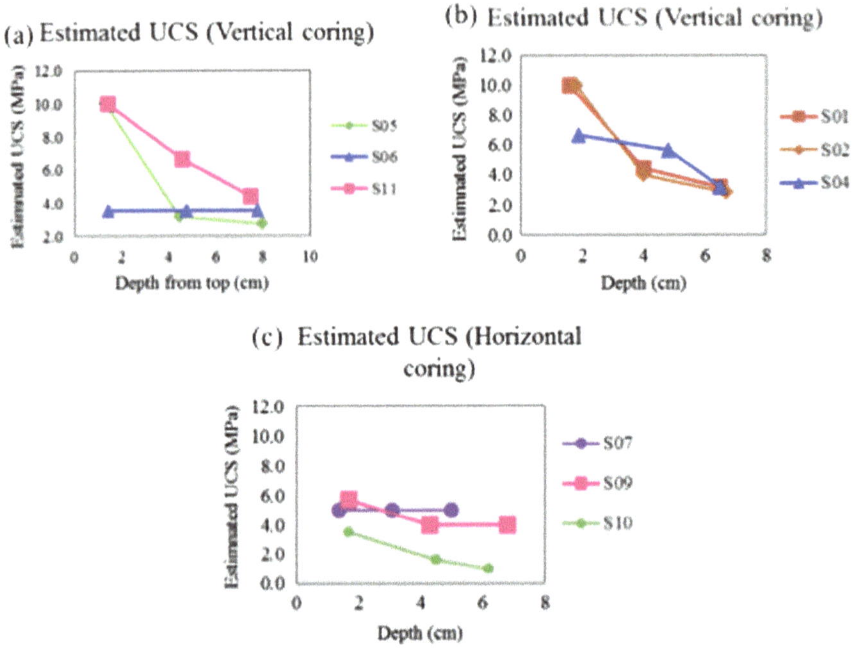

Fig. 9 Relationship between estimated UCS and depth of the sample for **a** vertically cored samples—test case 1, **b** vertically cored samples—test case 2 and **c** horizontally cored samples—test case 2

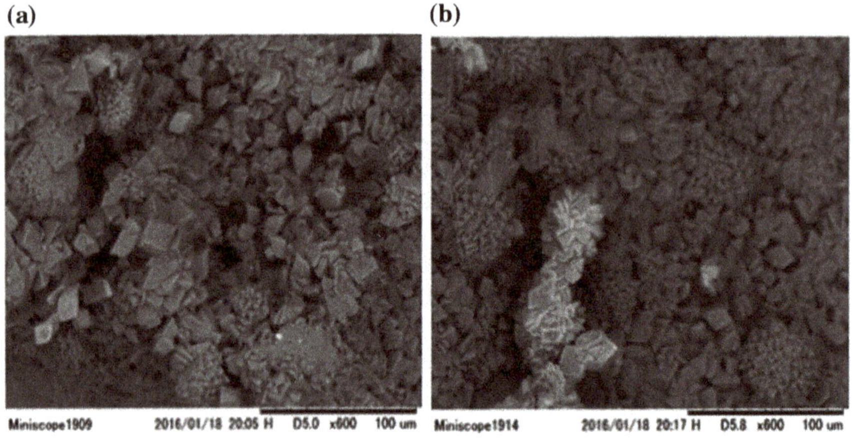

Fig. 10 SEM images **a** test case 1 and **b** test case 2

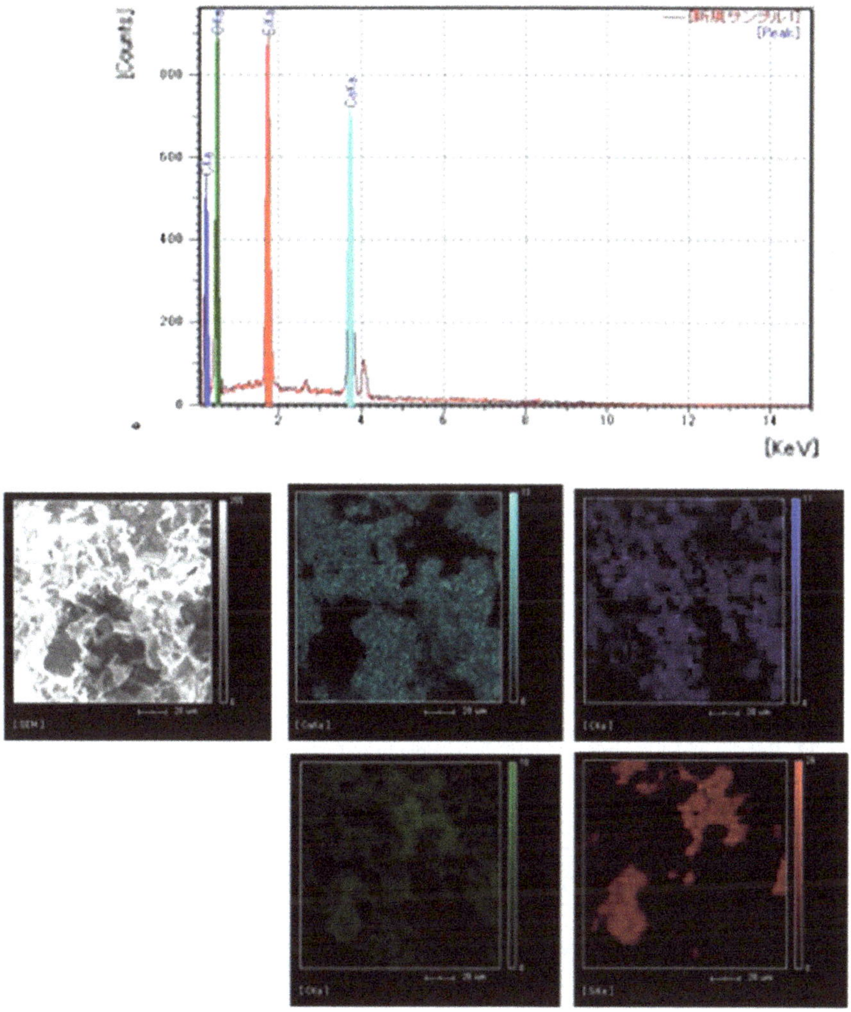

Fig. 11 EDX analysis for testing case 1

3.3.5 XRD Results

The XRD observation results (Fig. 12) indicate that the calcite precipitation was larger at the top of the sample than bottom of the sample (SiO_2—91.5% and $CaCO_3$—8.5% at the top and SiO_2—96.5% and $CaCO_3$—3.5% at the bottom). However, the difference of the distribution of calcite precipitation showed little variation in the sample which cored horizontally.

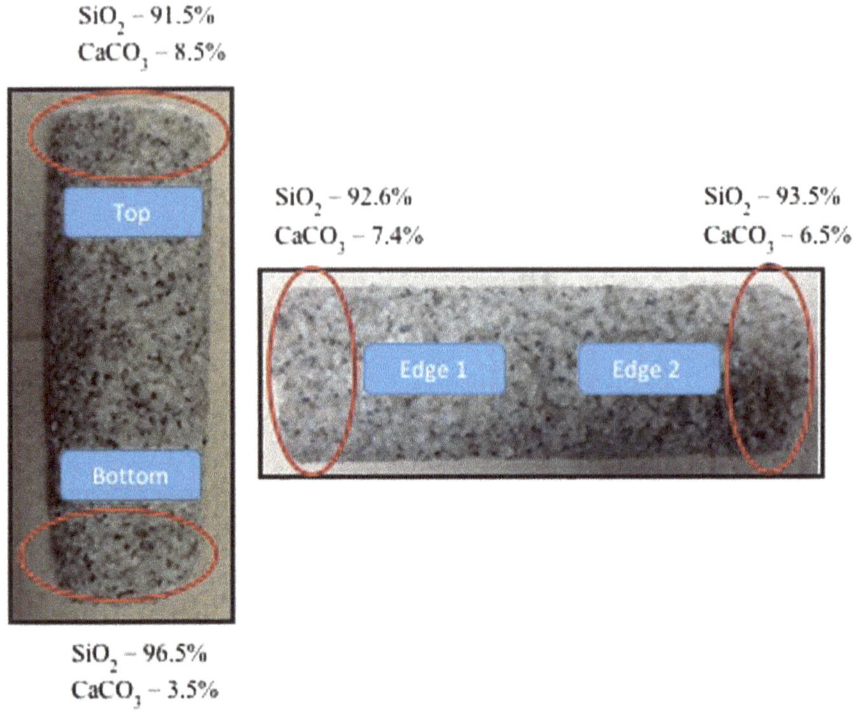

Fig. 12 Results of XRD observation of the sample in test case 2 vertically and horizontally

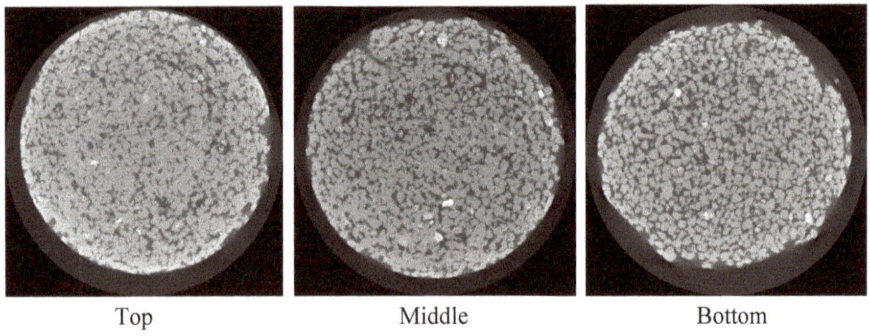

Fig. 13 Results of X-CT at top, middle, and bottom of the sample

3.3.6 X-CT Results

The X-CT results show clearly that the sand particles bonded with $CaCO_3$ precipitation were large in the top of the sample than middle and bottom (Fig. 13). Therefore, the strength decreased with the depth of the sample. Further examination of getting a homogeneous sample is needed in future.

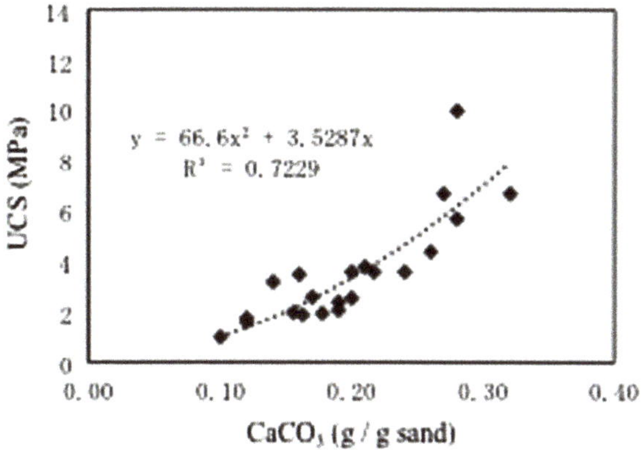

Fig. 14 Relationship between UCS and $CaCO_3$ content of samples, which were taken from three laboratory model samples

3.4 Discussion

3.4.1 Comparison of Relationship Between UCS and $CaCO_3$ Content with Previous Studies

To determine precipitated $CaCO_3$ in the soil specimens, specimens were crushed using a mortar and oven-dried. The dry soil was washed in HCl solution (1.0 M) to dissolve precipitated carbonates, rinsed, drained, and oven-dried. The difference between the two weights is considered the weight of the carbonates that were precipitated in the specimen (Rebata-Landa 2007). Figure 14 shows that the strength of the sample increased with an increase in the $CaCO_3$ precipitation.

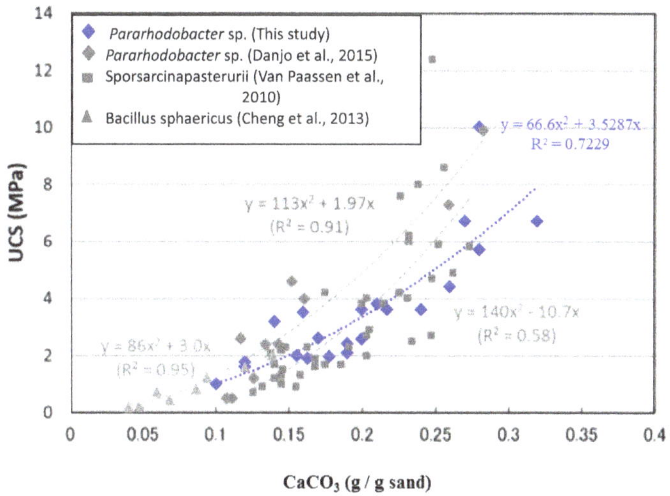

Fig. 15 Relationship between UCS and total $CaCO_3$ precipitation content for previous studies

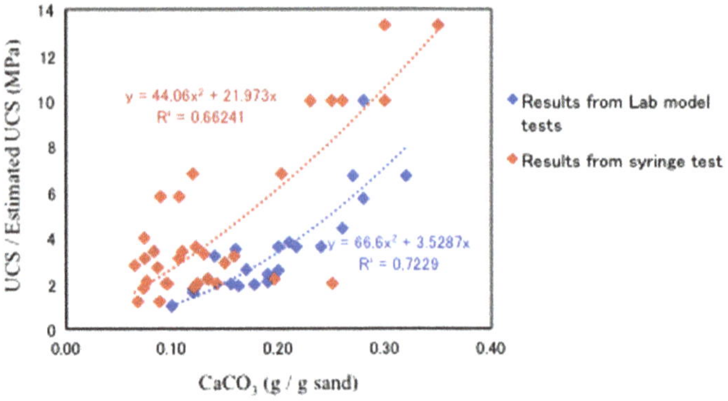

Fig. 16 Relationship between estimated UCS/UCS and total $CaCO_3$ precipitation content for syringe test and model test

The results for other bacteria are shown in Fig. 15 (Van Paassen et al. 2010; Cheng et al. 2013; and Danjo 2015). *Sporosarcina pasteurii* has been most widely used in investigations of sand improvement using bacteria, while *Bacillus sphaericus* was isolated by Al-Thawadi and Cord–Ruwisch (2012). Danjo (2015) found that the UCS of the specimen prepared using *Pararhodobacter* sp. was higher than that of the specimen generated using *Sporosarcina pasteurii*, even though both specimens contained the same amount of total $CaCO_3$ precipitation (Fig. 15).

According to this study, we found that the UCS of the silica sand specimen prepared using *Pararhodobacter* sp. (Amarakoon 2016) was less than the UCS of the coral sand specimen prepared using *Pararhodobacter* sp. (Danjo and Kawasaki 2016). However, because these specimens contained different kinds of sand and were cured under different conditions, it is unclear which bacteria are better for sand cementation. Conversely, the different amounts of total precipitation could explain the different UCS values of the specimens produced using *Bacillus sphaericus* or the other two bacteria.

The results from syringe test measured the estimated UCS for the syringe solidification test samples. In this study direct UCS value was obtained for the cored samples. Figure 16 shows the results obtained by syringe and model test samples. The graphs show that the rates of increase in UCS value are nearly parallel to each other. However, there is a gap between two graphs. This may happen due to an error coefficient between UCS and estimated UCS value. Future experiment is needed to identify the relationship between UCS and estimated UCS value.

3.4.2 Usefulness and Applicability of Model Test in Practical Stage

The sand solidification using local ureolytic bacteria (*Pararhodobacter* sp.) can be mainly used as a construction material for preserving coastal regions and/or healing of coastal concrete structures. Presently breakwater construction is used for prevention of coastal erosion. However, production of cement, which is a major construction material for breakwater construction, is energy-consuming and environmentally unfriendly. Therefore, this MICP-treated method offers a promising technique that can be used to solve coastal erosion.

In addition, this method can be used for soil improvement in land usage also. However, according to this study, the strength was decreased with the depth. Therefore, there is a limitation for applying in practical stage. Further investigation is needed to identify the efficient injection system for getting the uniform solidified sample before applying for a flat area practically. However, this study method can be used in a slope area by injecting vertically. From this technique, it can improve the soil in a slope area.

3.4.3 Comparison Between Syringe Solidification Test and Model Test

Most studies on MICP soil improvement used cylindrical columns or syringes for sample preparation by pumping or injections methods. In Sect. 2, it described syringe solidification test. However, it faced many practical issues. In the syringe test, there can be a possibility of blocking the bacterial solution and the consolidation solution penetrating in to the soil sample. This was caused because of the syringe size was very small and which lead easy bio-clogging. To avoid this problem, the large size of sample preparation was used. In Sect. 3, the small size of laboratory model test was described. The methodology was same as syringe solidification test, and here, materials are needed more than compared to the syringe test. In addition, unconfined compressive strength (UCS), X-ray diffraction (XRD), X-CT, and color measurement tests were conducted, which could not be conducted for syringe samples, because the sample size is not matched with the required size for conducting previously mentioned tests. Therefore, sufficient soil parameters were obtained from the model test.

4 Future Improvements

This study shows that the optimization of the MICP process is possible in controlled laboratory-scale experiments. The applicability of the MICP on the field scale still requires further investigation, as common natural conditions like high pore-water pressure, non-uniform flow field, and soil heterogeneity have not been sufficiently investigated yet.

In this study, the strength decreased with the depth of the sample. Therefore, further studies that achieve a uniformly solidified sample using *Pararhodobacter* sp. are needed. Furthermore, the cementation media and bacterial solution were added using the injection method, the sample was saturated throughout the testing period, and the cementation media was kept around 2 cm above from the top of sample. Ca^{2+} and bacteria were present in the sample, as well as the solution that was kept for saturation. Due to the presence of Ca^{2+} and bacteria, calcite precipitation might have occurred at the top of the sample. Therefore, the top layer might have hardened. Furthermore, the cementation media might have penetrated into the soil pores that were under pressure to some extent. Moreover, the effluent also reduces the number of bacteria and a portion of urease produced by bacteria, and the samples may not be uniform along the flow. However, to clarify this matter, further investigations with a reduced cementation media level at the top of the sample should be conducted in future.

MICP is a sustainable and environmentally friendly technique that must be improved both at laboratory and field scales. This technique must be optimized to find the best conditions (pH, soil, temperature, concentration of cementation media, etc.) for bacterial activity and to achieve a homogeneous distribution in the soil. It is believed that the conditions for bacterial activity were achieved, and therefore, future research must focus on finding efficient injection systems for both bacteria and cementation media. Only after solving these problems, it will be possible to compare the UCS from this treatment with that of soil–cement mixtures; however, it is expected that UCS of cement may be larger.

5 Conclusions

Microbial-induced calcite precipitation utilizing urea hydrolysis is a complex biochemical process, especially when it takes place between sand particles for improvement of soil engineering properties. To cement sand specimens to a UCS of several MPa using local ureolytic bacteria from silica sand and to consider the influences of various test conditions on the UCS, we conducted silica sand cementation tests. The main findings of our study are as follows:

(1) The model test specimens were cemented up to 10 MPa estimated UCS after 14 days under a curing temperature of 25 °C, an injection interval of the cementation solution of 1 day and in the presence of a cementation solution containing 0.5 M Ca^{2+}.

(2) Multiple regression analysis revealed that the relevant test conditions influencing the estimated UCS of the specimen were test period (*D* (days)) and Ca^{2+} concentration of the cementation solution (C_{ca}(M)). The formula for predicting the estimated UCS (q_{eu}(MPa)) was $q_{eu} = 13.99\ C_{ca} + 0.37\ D - 0.09$.

The results of this study will contribute to the development of new techniques for bio-stimulation and silica sand improvement. In future, environmental impact assessments, tank cementation tests, and field tests are needed to enable widespread application of this soil improvement technique using bacteria.

Acknowledgements This work was partly supported by JSPS KAKENHI Grant Number JP16H04404 and Grant for Environmental Research Projects from the Sumitomo Foundation.

References

Al-Thawadi S, Cord-Ruwisch R (2012) J Adv Sci Eng Res 2:13–26

Amarakoon GGNN (2016) Doctoral thesis. Hokkaido University, Japan

Burbank MB, Weaver TJ, Green TL, Williams BC, Crawford RL (2011) Precipitation of calcite by indigenous microorganisms to strengthen liquefiablesoils. J Geomicrobiol 28(4):301–312

Cheng L, Cord-Ruwisch R, Shahin MA (2013) J Can Geotech 50:81–90

Danjo T, Kawasaki S (2016) Microbially induced sand cementation method using Pararhodobacter sp. strain S01, inspired by beachrock formation mechanism. J Mater Trans 57(3):428–437

Danjo T (2015) Doctoral thesis. Hokkaido University, Japan

DeJong JT, Fritzges MB, Nusslein K (2006) Microbial induced cementation to control sand response to undrained shear. J Geotech Geoenviron Eng 132:1381–1392

DeJong JT, Mortensen BM, Martinez BC, Nelson DC (2010) Bio-mediated soil improvement. Ecol Eng 36:197–210

Ferris FG, Wiese RG, Fyfe WS (1994) Precipitation of carbonate minerals by microorganisms: implications for silicate weathering and the global carbon dioxide budget. Geomicrobiol J 12:1–13

Foesel BU, Darke HL, Schramm A (2011) Defluviimonasdenitrificans gen. nov., sp. nov., and Pararhodobactor aggregans gen. nov., sp. nov., non-phototropic Rhodobacteraceae from the biofiter of a marine aguaculture. J Syst Appl Microbiol 34:498–502

Fujita Y, Ferris FG, Lawson RD, Colwell FS, Smith RW (2000) Calcium carbonate precipitation by ureolytic subsurface bacteria. J Geomicrobiol 17(4):305–318

Khanafari A, Khans FN, Sepahy AA (2011) An investigation of biocement production from hardwater. Middle-East J Sci Res 7:1990–9233

Mitchell JK, Santamarina CJ (2005) Biological considerations in geotechnical engineering. J Geotech Geoenviron Eng 131(10):1222–1233

Mitchell AC, Dideriksen K, Spangler LH, Cunningham AB, Gerlach R (2010) Microbially enhanced carbon capture and storage by mineral-trapping and solubility-trapping. Environ Sci Technol 44:5270–5276

Mortensen BM, Haber MJ, DeJong JT, Caslake LF, Nelson DC (2011) Effects of environmental factors on microbial induced calcium carbonate precipitation. J Appl Microbiol 111(2):338–349

Qabany, AA, Soga, K, Santamarina, C (2012) Factors affecting efficiency of microbially induced calcite precipitation. J Geotech Geoenviron Eng 138(8), 992–1001

Rebata-Landa, V (2007) Microbial activity in sediments: effects on soil behavior. PhD thesis, Georgia Institute of Technology, Atlanta, GA

Van Paassen LA, Daza CM, Staal M, Sorokin DY, van der Zonb W, van Loosdrecht MC (2010) Potential soil reinforcement by biological denitrification. Ecol Eng 36(2):168–175

Van Paassen LA, Harkes MP, Van Zwieten GA, Van der Zon WH, Van der Star WRL, Van Loosdrecht MCM (2009) Scale up of BioGrout: a biological ground reinforcement method. In: 17th international conference on soil mechanics and geotechnical engineering, pp 2328–2333

Whiffin VS, Van Paassen LA, Harkes MP (2007) Microbial carbonate precipitation as a soil improvement technique. J Geomicrobiol 24:417–423

Effect of Plant-Derived Urease-Induced Carbonate Formation on the Strength Enhancement of Sandy Soil

R. A. N. Dilrukshi and Satoru Kawasaki

Abstract The formation of the carbonate biomineral using the byproducts of enzymatic urea hydrolysis is widely explored within different areas of the field of engineering. Various types of urease sources such as ureolytic bacteria and microalgae, soil urease, and plant urease have been applied for the formation of carbonate. At present, the potential of using plant-derived urease enzyme-induced carbonate (mainly calcium carbonate (CC)) precipitation has been demonstrated at laboratory scale. Biominerals are formed by byproducts resulting from a series of chemical reactions. Those reactions are managed by biological activities. Sand cementation using biomineral precipitation helps to improve mechanical properties and hence to improve bearing capacity and to resist liquefaction, control of soil erosion by surficial stabilization, and to reduce the hydraulic conductivity of sand. In this study, plant-derived urease (crude extract of watermelon (*Citrullus lanatus*) seeds)-induced CC precipitation was considered. A significant strength improvement ranging from estimated UCS of 500 kPa to 4.0 MPa was observed in urea–urease-treated sand specimens under different concentrations of $CaCl_2$–urea, urease, and different curing periods (days) rather than the non-treated sand specimens. The carbonate ions (hereafter, CO_3^{2-}) resulting from urea hydrolysis play a major role to precipitate CC. The use of the biominerals that cause minimal negative impact on the ecology and the environment contributes to the sustainable development.

Keywords Biominerals · Soil · Urease · Sandstone · Calcium carbonate
Sustainable development

R. A. N. Dilrukshi (✉)
Graduate School of Engineering, Hokkaido University, Sapporo 060-8628, Japan
e-mail: nadilrukshi@yahoo.com

S. Kawasaki
Faculty of Engineering, Hokkaido University, Sapporo 060-8628, Japan

V. Achal and A. Mukherjee (eds.), *Ecological Wisdom Inspired Restoration Engineering*, EcoWISE, https://doi.org/10.1007/978-981-13-0149-0_5

1 Introduction

Currently, the quantity of lands that can be used for human activities is limited due to weak properties of subsurface soil, topography, and different climate conditions. Weak soil characteristics can be modified by removing/replacing unsuitable soils and modifying existing ground using a suitable ground improvement method. Ground improvement is a technique used to change the engineering properties of subsurface soil to enhance bearing capacity in order to use as a countermeasure against natural disasters, including liquefaction of saturated loose sand. Widely practicing ground improvement techniques are cement, chemical, compaction, fracture and jet grouting, micropiles, ground anchors, soil nailing, vibro compaction, stone columns, concrete columns, piers, etc. (Karol 2003). All of the above ground improvement techniques require cement and/or chemicals. Widely used chemical grouts are sodium silicate, acrylate, urethane, lignin, and resin grouts (US Army Corps of Engineers 1995). All the above chemical grouts are hazardous, and hence, the groundwater can be polluted. Furthermore, the cement-based concrete system is not favored to the environment due to CO_2 emission during cement production (Gerilla et al. 2007; Mora 2007). Therefore, a proper planning is needed to improve the soil characteristics in order to assure the least damage to the environment.

In 2005, the idea of the use of biotechnology for ground improvement started spreading through the field of geotechnical engineering, and National Research Council identified as biogeotechnical engineering is an important research area (Dejong et al. 2013). The grout forms by biological actions are named as "biogrouts" (Van Paassen et al. 2009). Some of biogrout formation mechanisms are listed as carbonate precipitation using urea and ureolytic bacteria (Harkes et al. 2010) or urea and purified/crude extracts of plant urease enzyme (Park et al. 2014; Nam et al. 2014; Neupane et al. 2013) or using glucose and yeast (Kawasaki et al. 2006), iron/manganese compound precipitation using iron-oxidizing bacteria (Weaver et al. 2011), siloxane bond formation using glucose and yeast (Terajima et al. 2009), calcium phosphate compound (CPC)-based chemical grouts (CPC-chem) formation by its self-setting mechanism (Akiyama and Kawasaki 2012a) and CPC biogrout (CPC-bio) formation by using ureolytic microorganisms, an ammonia source along with CPC-chem (Akiyama and Kawasaki 2012b). Although a temperature about 1500 °C is required for the production of ordinary cement, relatively low temperatures are required for the production of reagents for biogrouts. Therefore, biogrout needs less energy, and it releases fewer amounts of greenhouse gases (Kanafari et al. 2011). Various biominerals such as calcium carbonate (CC), calcium phosphate, calcium oxalate, calcium sulfate, silicate, and iron oxide can be found in nature. The biomimetic in which lessons learned from the nature forms the basis for the evolution of novel biogrouting materials for the ground improvement, and hence, the use of these biogrouts do not permanently change the subsurface condition after the ground improvement. Furthermore, the negative effect on groundwater can also be minimized from novel biogrouting materials rather than synthetic chemical grouts. Therefore, above-mentioned advantages of biotechnology-based ground improvement practices may help to maintain the ecological balance of the environment.

The objective of this study was to strengthen the small-scale sand specimens to an estimated unconfined compressive strength (UCS) ranging from several kPa to MPa using crude extract of watermelon seeds urease-induced CC precipitation technique and identify the range of applicability of this method using obtained results such as to mitigate the liquefaction and to create artificial soft rocks.

1.1 Plant-Derived Urease Enzyme and its Activity

Particularly, leguminosae and cucurbitaceae families are rich sources for plant ureases (http://www.ufrgs.br/laprotox/en/what-we-do/research-lines/plant-ureases). The enzyme urease is a protein, and it may occur in various plant tissues and concentrate in higher amounts in seeds and roots. Key examples for leguminosae family are jack bean (*Canavalia ensiformis*) and soybean (*Glycine max*) and watermelon (*Citrullus vulgaris*) and pumpkin (*Cucurbita maxima*) for cucurbitaceae family. In addition to that, white mulberry (*Morus alba*), cotton (*Gossypium hirsutum*), potato (*Solanum tuberosum*), tomato (*Lycopersicon esculentum*), and pigeon pea (*Cajanus cajan*) belong to other families. Urea hydrolysis is catalyzed by enzyme urease to form ammonium ions (NH_4^+) and carbonate ions (CO_3^{2-}). One mole of ammonia (NH_3) and one mole of carbamate resulted from the hydrolysis of one mole of urea (Eq. (1)) are further hydrolyzed to one mole of NH_3 and one mole of carbonic acid (Eq. (2)). Afterward, NH_3 and carbamate equilibrate in water to form bicarbonate and 2 mol of NH_4^+ and hydroxide ions (Eqs. (3) and (4)) (De Muynck et al. 2010).

$$CO\,(NH_2)_2 + H_2O \rightarrow NH_2COOH + NH_3 \tag{1}$$

$$NH_2COOH + H_2O \rightarrow NH_3 + H_2CO_3 \tag{2}$$

$$2NH_3 + 2H_2O \rightarrow 2NH_4^+ + 2OH^- \tag{3}$$

$$2OH^- + H_2CO_3 \rightarrow CO_3^{2-} + 2H_2O \tag{4}$$

Total reaction:

$$CO\,(NH_2)_2 + 2H_2O \rightarrow 2NH_4^+ + CO_3^{2-} \tag{5}$$

$$Ca^{2+} + CO_3^{2-} \rightarrow CaCO_3 \tag{6}$$

Calcium ions (Ca^{2+}) in the system help to precipitate CC after a definite level of supersaturation. When the NH_3 reacts with water, it creates OH^- ions and it helps to raise the pH in the system. The increase in pH provides an optimum condition for the carbonate precipitation. Hydrolysis of urea is accelerated by urease by a factor of 10^{14} in comparison with the spontaneous reaction (Hausinger 1993).

Jack bean urease is the most studied plant urease source (Andrews et al. 1984) and the first enzyme crystallized (Sumner 1926). Molecular weight of jack bean urease is 480 kDa (Goring et al. 1962). Out of 47 SH groups in the urease species, four to eight of these groups are needed for the activity (Goring and Chin 1965). Maximum urease activity of jack bean occurs at 65 °C (Sumner 1951), and it is totally inactive at temperature above 70 °C (Frankenberger and Tabatabai 1982), and optimum pH range of jack bean urease is 6.0–7.0 (Lai and Tabatabai 1992; Boyd and Mortland 1985).

Maximum activity of chickpea seed (*Cicer arietinum* L.) occurred at pH 7.2 (Pervin et al. 2013), and this result is similar to the results obtained from urease from jack bean (Sung et al. 1989) and pigeon pea (Das et al. 2002) but different from urease obtained from mulberry leaves (Hirayama et al. 2000). Maximum activity of chickpea seeds can be obtained at temperature of 40 °C, and beyond that, the enzyme activity is loosed (El-Hefnawy et al. 2014). This result is closely related to the results obtained for pigeon pea urease (Das et al. 2002 and Srivastava et al. 2001) and differs from the results obtained by El-Shora (2001). Urease activity is closely related to substrate concentration. Urease activity increases with the increase in substrate concentration. However, after certain value, the urease activity decreases with the increase in urea concentration (Pervin et al. 2013). The rate of urea hydrolysis is increased with the increase in urea concentrations until a certain level is achieved. After that, the urea hydrolysis is decreased due to substrate inhibition at high urea concentrations (Singh and Nye 1984; Fidaleo and Lavecchia 2003). Muskmelon (*Cucumis melo*), which is also named as Kharbooza and Banggi, is another plant urease source (Wright et al. 2007). Furthermore, urease from mulberry leaves (*Morus alba*) has newly subjected to detailed investigation (Hirayama et al. 2000).

1.2 Importance of Plant-Derived Urease Enzyme Over the Microbial Urease

The extraction of plant urease enzyme from urease containing plant species is simple, and some purified plant urease enzyme is available from laboratory suppliers. The size of the plant enzyme is approximately 12 nm in dimension (Blakely and Zerner 1984), and ureolytic microbes are in the range of 500–5000 nm. Therefore, there is a possibility to apply plant-derived urease to much finer soils. Furthermore, plant urease enzyme can be extracted as a free enzyme, and therefore, its lifespan is short, and activity and function decrease with time (Marzadori et al. 1998; Pettit et al. 1976). This limited lifespan of the enzyme reduces the long-term effects to the ecosystem. However, in case of microbial urease, the organisms leave behind. Therefore, the approvals and licenses from the government are required to use microbial urease, and continuous monitoring of microbial ecology for safety is needed (Akiyama et al. 2011). Microbial-induced carbonate precipitation (MICP) may be limited in deeper areas in the subsurface due to limited growth of bacteria and movement in subsoil in deeper areas. MICP may also be limited in fine soils. Therefore, the applicability of MICP is limited to well-graded gravel (GW), poorly graded gravel (GP), well-graded sand (SW), poorly graded sand (SP), silt (ML), and organic soils (Mitchell and Santamarina 2005). Moreover, the production of ureolytic bacteria and isolation of pure bacteria are costly processes for applying to the biocementation technology (Van et al. 2011). Although the commercially available plant urease enzyme is little expensive, laboratory grade enzyme is very effective (Knorr 2014). Crude extracts of jack bean (*Canavalia ensiformis*) have the potential to be used as an alternative to the commercially available urease (Nam et al. 2014). The advantages of plant-derived urease are very important to the geotechnologists and/or biotechnologists to a deep investigation about the biocementation using plant-derived urease instead of microbial urease.

1.3 Existing Applications of Plant-Derived Urease-Induced CC Precipitation

Plant-derived urease enzyme has been applied to precipitate CC , hence, to reduce the permeability of soils, to strengthen loose soils, and to prevent soil erosion. However, any kind of large-scale application in the field has not been reported. A UCS of 317 kPa has been achieved using jack bean urease and calcium chloride stock solution as the calcium source (Park et al. 2014). A decrease in the permeability has also been investigated using plant urease-induced calcite formation (Nemati and Voordouw 2003). At a low enzyme concentration (0.03 g/L), the permeability reduction increases with temperature from 22 to 30 °C. However, the influence of temperature is less with high concentration of enzyme used in the above study. CC has been precipitated with the aid of jack bean (*Canavalia ensiformis*) urease (specific activity is 15,000–50,000 units/g solid) to make a building material (Bull 2014). Maximum UCS of 319 kPa and an elastic modulus of nearly 10 MPa have been obtained after curing and air drying for 2 weeks. High concentrations of $CaCl_2$ and urea have not been successful for CC cementation. Hence, it is useful to study allowable limits of concentrations of each solution for a better cementation. Another study has been done to precipitate CC using jack bean urease and, hence, to observe the change in permeability as well as strength characteristics of soil (Yasuhara et al. 2012). Permeability reduction of more than one order of magnitude and maximum UCS of 1.6 MPa has been achieved. Neupane et al. (2014) have also obtained same UCS mentioned above for oven-dried specimens made using commercially available urease, with the activity of 2970 units/g. The strength improvement and permeability reduction of soil are important features in geotechnical engineering. As an example, UCS value of 100 kPa is needed to avoid ground liquefaction during earthquakes (Yamazaki et al. 1998). All of the past research studies based on plant-derived urease-induced CC precipitation have successfully achieved this target value.

Although the plant urease and bacterial urease hydrolyze urea in the same way, the formation mechanism is slight different. In the MICP process, negatively charged groups on bacterial cell surfaces combine with divalent cations like Ca^{2+} and Mg^{2+} and bind to their cell surfaces easily at neutral pH and the bound cation (metal ions) reacts with anions (CO_3^{2-}) to form CO_3^{2-} compound. Therefore, it makes ideal nucleation sites for carbonate deposition (Stocks-Fischer et al. 1999 and Ramachandran et al. 2001). However, in case of plant-derived urease, it does not provide a nucleation site and just acts as a catalyst for urea hydrolysis and liberates NH_4^+ and CO_3^{2-}. The resulted NH_4^+ helps to provide a neutral pH environment, and CO_3^{2-} helps to precipitate CC. The primary structures (amino acid sequences) of plant and bacterial urease are different, and therefore, the impact of each urease on phase formation, crystallization, and the morphology of CC particles is different (Sondi I and Sondi BS 2005). Some factors influencing for optimal urease activity such as temperature, pH, and substrate concentration may also differ from one urease source to another urease source. Therefore, the UCS value obtained for same concentration of $CaCl_2$, urea, and urease under same temperature may be different.

2 Methodology of the Current Study

2.1 *Selection of a Plant Species Having Urease Activity and Extraction of Urease Enzyme from Seeds*

Different plant species belong to cucurbit family (*Cucurbitaceae*) such as pumpkin (*Cucurbita species*), watermelon (*Citrullus lanatus*), and melon (*Cucumis melo*) which have urease activity were selected. After considering the results of the rate of change of pH in different urea–urease solutions presented by Dilrukshi et al. (2015), watermelon seed urease–urea solution has shown higher pH rather than other two solutions. Therefore, watermelon seeds were selected for the current study.

2.2 *Quantitative Determination of Urease Activity*

A quantitative value for urease activity was obtained using indophenol method (Natarajan 1995). In this case, the enzyme urease is used to catalyze the hydrolysis of urea into CO_2 and NH_4^+. The resultant NH_4^+ reacts with phenol consists of hypochlorite, and blue color indophenol resulted (Eq. (7)). The intensity of the color is proportional to the NH_4^+ concentration in the sample, and it is measured at 630 nm wavelength. Amount of NH_4^+ released is determined by using a previously prepared standard curve with relating absorbance at 630 nm for 0.1, 1, and 10 mg/L NH_4^+ solutions prepared from 1000 mg/L NH_4Cl solution.

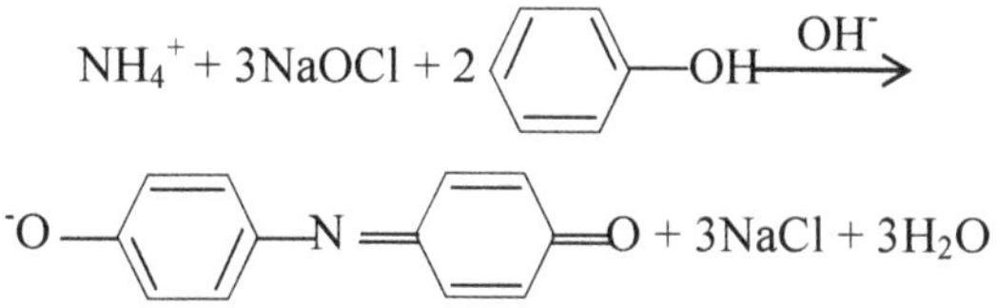

(7)

Indophenol (Blue)

In this study, watermelon seeds were finely ground using mortar and pestle. The crushed seeds (0.5 g) were stirred at 500 rpm with 10 mL of distilled water for 1 h. The crude extract was obtained after filtering and collecting the filtrate. One part of the collected crude extract was directly used to measure the urease activity, and the other part of the extract was centrifuged at a rate of 10000 rpm for 3 min under 25 ° C, and the urease activity in the supernatant was measured. The purpose of measuring urease activity before and after centrifuge was to identify the effect of solid part in the extract on urease activity. Furthermore, the temperature effect on urease activity was examined by measuring urease activity at the range of temperatures varying from 25 to 70 °C. Freshly prepared crude extract was used throughout this study. Urea solution (0.1 M) in phosphate buffer with pH around 7.01 was used as the substrate.

2.3 *Syringe Solidification Test*

Commercially available Mikawa sand-no. 4 (43.5 g) (particle size distribution is shown in Fig. 1) was placed in a 35-mL syringe (diameter, $\varphi = 2.3$ cm, height h = 7.1 cm) under three equal layers, and each layer was compacted, and final density of the specimen was set to the maximum density of Mikawa sand-no. 4. The sand was used after drying at 110 °C about 1 day. Some physical properties of Mikawa sand-no. 4 are shown in Table 1. The syringe specimens were kept in an incubator throughout the experiment to control the temperature at 25 °C. The setting of syringe solidification test is shown in Fig. 2. Subsequently, 13.1 mL of $CaCl_2$–urea–urease solution was added to the syringe (calculated initial void volume was 13.1 mL) and kept the final level of the solution as top surface level of the sand. The centrifuged crude extract was used to dissolve $CaCl_2$ and urea, and finally, $CaCl_2$–urea–urease solution was prepared. The purpose of selecting centrifuged crude extract was to prevent the clogging of specimen with solid particles in the crude extract. However, centrifuged crude extract was not a fully transparent

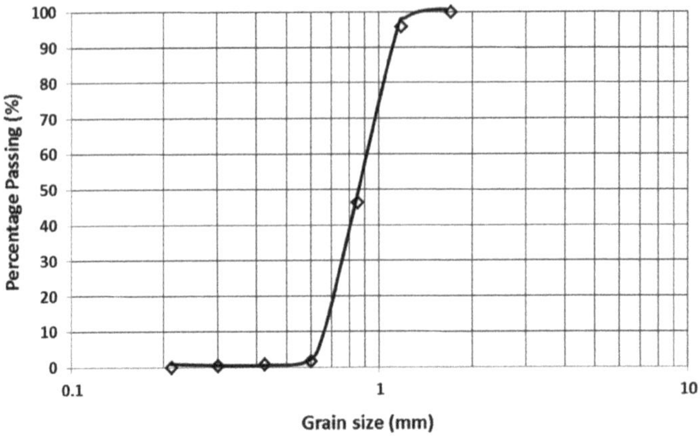

Fig. 1 Particle size distribution of Mikawa sand-no.4

Table 1 Physical properties of Mikawa sand-no. 4

Physical property	Value
Minimum density (g/cm^3)	1.256
Maximum density (g/cm^3)	1.476
Mean diameter (μm)	870
Particle density (g/cm^3)	2.66

solution and it also consisted of some suspended particles. After curing 1 day, the solution with same concentration and volume was added and, at the same time, drained the previously injected solution. While draining out the solution, initially, the rate was maintained at 0.05 mL/s. However, this rate may change (decrease) with injection period due to precipitated CC within the specimen. The Ca^{2+} concentration and pH of the outlet were measured once every 2 days to determine temporal variations of these parameters within the specimen. At the same time, two control test specimens were prepared only using $CaCl_2$–urea and $CaCl_2$–urease. After required curing days (7, 14, 21, and 28 days in this study), each specimen was carefully removed from the syringe and needle penetration (NP) test was conducted using a needle penetration device mentioned by Danjo and Kawasaki (2016).

An estimated UCS was obtained from the NP based on the following regression equation (Eq. (8)) (correlation coefficient: 0.941, x: NP (N/mm), y: UCS (MPa))

$$\log(y) = 0.978 \log(x) + 2.621 \tag{8}$$

In addition, samples were observed by scanning electron microscopy (SEM) (TM 3000 Miniscope, HITACHI).

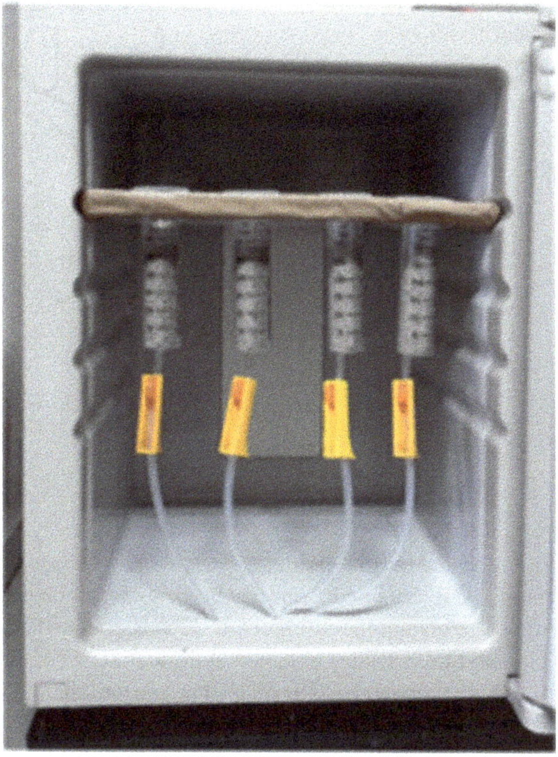

Fig. 2 Setting of the syringe solidification test

3 Results and Discussion

3.1 Quantitative Determination of Urease Activity

According to the results of the urease activity, U/mL (μmol urea hydrolyzed/min./mL), the crude extracts before and after centrifuge showed different activity values. The urease activity values before and after centrifuge were 4.391 and 3.912 U/mL, respectively, at 25 °C for the crushed seeds solution with 50 g/L. The urease activity before centrifuge was little higher than after centrifuge. This may be due to the presence of some urease activity in the fine solid particles in the crude extract that cannot be dissolved in distilled water during 1-hour soaking period. Furthermore, the urease activity values at different temperatures ranging from 25 to 70 °C were examined and the maximum activity was obtained around 50 °C (Fig. 3). Afterward, the activity abruptly declined with further increase in temperature. That means they are no longer active and cannot function properly and reach the denaturation.

Fig. 3 Variation of urease activity, (U/mL) with temperature (°C)

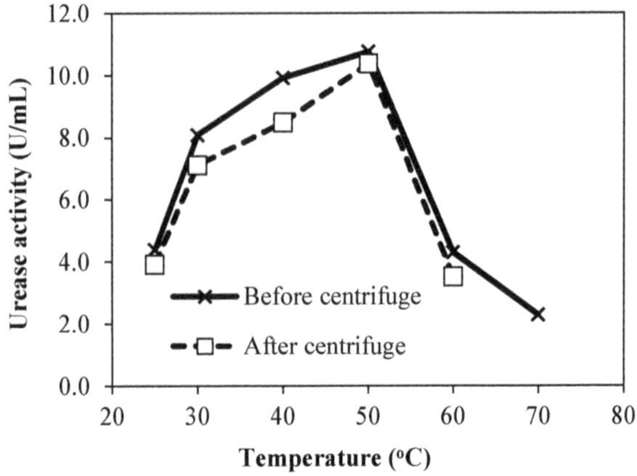

3.2 *Syringe Solidification Test*

After decided curing period, the needle penetration inclination (Np) values of each sample were obtained from needle penetration device as described in the methodology section and the estimated UCS was obtained from NP value by using Eq. (8). Estimated UCS values obtained at different test conditions are shown in Fig. 4. The minimum estimated UCS value that can be obtained from the needle penetration device was 200 kPa. Therefore, the estimated UCS values less than 200 kPa were reported as 0 kPa for the ease of representing it graphically. The appearance of some syringe test specimens just before the needle penetration test is shown in Fig. 5. According to Fig. 4a and c, estimated UCS values were increased with curing time (days). It can be simply explained using the amount of precipitated CC. The amount of precipitated CC increases with the increase in curing time. The total

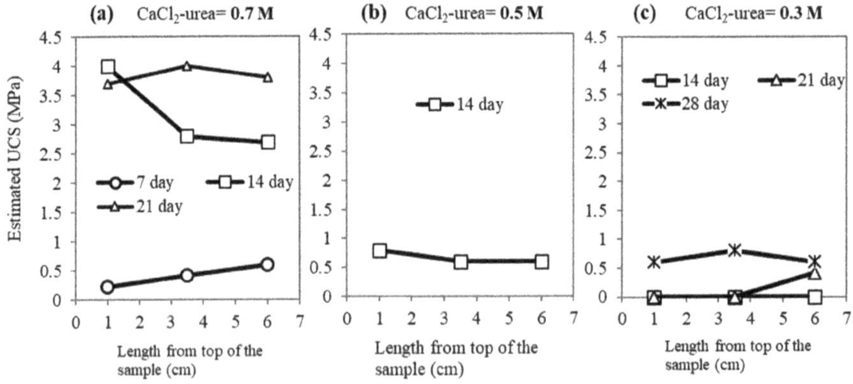

Fig. 4 Effect of curing time on estimated UCS (MPa) with different concentrations of $CaCl_2$–urea: **a** 0.7 M, **b** 0.5 M, and **c** 0.3 M with constant urease activity, (U/mL) of 3.912 U/mL

Fig. 5 Appearance of syringe test specimens after 14 days of curing time: **a** 0.7 M $CaCl_2$+ urea, no urease **b** 0.7 M $CaCl_2$+ urease, 3.912 U/mL, no urea, and **c** 0.7 M $CaCl_2$+ urea +urease, 3.912 U/mL

CC precipitation within the syringe specimen relevant to $CaCl_2$–urea concentration of 0.7 M as in Fig. 4a was calculated using the method described by Danjo and Kawasaki (2016), and it was 0.11, 0.21, and 0.23 g/g sand after 7, 14, and 21 days of curing time, respectively. The precipitated CC relevant to $CaCl_2$–urea concentration of 0.3 M as in Fig. 4c was 0.10, 0.14, and 0.18 g/g sand after 14, 21, and 28 days of curing time, respectively. The amount of precipitated CC after 7 days of injections with 0.7 M $CaCl_2$–urea is nearly same to the precipitated CC after 14 days of injections with 0.3 M $CaCl_2$–urea. However, estimated UCS values were different in those two testing cases as shown in Fig. 4a and c. The low concentration of urea (0.3 M) was rapidly hydrolyzed by the selected urease concentration rather than the high concentration of urea (0.7 M). Therefore, CC started to precipitate before injecting the solution into the syringe specimen in 0.3 M $CaCl_2$–urea–urease solution. Furthermore, some of the pre-precipitated CC remained at the top surface of the specimen while injecting, and hence, the total amount of calculated CC was not precipitated within the specimen in 0.3-M $CaCl_2$–urea–urease solution. This may be a possible reason to get low estimated UCS value for the specimen treated with 0.3 M $CaCl_2$–urea. Therefore, it is needed to select the best urease concentration appropriate to each concentration of $CaCl_2$–urea solution. To investigate the effect of urease concentration on estimated UCS with different concentrations of $CaCl_2$–urea, concentration of urease was reduced to 0.877 U/mL. The results are graphically shown in Fig. 6. According to the results, a higher estimated UCS was obtained at low concentration of urease rather than high concentration of urease for 0.5 M and 0.3 M $CaCl_2$–urea. The rate of hydrolysis of the urea tends to be decreased with decreasing the concentration of urease, and it may be favorable for precipitating CC within the specimen in the test cases relevant to 0.5 and 0.3 M concentrations of $CaCl_2$–urea. Conversely, in the case of 0.7 M $CaCl_2$–urea, the estimated UCS was decreased with decreasing the concentration of urease, which may be due to an insufficient amount of urease to hydrolyze 0.7 M urea and hence insufficient amount of CO_3^{2-} to precipitate CC within 1-day injection interval. This can be clearly understood using the Ca^{2+} concentrations in the outlet solutions as shown in Fig. 7a. At low concentration of urease, the Ca^{2+} concentration in the outlet was higher than the high concentration of urease. That

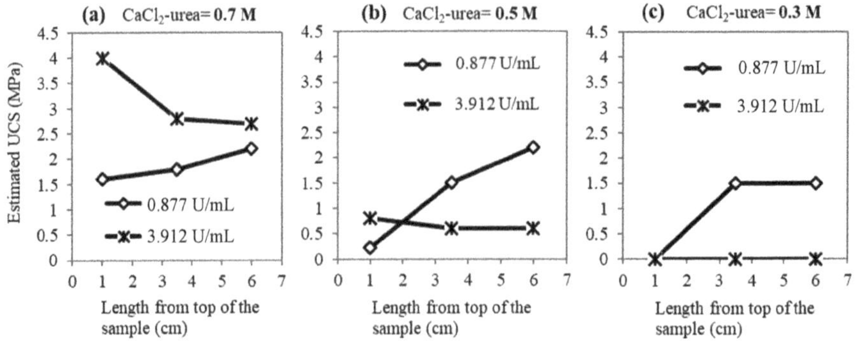

Fig. 6 Effect of urease activity on estimated UCS (MPa) with different concentrations of $CaCl_2$–urea: **a** 0.7 M, **b** 0.5 M, and **c** 0.3 M (all testing cases are after14 days of curing time)

means the rate of urea hydrolysis is low at low concentration of urease and the produced CO_3^{2-} is not sufficient to bind with free Ca^{2+} in the solution. Furthermore, the Ca^{2+} concentration in the outlet solution was decreased and becomes constant with curing time in all testing cases. The fine particles remain in the crude extract showed some urease activity at urease activity determination. After injecting the solution, those particles may remain within the specimen and it may also help to hydrolyze the urea and hence to precipitate Ca^{2+} as CC. According to the SEM images (Fig. 8), we could observe most of the fine particles remain in the upper portion of the sample. In addition to the fine particles, the urease adsorbed by sand particles may also help to hydrolyze the urea.

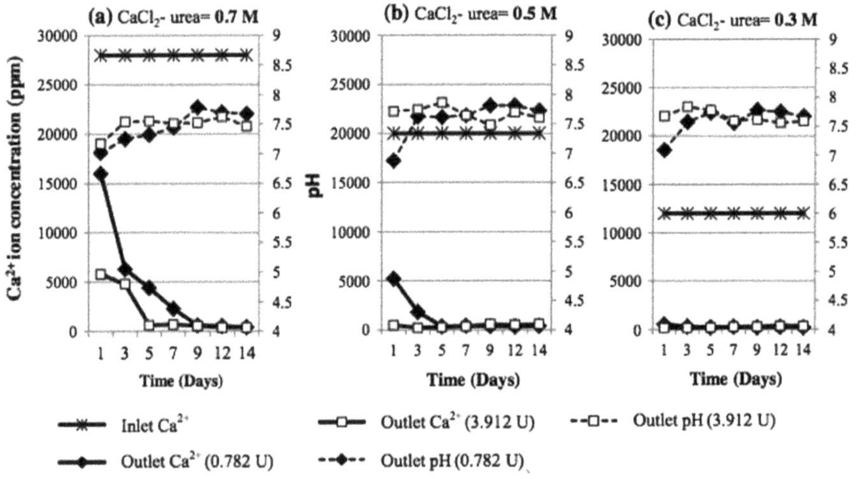

Fig. 7 Temporal variation of Ca^{2+} and pH with different concentrations of $CaCl_2$–urea: **a** 0.7 M, **b** 0.5 M and **c** 0.3 M solutions

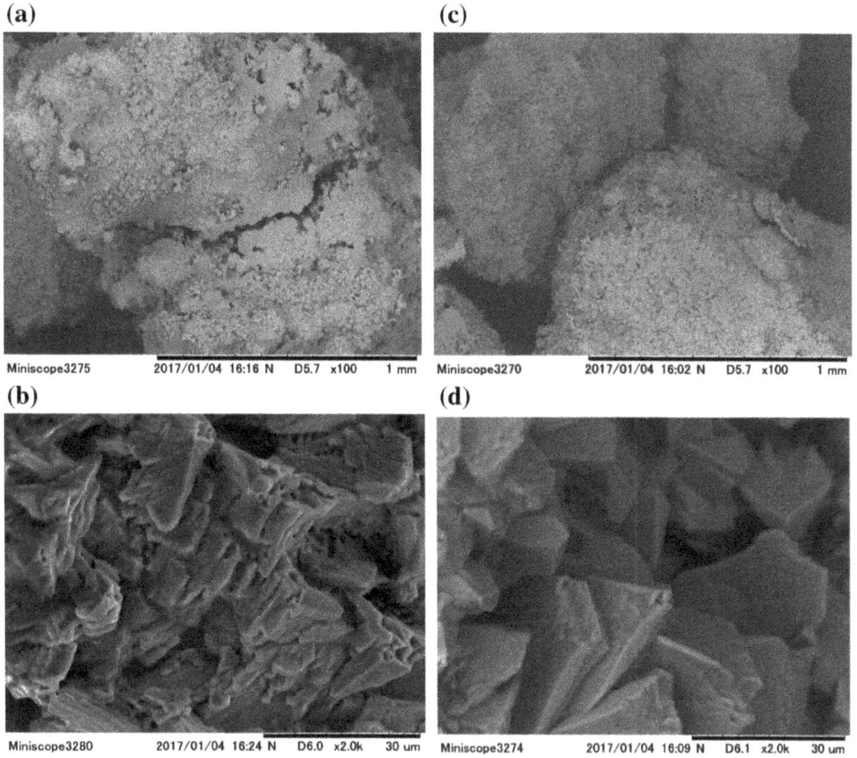

Fig. 8 SEM images of CC precipitated Mikawa sand test pieces. (**a** and **b**) upper portion of the specimen and (**c** and **d**) bottom portion of the specimen, **a** x 100, **b** x 2000, **c** x 100, and **d** x 2000

4 Conclusions

This chapter presents the use of plant-derived urease-induced carbonate formation as an alternative method for MICP. The crude extract of crushed watermelon seeds was successfully applied for the current study, and small-scale test specimens made from commercially available Mikawa sand-no. 4 were cemented and achieved satisfactory estimated UCS values. According to the results of this study, the following conclusions were summarized.

- Estimated UCS of several kPa to MPa was obtained by changing the concentration of $CaCl_2$–urea, urease as well as curing time. The increase in curing time caused to increase in estimated UCS value. The increase in $CaCl_2$–urea concentration from 0.3 M to 0.7 M also caused to increase the estimated UCS value. However, the effect of urease activity on estimated UCS is different for different concentrations of $CaCl_2$–urea solutions. It is better to keep low urease activity values for low concentrations of $CaCl_2$–urea solutions and high urease activity values for high concentrations of $CaCl_2$–urea solutions to maintain the rate of urea hydrolysis and finally to maintain the rate of carbonate precipitation. In this study, the urease activity of 3.912 U/mL is suitable for 0.7-M $CaCl_2$–urea solutions and 0.877 U/mL for 0.3-M and 0.5-M $CaCl_2$–urea solutions.

- By changing each and every parameter mentioned above, there is a possibility to apply this method for some applications such as strength improvement of weak unconsolidated soil, especially saturated loose sand to mitigate liquefaction, to protect limestone monuments, and to create artificial soft rock (like sandstone).
- The urease obtained from the crude extract of crushed dry watermelon seeds has a potential to be applied as an alternative to the commercially available plant urease for carbonate precipitation. Further laboratory investigations should be carried out at different temperatures to determine the temperature effect on carbonate precipitation. It is very important because urease activity is highly depending on temperature. A uniform distribution of $CaCO_3$ should also be achieved prior to the large-scale applications. Furthermore, some properties of subsurface soil such as soil type, pH of the soil, mineral content, and their interaction with groundwater should be examined before applying this method to the real fields in order to get maximum output from this technique. Durability and reversible process of the precipitated carbonate should also be investigated under economical point of view. Since biominerals have brought a new revolution in different engineering applications, it is needed to explore this novel area deeply in order to bring this method to an environmental-friendly and cost-effective method and to develop this technology from laboratory to large-scale field applications.

References

Akiyama M, Kawasaki S (2012a) Novel grout material using calcium phosphate compounds: in vitro evaluation of crystal precipitation and strength reinforcement. Eng Geol 125:119–128

Akiyama M, Kawasaki S (2012b) Microbially mediated sand solidification using calcium phosphate compounds. Eng Geol 137–138:29–39

Akiyama M, Kawasaki S, Shimeno A (2011) Fundamental study on new grouting material using calcium phosphate compounds-Effect of addition of ammonia sources and soil microorganisms on unconfined compression strength. Japanese Geotech J 6(4):513–524

Andrews RK, Blakeley RL, Zerner B (1984) Urea and urease. Adv Inorg Biochem 6:245–283

Blakely RL, Zerner B (1984) Jack bean urease: the first nickel enzyme. J Mol Catal 23:263–292. https://doi.org/10.1016/0304-5102(84)80014-0

Boyd SA, Mortland MM (1985) Urease activity on a clay-organic complex. Soil Sci Soc Am J 49:617–622. https://doi.org/10.2136/sssaj1985.03615995004900030018x

Bull M (2014) Bio inspired cementation of soil using plant enzyme. The thesis, Barrett, The Honors College at Arizona State University

Danjo T, Kawasaki S (2016) Microbially induced sand cementation method using *Pararhodobacter* sp. strain SO1, inspired by beachrock formation mechanism. Mater Trans 57(3):428–437. https://doi.org/10.2320/matertrans.M-M2015842

Das N, Kayastha AM, Srivastava PK (2002) Purification and characterization of urease from dehusked pigeonpea (*Cajanus cajan* L) seeds. Phytochemistry 61:513–521. https://doi.org/10.1016/S0031-9422(02)00270-4
De Muynck W, De Belie N, Verstraete W (2010) Microbial carbonate precipitation in construction materials: a review. Ecol Eng 36:118–136
DeJong JT, Soga KS, Kavazanjian E et al (2013) Biogeochemical processes and geotechnical applications: progress, opportunities, and challenges. Geotechnique 63:287–301
Dilrukshi RAN, Watanabe J, Kawasaki S (2015) Sand cementation test using plant-derived urease and calcium phosphate compound. Mater Trans 56(9):1565–1572
El-Hefnawy ME, Sakran M, Ismail AI et al (2014) Extraction, purification, kinetic and thermodynamic properties of urease from germinating *Pisum Sativum* L. seeds. J BMC Biochem 15(1):15. https://doi.org/10.1186/1471-2091-15-15
El-Shora HM (2001) Properties and immobilization of urease from leaves of *Chenopodium album* (C3). Botan Bull Acad Sinica 42:251–258
Fidaleo M, Lavecchia R (2003) Kinetic study of enzymatic urea hydrolysis in the pH range 4–9. Chem Biochem Eng Q 17:311–318
Frankenberger WT.jr, Tabatabai MA (1982) Amidase and urease activity in plants. Plant Soil 64 (2):153–166. https://doi.org/10.1007/bf02184247
Gerilla GP, Teknomo K, Hokao K (2007) An environmental assessment of wood and steel reinforced concrete housing construction. Build Environ 42(7):2778–2784
Goring G, Chin CC (1965) Urease. IV. Its reaction with N-ethylmaleimide and with silver ion. Biochim Biophys Acta 99:418–426. https://doi.org/10.1016/S0926-6593(65)80196-5
Goring G, Fuchs E, Butler LG et al (1962) Some properties of urease. Biochemistry 1:911–916. https://doi.org/10.1021/bi00911a026
Harkes MP, Van Paassen LA, Booster JL et al (2010) Fixation and distribution of bacterial activity in sand to induce carbonate precipitation for ground reinforcement. Ecol Eng 36:112–117. https://doi.org/10.1016/j.ecoleng.2009.01.004
Hausinger RP (1993) Biochemistry of nickel. Plenum Press, New York Chapter 3
Hirayama C, Sugimura M, Saito H et al (2000) Purification and properties of urease from the leaf of mulberry, *Morus alba*. Phytochemistry 53:325–330. https://doi.org/10.1016/S0031-9422(99)00521-X
http://www.ufrgs.br/laprotox/en/what-we-do/research-lines/plant-ureases
Kanafari A, Khams FN, Sepahy AA (2011) An investigation of biocement production from hard water. Middle-East J Sci Res 7:964–971
Karol RH (ed) (2003) Chemical grouting and soil stabilization. M. Dekker, New York
Kawasaki S, Murao A, Hiroyoshi N et al (2006) Fundamental study on novel grout cementing due to microbial metabolism. J Jpn Soc Eng Geol 47:2–12 (in Japanese with English abstract)
Knorr B (2014) Enzyme-induced carbonate precipitation for the mitigation of fugitive dust. A thesis for the degree Master of Science, Arizona State University
Lai CM, Tabatabai MA (1992) Kinetic parameters of immobilized urease. Soil Biol Biochem 24:225–228. https://doi.org/10.1016/0038-0717(92)90222-J
Marzadori C, Miletti S, Gessa C et al (1998) Immobilization of jack bean urease on hydroxyapatite: urease immobilization in alkaline soils. Soil Biol Biochem 30:1485–1490. https://doi.org/10.1016/S0038-0717(98)00051-0
Mitchell JK, Santamarina JC (2005) Biological considerations in geotechnical engineering. J Geotech Geoenviron Eng 131:1222–1233. https://doi.org/10.1061/(ASCE)1090-0241(2005)131:10(1222)
Mora EP (2007) Life cycle, sustainability and the transcendent quality of building materials. Build Environ 42(3):1329–1334
Nam IH, Chon CM, Jung KY et al (2014) Calcite precipitation by ureolytic plant (*Canavalia ensiformis*) extracts as effective biomaterials. KSCE J Civil Eng, (0000) 00(0):1–6. https://doi.org/10.1007/s12205-014-0558-3
Natarajan KR (1995) Kinetic study of the enzyme Urease from *Dolichos biflorus*. J Chem Educ 72 (6):556–557. https://doi.org/10.1021/ed072p556

Nemati M, Voordouw G (2003) Modification of porous media permeability, using calcium carbonate produced enzymatically in situ. Enzyme Microbial Technol 33:635–642. https://doi.org/10.1016/S0141-0229(03)00191-1

Neupane D, Yasuhara H, Kinoshita N (2014) Optimization of treatment area of enzyme mediated calcite cementation: a novel soil improvement technique. In: 4th international conference on geotechnique, construction materials and environment, Brisbane, Australia

Neupane D, Yasuhara H, Kinoshita N et al (2013) Applicability of enzymatic calcium carbonate precipitation as a soil-strengthening technique. J Geotech Geoenviron Eng 139:2201–2211. https://doi.org/10.1061/(ASCE)GT.1943-5606.0000959

Park SS, Choi SG, Nam IH (2014) Effect of plant induced calcite precipitation on the strength of sand. J Mater Civ Eng 06014017:1–5. https://doi.org/10.1061/(ASCE)MT.1943-5533.0001029

Pervin MS, Jahan MGS, Masud Rana AYKMd et al (2013) Effects of some environmental variables on urease in germinating chickpea (*Cicer arietinum L*) seed. J Stress Physiol Biochem 9:345–356

Pettit NM, Smith ARJ, Freedman RB et al (1976) Soil urease: activity, stability, and kinetic properties. Soil Biol Biochem 8:479–484. https://doi.org/10.1016/0038-0717(76)90089-4

Ramachandran SK, Ramakrishnan V, Bang SS (2001) Remediation of concrete using micro-organisms. J ACI Mater 98:3–9

Singh R, Nye PH (1984) The effect of soil pH and high urea concentrations on urease activity in soil. J Soil Sci 35:519–527. https://doi.org/10.1111/j.1365-2389.1984.tb00609.x

Sondi I, Sondi BS (2005) Influence of the primary structure of enzymes on the formation of $CaCO_3$ polymorphs: a comparison of plant (*Canavalia ensiformis*) and bacterial (*Bacillus pasteurii*) ureases. Langmuir 21:8876–8882. https://doi.org/10.1021/la051129v

Srivastava PK, Kayastha AM, Srinivasan (2001) Characterization of gelatin immobilized pigeonpea urease and preparation of a new urea biosensor. Biotechnol Appl Biochem 34:55–62. https://doi.org/10.1042/ba20010016

Stocks-Fischer S, Galinat JK, Bang SS (1999) Microbiological precipitation of $CaCO_3$. Soil Biol Biochem 31:1563–1571. https://doi.org/10.1016/S0038-0717(99)00082-6

Sumner JB (1926) The isolation and crystallization of the enzyme urease. J Biol Chem 69:435–441

Sumner JB (1951) Urease. In: The enzymes, vol 1. Academic Press, New York, pp 873–892

Sung HY, Lee WM, Chiou MJ et al (1989) A procedure for purifying jack bean urease for clinical use. Proc Natl Sci Counc Repub China B 13:250–257

Terajima R, Shimada S, Oyama T et al (2009) Fundamental study of siliceous biogrout for eco-friendly soil improvement. J Jpn Soc Civil Eng Ser C (Geosph Eng) 65:120–130 (in Japanese with English abstract)

US Army Corps of Engineers (1995) Engineering and design: Chemical grouting, Engineer Manual 1110-1-3500

Van MA, van den Ham GA, Blauw M et al (2011) Preventing internal erosion phenomena with the BioGrout process. In: Proceedings the 15th European conference on soil mechanics and geotechnical engineering-geotechnics of hard soils—weak rocks, pp 1079–1084. https://doi.org/10.3233/978-1-60750-801-4-1079

Van Paassen LA, Harkes MP, Van Zwieten GA et al (2009) Scale up of BioGrout: a biological ground improvement method. In: 17th international conference on Soil Mechanics and Geotechnical Engineering, pp 2328–2333

Weaver TJ, Burbank M, Lewis A et al (2011) Bio-induced calcite, iron and manganese precipitation for geotechnical engineering applications. Geo-Frontiers:3975–3983. https://doi.org/10.1061/41165(397)406

Wright CI, Van-Buren L, Kroner CI et al (2007) Herbal medicines as diuretics: a review of the scientific evidence. J Ethnopharmacol 114:1–31. https://doi.org/10.1016/j.jep.2007.07.023

Yamazaki H, Maeda K, Takahashi K et al (1998) Development of counter-measure against liquefaction by using solution type grout. Tech Note Port Harb Res Inst 905:29 (in Japanese)

Yasuhara H, Neupane D, Hayashi K et al (2012) Experiments and predictions of physical properties of sand cemented by enzymatically-induced carbonate precipitation. Soils Found 52:539–549. https://doi.org/10.1016/j.sandf.2012.05.011

Material Properties of Agriculture Straw Fibre-Reinforced Concrete

Chee Seong Chin and Bhooma Nepal

Abstract China is the world largest agricultural country with the sown grain area of 113.34 million hectares recording a total grain output of 576.35 million tonnes in 2015 (NBS China 2016). The high production of agricultural products has resulted in a significant amount of straw from crops (e.g. wheat, rice and corn) where burning has been the most convenient way of disposal. The consequence has caused serious waste of natural resources, high carbon emissions, air pollution and environmental issues (haze) particularly in the central part of China such as Hunan, Hubei, Anhui, Jiangxi and Jiangsu provinces. The Chinese Government reacted to this problem by introducing environmental regulation since 2000 with an aim to diminish the straw-burning activities. However, this has not been shown to be a viable solution due to the lack of robust, cost-effective and sustainable recycling/reuse technology. The common research dealing with agricultural straw has been the substitution for fertilizer, conversion to bioenergy and making recycled composites but these techniques are still at their early stage of development and lack of economic advantage. The combination of straw with cement-based materials would have the greatest potential for its widespread application due to the rapid development of China in construction industry. However, recent research has been primarily restricted to adopt straw fibres as reinforcements in concrete masonry blocks where its fundamental material properties are still not fully understood. Fibre-reinforced concrete (FRC) has been a widely used construction material because of its exceptional performance, efficiency and cost-saving features. In order to save carbon emission and reduce environmental impact, by making an effective use of straw fibres in concrete is no doubt a significant step towards low carbon and sustainable construction in the era of rapid economic growth and industrialisation in China. The main aim of this research is to advance knowledge on the mechanical characteristics of agricultural straw fibre-reinforced concrete (ASFRC) that involves the investigation of its compressive and tensile performance.

C. S. Chin (✉) · B. Nepal
Xi'an Jiaotong-Liverpool University, Suzhou, China
e-mail: Chee.Chin@xjtlu.edu.cn

V. Achal and A. Mukherjee (eds.), *Ecological Wisdom Inspired Restoration Engineering*, EcoWISE, https://doi.org/10.1007/978-981-13-0149-0_6

Keywords Agricultural straw · Fibre-reinforced concrete · Uniaxial compressive strength · Residual flexural tensile strength

1 Introduction

The significant production of agricultural outputs to accommodate burgeoning population growth in China has resulted in a substantial amount of agricultural waste like wheat, rice and corn straws where the main method of disposal has been by means of uncontrolled burning which has caused a massive impact on carbon emissions and environmental pollution. Therefore, by making a proper utilisation of agricultural straw fibres is no doubt a significant step towards low carbon and sustainable construction.

The technology of employing natural fibres to reinforce cementitious materials can be traced back to prehistoric times where ancient Egyptian used straw fibres to strengthen mud bricks approximately 3000 BC (RTA 2006). The process of making straw fibre-reinforced clay bricks is vividly depicted in paintings on limestone in the eighteenth dynasty tomb of Rekhmire where ancient Egyptians were shown mixing the piles of clay with straw fibre and water, which was then compacted into oblong forms using a wooden frame before being laid out to dry in the sun.

Concrete is brittle when the confining pressure is low and contains numerous microcracks where rapid propagation of these cracks under applied loads is considered responsible for its weakness in tension. Therefore, it is reasonable to employ randomly distributed fibres to bridge the onset of the cracks to increase the tensile strength and ductility of the material. ACI Committee 116 (2000) has defined the term FRC as concrete containing dispersed, randomly oriented fibres.

Over few decades have passed since the initiation of the modern era of research and development on FRC where the use of fibres is a well-known method to enhance the ductility, cracking resistance and long-term integrity of the conventional concrete matrix. FRC has been a widely used construction material because of its exceptional performance, efficiency and cost-saving features. However, the potential use of agricultural fibres in modern construction is less well researched particularly when compared to steel and polymeric fibres. This research project is targeted to perform a detailed experimental investigation of the mechanical behaviour of concrete with the incorporation of rice and wheat straw fibres at varying fibre volume fractions.

2 Agricultural Fibres

Fibres from agricultural by-products or waste, such as wheat straw, soy stalk and rice husk, have received considerable attention due to their nature as a by-product of food. Wheat and rice straw, rice husk, rice husk ash, coir, corn stalk, banana fibre and pineapple leaf fibres are lignocellulosic biomass that have been successfully utilised to reinforce polymeric matrices.

Rice is grown all over the world but primarily in tropical and sub-tropical climates. Some rice production also occurs in Mediterranean climates. Wheat is produced mostly in moderate and Mediterranean climates. Combined harvest of both rice and wheat is also in practice in a few regions and climates. Depending on location, temperature and availability of water, rice production is combined with wheat production or other crops in some regions. In tropical climates and with sufficient water availability, it is found that more than one crop of rice can be grown per year. It also signifies that straw is generated more than once a year. Wheat is often grown in rotation with other crops, especially sugar beets. For rice, the main producers are located in South and East Asia: China, India, Pakistan, Indonesia, Bangladesh and Vietnam. Wheat is produced in large quantities in Southern Asia, Eastern Europe, Northern America and Eastern and Central Asia. For the countries of the Europe, wheat is by far the more dominant crop in comparison to rice (Bakker et al. 2013).

Rice grain, often referred to as paddy rice or rough rice is primarily used for human food, with the remainder used for animal feed and other uses. Most wheat (more than two-thirds) produced is consumed by humans, generally in the form of flour to produce bread. About 17% of global production is used for animal feed, although this varies from country to country (in Europe and North America, more wheat is used for feed) (Bakker et al. 2013). Globally, approximately 740 million tonnes of rice paddy and a similar amount of wheat is produced each year (FAOSTAT 2014). On average 20% of the product is husk, giving an annual total production of 300 million tonnes. In most of the rice- and wheat-producing countries, much of the husk produced from processing is either burnt or dumped as waste. Only a minor portion of the agricultural residues is reserved as animal feed or household fuel. However, a huge quantity of the remaining straw and stems is not used as industrial raw materials and is burnt in the fields or disposed (Nepal et al. 2015). The burning of husk leaves a residue called husk ash. For every 1000 kg of paddy milled, about 220 kg (22%) of husk is produced (Kumar et al. 2012).

In the context of international research, a considerable effort is going on in the exploitation of fast growing, renewable, cheap agricultural crops and crop residues as possible fibre reinforcement in concrete. The basic advantage of natural fibres is that they are a low cost and widely available resource in many agricultural areas. Natural fibre-reinforced materials are environmental friendly materials producing less greenhouse gas emissions and pollutants. The use of natural fibres as reinforcement is a way to recycle these fibres and to produce a high-performance construction material. But the disadvantage of natural fibres is that they have a high variation of their properties, which may lead to unpredictable concrete properties. These challenges can be resolved by optimum use of fibre with respect to its quantity and aspect ratio. Minimal research has been carried out and the results of agricultural fibres have been compared with that of synthetic fibres (Nepal et al. 2017a, b). The fibres should be used in a manner so as not to degrade the strength of the concrete and also to make the concrete sustainable. But the disadvantage of natural fibres is that they have a high variation of their properties, which can lead to unpredictable concrete properties.

3 Characteristics of the Adopted Agricultural Straw Fibres

Several material properties of rice and wheat straws including both chemical and physical properties of fibres were investigated and they are presented in Table 1. Vario MICRO cube was used for the simultaneous CHNS analysis of different chemicals and in our experiments. Prior to the analysis, straw was dried at 105 °C for 24 h in an oven. For each sample, 10–15 mg are weighed into tin boats with a microbalance and analysed with a standard method. The combustion process occurred at high temperature in the combustion chamber with pure oxygen. For this analysis, helium is used as carrier gas and four measurements were conducted for each sample. Oxygen was the prime element in both the straw fibres at 51.61% for rice and 48.74% for wheat followed by carbon, nitrogen and hydrogen were in the trace amount.

Fibertec™ 2010 was used in our experiments to determine the chemical composition of fibres according to the Van Soest method. Solid samples were dried at 105 °C for 4 h and then subjected to defeat using acetone in cold extraction unit. After removal of fat, the sample was kept in hot extraction unit. Hemicellulose is extracted from the Neural Detergent Fibre (NDF). The similar process is repeated to extract cellulose and finally, in cold extraction unit to extract Acid Detergent Lignin (ADL) using 72% H_2SO_4. After that, it is cooled to room temperature in the desiccator and ignited at 525 °C in furnace to find the quantity of lignin.

The water absorption coefficient is an important parameter in the mix design of concrete based on plant fibre. The amount of water absorbed by a material under specified test conditions is determined by the water absorption test. It is commonly

Table 1 Material properties of agricultural straw fibres

Types of properties		Rice straw	Wheat straw
Chemical properties	Nitrogen (%)	0.44	0.17
	Carbon (%)	42.00	45.05
	Hydrogen (%)	5.95	6.04
	Oxygen (%)	51.61	48.74
	Cellulose (%)	47.48	35.00
	Hemicellulose (%)	14.14	21.30
	Lignin (%)	6.38	17.50
Physical properties	Water absorption capacity (24 h) (%)	332.33	269.55
	Density (g/cm^3)	1.10	0.60
	Tensile Test (N/mm^2)	198.54	278.00

expressed as a weight per cent of the test specimen. The fibres are oven dried for 24 h at 60 °C and first soaked in water, and brought to the saturated condition, weighed and brought to an oven dry condition. The water absorption coefficient was determined by submerging the samples in the water for several durations (1, 2, 5, 10, 30, 60 min and 24 h). The samples are weighed before and after wetting and water absorption was determined. Water absorption was higher in the rice fibres at 332.33%, which exceeded the wheat water absorption by 23.30%.

For fibre density, several aspects of its properties were put into consideration. Density was determined using a true density analyser in Quantachrome's material characterisation laboratory, which works on the principle of Helium pycnometer. Since helium, which can enter even the smallest voids or pores, is used to measure the unknown volume of a material with a known weight, the final result is often referred to as the true density. The principle of this method is based on putting a sample of known mass into a cell of known volume. Then helium is introduced into the cell, which is at a vacuum, so helium occupies the entire volume of the cell that is not occupied by the sample. The actual volume of the sample can be determined since the volume of the cell is known. The actual density of the sample material then can be accurately determined by this technique.

The tensile strength of the straw can be determined from a simple grip mechanism. A machine is to be used to grip the specimen around its two grips and the tension force was generated on the straw (Nepal et al. 2017a, b). The cross-sectional dimension of the fibres was determined using a digital vernier calliper. Tensile test showed that rice straw fibres and wheat straw fibres had a noteworthy strength of 198.54 MPa and 278.00 MPa, respectively. In comparison, the tensile strength of rice was 71% of that of wheat straw.

4 Concrete with Agricultural Straw Fibres

Using fibre reinforcements is an effective and economical technique to improve the toughness of cementitious materials. A clear understanding of the mechanical properties of ASFRC is of major importance before its industrial application can be fostered. Some of these properties, such as compressive strength, can be measured by standard testing methods used for plain concrete. Other material properties such as tensile strength and energy absorption capacity, however, are quite different from those of the conventional concrete, and these must be evaluated by special testing method. In order to obtain the useful design parameters to facilitate the practical application of ASFRC, compressive strength (including stress–strain relation) and tensile behaviour (in terms of residual flexural tensile strength) will be performed with reference to BS EN 12390-3 (2009) and BS EN 14651 (2005), respectively. The plain concrete is designed to have 37 MPa of compressive strength using the BRE method (Teychenné et al. 1997). The proportion of mix was determined as 1.00: 0.55: 2.14: 3.35 (cement:water:sand:coarse aggregate). The aspect ratio is conventionally defined by the fibre length over fibre diameter. For the fibres used in

Table 2 Description of concrete mixes

Label	Fibre type	Fibre volume fraction (%)	Fibre length	Fibre diameter (Equivalent)
P	Plain concrete	0.00	–	–
0.25R	Rice straws	0.25	45 mm	0.47 mm
0.50R		0.50		
1.00R		1.00		
0.25W	Wheat straws	0.25		
0.50W		0.50		
1.00W		1.00		

Table 3 Summary of testing results

Label	Cube strength (MPa)	Cylinder strength (MPa)	Flexural load (kN)	Limit of proportionality (MPa)	Energy absorption up to 3.02 mm (KNmm)
P	47.10	39.75	15.80	5.06	9.07
0.25R	38.33	30.03	14.47	4.63	5.35
0.25W	47.11	42.43	17.55	5.61	7.97
0.50R	37.50	26.17	15.22	4.87	4.90
0.50W	42.33	32.75	15.79	5.05	6.34
1.00R	26.08	22.58	13.20	4.22	5.17
1.00W	36.98	24.55	14.46	4.63	4.07

our experiments, the aspect ratio was around 96. All the fibres were straight and monofilament fibres. The detailed descriptions of the concrete mixes and the fibres adopted can be found in Table 2. The summary of testing results for all concrete mixes is shown in Table 3.

4.1 Uniaxial Compressive Strength

The compressive properties of FRC is not much different from plain concrete where this property can be measured using the same testing methods as those specified for ordinary hardened concrete (Bentur and Mindess 1990; Mindess et al. 1995). The uniaxial compression tests will be performed according to BS EN 12390-3 (2009) where a loading rate of 0.1 mm/min was adopted. The cubes and cylinders used for compression test were, respectively, 150 mm × 150 mm × 150 mm and 150 mm (diameter) × 300 mm (height) in size. When casting the specimen, external vibration is preferred since an internal vibrator may adversely influence the random distribution and alignment of fibres. The frictional forces which develop between the platen plates of the testing machine and the contact faces of the test specimen produce a multiaxial stress state which would increase the compressive strength of the concrete (Kong and Evans 1992). In order to ensure parallel loading faces, constant height and avoid stress concentration due to rough casting surface, adjustment of the cylindrical specimen by grinding has been carried out. Adjustment of cube specimen is unnecessary since the smooth faces perpendicular to the direction of casting would be placed in contact with the plates of the testing machine.

Figure 1 shows the stress versus strain comparison of plain concrete, as well as 0.25%, 0.5% and 1% rice, and wheat straw-reinforced concrete cubes and cylinders. Wheat straw has increased stress, which indicates it has better compressive strength capacity compared to plain concrete and rice straw fibres for 0.25% volume fraction. The cylinder crushing compressive strength for plain concrete was 47.10 KN for cubes and 39.75 KN for cylinders. The maximum stress for cylinder for 0.25% volume fraction was 29.57 for rice straw and 41.67 for wheat straw. For wheat straw, this is an increase of 6% when compared to plain concrete. For 0.25% cube specimens, the increase in compressive strength were seen in wheat straw at 47.79 MPa when compared to 47.02 MPa for plain concrete. Rice straw fibre has good post crack elongation similar to that of plain concrete. It is also evident that for 0.5% rice and wheat straw fibres, wheat has a better compressive stress when compared to rice but the strength has decreased in this volume in comparison to plain concrete. There was a decrease of 21% for 0.5% rice straw cube strength and 10.8% for wheat straw. For cylindrical compressive strength, there was a decrease of 33% for rice and 17.8% for wheat straw with 0.5% fibres.

The maximum decrease in strength was seen for 1% volume fraction for rice straw for both cube and cylindrical specimens. However, wheat straw fibres had comparatively better strength than rice straw. Thus, from the above test results, it can be seen that with the increase in fibre reinforcement, there is a decrease in compressive stress but with 0.25% fibre volume, the compressive stress is significantly improved for wheat straw fibres. The test setup of beam, cylinder and cubes are shown in the Fig. 2. Figure 3 presents the failure pattern of each specimen.

4.2 Residual Flexural Strength

The term toughness can be generally defined as the area under a load-deflection curve which is a measure of the energy absorption capacity of a material. It can also

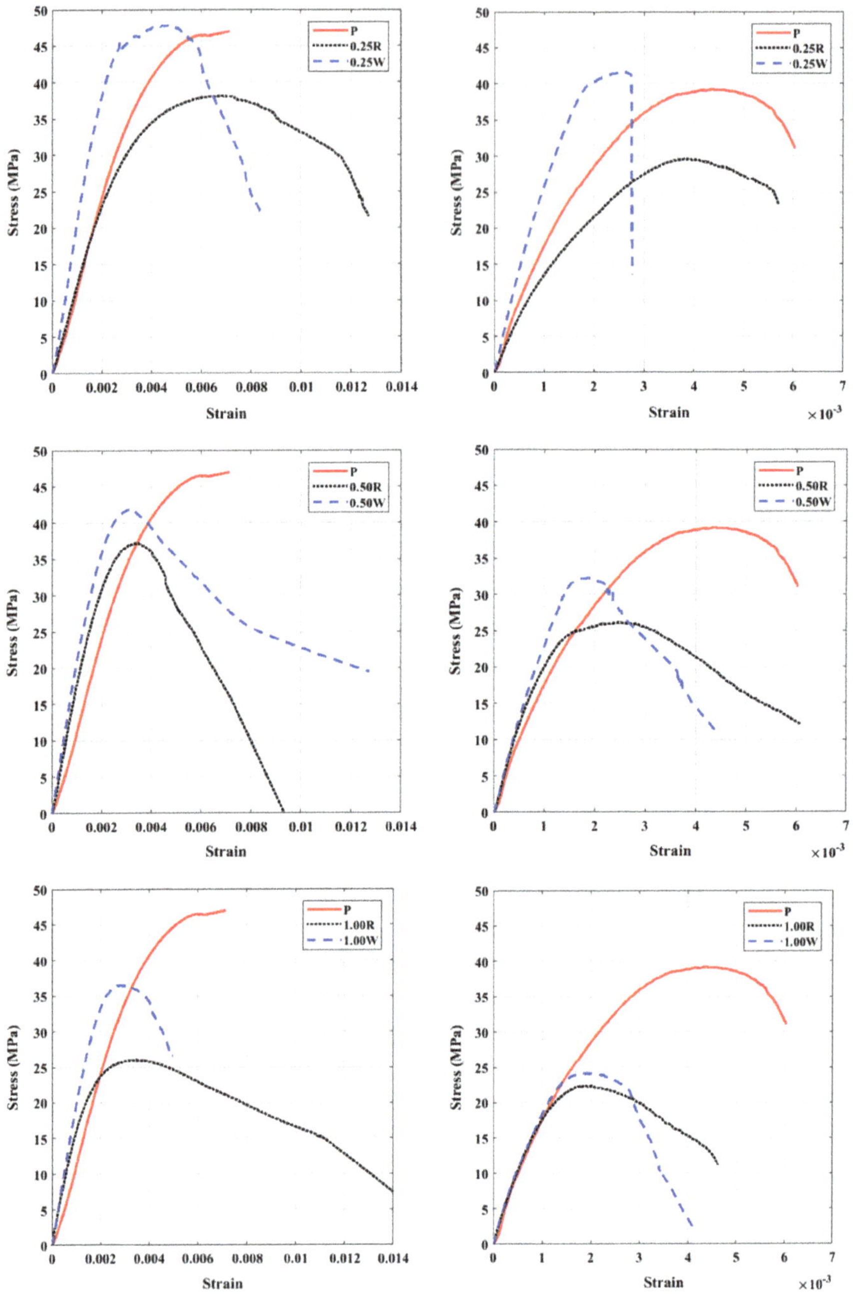

(a) Cubical specimens (b) Cylindrical specimens

Fig. 1 Stress–strain responses of all concrete mixes

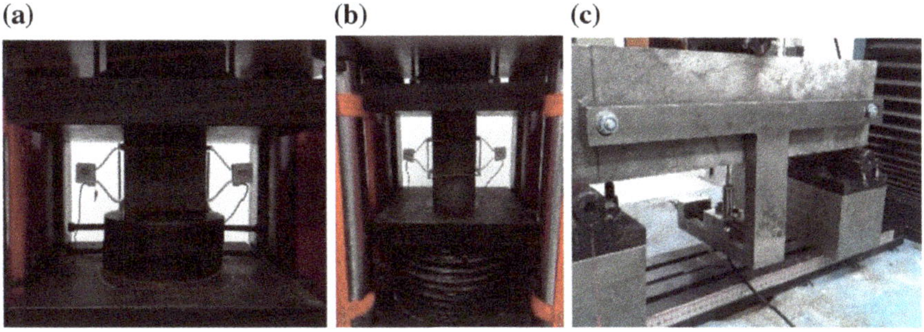

Fig. 2 Test setup for **a** Cubes, **b** Cylinders and **c** Beams

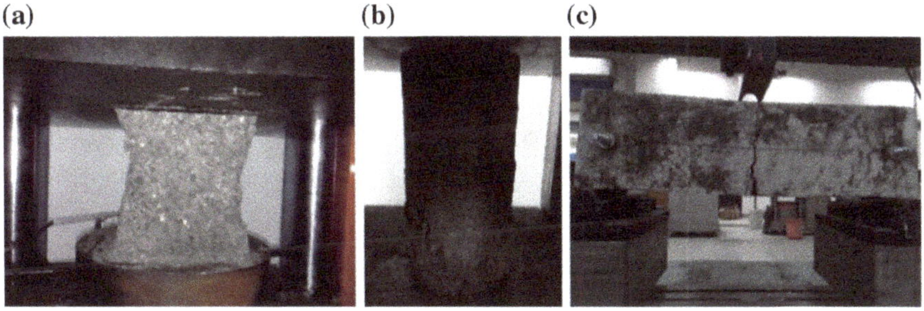

Fig. 3 Failure pattern for **a** Cubes, **b** Cylinders and **c** Beams

be used to characterise the material's ability to resist fracture when subjected to certain loads (e.g. dynamic or impact loading). BS EN 14651 (2005) has been one of the most commonly used methods to investigate the flexural toughness of cementitious materials containing fibres since it is easy to perform and it simulates the loading conditions for many practical applications. It can be used to appropriately characterise the cracking resistance and ductility of the material due to the addition of fibres. The flexural behaviour of ASFRC has been evaluated by providing the beam specimens with centre point loading (span = 500 mm) where the specimen was a notched prism with a square cross section of 150 × 150 mm (Fig. 2). The specimen length was fixed at 550 mm. The width of the notch was 5 mm whilst the height was 25 mm located at the bottom centre of the beam. Linear variable differential transformer (LVDT) to an accuracy of 0.01 mm was used with the help of a 1 mm thick aluminium plate to measure the deflection. A frame was used to hold the LVDT(s) to ensure accurate measurement of the mid-span deflection to avoid errors due to seating or twisting of the specimen on its support (see Fig. 2). The testing machine should be operated so that the deflection increases at a constant rate of 0.05 mm/min. The data acquiring system should be set to have a recording rate not lower than 5 Hz. When deflection reaches 0.125 mm, a constant loading rate of 0.2 mm/min should be used.

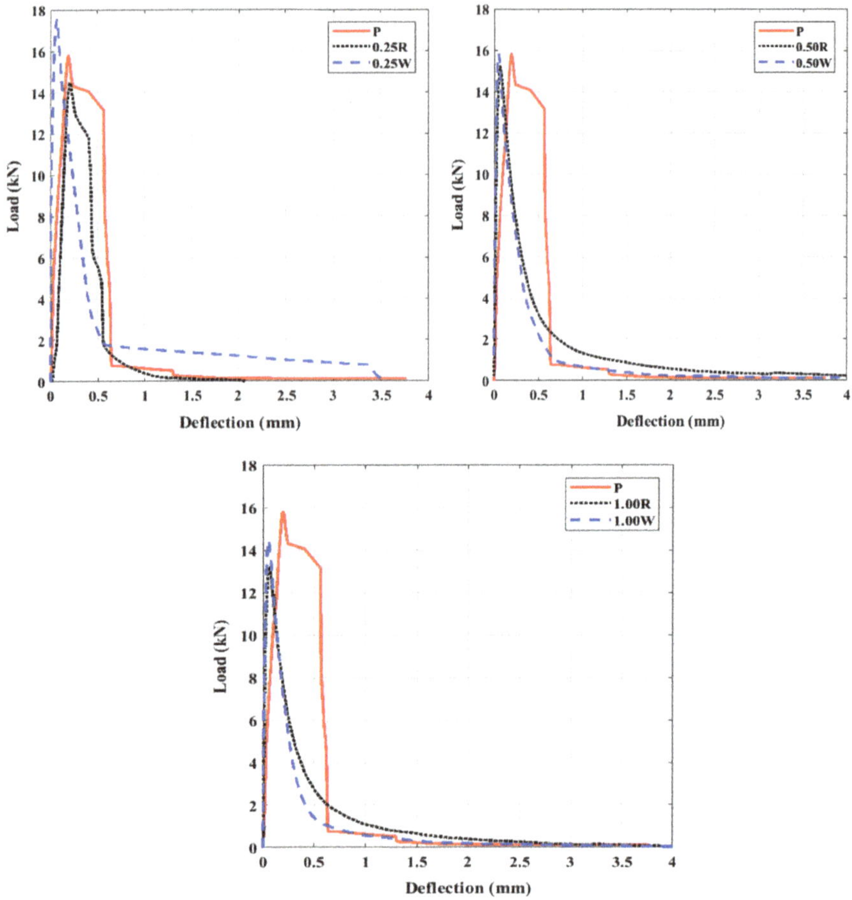

Fig. 4 Load-deflection responses of all concrete mixes

Figure 4 shows the load-deflection comparison of plain concrete beams and those reinforced with rice and wheat straw fibres at various fibre percentages. The flexural strength test results indicate a sudden brittle failure of plain concrete around 0.5 mm whereas the straw fibres have better post-yield ductility and brittle failure is not evident in the beams.

For 0.25% natural fibre, the maximum flexural load sustained by the beam was 14.47 kN for rice straw fibres and 17.54 kN for wheat straw. In comparison to plain concrete, the wheat straw fibre increased by 11% compared to plain concrete. Similarly, the LOP value was higher at 5.61 MPa for wheat straw compared to 5.05 MPa for plain concrete. The flexural load for higher volume fraction, however, decreases in comparison to the plain concrete.

The significant observation from this test data is the post-yield ductility of fibre reinforcement. Comparing the 2.17 mm CMOD load, the plain concrete sustains only 0.14 kN of load at that instant whereas, wheat straw recorded 1.16kN of load. This is also observed in rice straw for 0.5% volume fraction with rice straw at 0.5 kN flexural load. The important characteristic of fibre reinforcement is observed from the load-displacement responses from Fig. 4. Fibre reinforcement does not show a sudden drop in load-carrying capacity and implies a ductile behaviour with a gradual decrease of load and longer deflection.

5 Conclusion

From the experimental observations, several conclusions can be drawn:

(1) It can be observed that for 0.25% volume fraction, there is an improvement in compressive strength as well as flexural strength for wheat straw fibres. Wheat has also improved performance in 0.5 as well as 1% volume fraction compared to plain concrete.
(2) There is a substantial increase in compressive strength for 0.25% volume fraction for wheat straw for both cubic and cylindrical specimens. This indicates that wheat can be more advantageous over rice straw fibres. This is also consistent in all the other concrete properties as well.
(3) Rice and wheat fibres as observed from its chemical composition are basically made up of cellulose and other organic derivatives of cellulose. This needs to be treated to avoid decay and other detrimental effects when they are used with concrete.
(4) Wheat straw fibres also have high water absorption. Whilst mixing with concrete, this need to be carefully taken into account so that the desired mix can be achieved.
(5) Further investigation into material properties and interaction of fibres with concrete needs to be studied to arrive in conclusion regarding the optimum volume fraction of natural fibre for use in concrete.

References

ACI Committee 116 (2000) Cement and concrete terminology, ACI manual of concrete practice, ACI 116R-00, 116R-16

Bakker R, Elbersen W, Poppens R, Lesschen JP (2013) Rice straw and wheat straw. Potential feedstocks for the biobased economy. Wageningen UR, Food and Biobased Research. Netherlands programmes sustainable biomass, Wageningen, Netherlands

Bentur A, Mindess S (1990) Fibre reinforced cementitious composites. Elsevier Applied Science, London

BS EN 14651 (2005) Test method for metallic fibered concrete—measuring the flexural tensile strength (Limit of Proportionality (LOP), Residual). British Standard Institution

BS EN 12390-3 (2009) Testing hardened concrete—part 3: compressive strength of test specimens. British Standard Institution

FAOSTAT (2014) Production. http://faostat3.fao.org/download/Q/QC/E. Accessed 20 Dec 2015

Kong FK, Evans RH (1992) Reinforced and prestressed concrete, 3rd edn. Springer, Berlin

Kumar A, Mohanta K, Kumar D, Parkash O (2012) Properties and industrial applications of rice husk: a review. Int J Emerg Technol Adv Eng 2(10):86–90

Mindess S, Taerwa L, Lin YZ, Ansari F, Batson G (1995) Standard testing, high performance fiber reinforced cement composites 2 (HPFRCC 2). In: Proceedings of the second international RILEM workshop, 1995, pp 383–421

NBS China (2016) Statistical communiqué of the People's Republic of China on the 2016 National Economic and Social Development. National Bureau of Statistics China. http://www.stats.gov.cn/tjsj/ndsj/2016/indexeh.htm. Accessed 31 Aug 2017

Nepal B, Chin CS, Jones S (2015) A review on agricultural fibre reinforced concrete. In: Sustainable buildings and structures: proceedings of the 1st international conference on sustainable buildings and structures, Suzhou, PR China, 29 Oct–1 Nov 2015, p 125. CRC Press

Nepal B, Chin CS, Jones S (2017a) Cementitious material with bio-based recyled agricultural waste. In: ICBBM 2017, 2nd international conference on bio-based building materials, clermont ferrand, France, 21 June 2017. RILEM, pp 589–596

Nepal B, Chin CS, Jones S (2017b) Comparative study of the use of man-made and natural fibre reinforced concrete for ground floor slab applications. Int J Struct Civil Eng Res 6(3):182–185. https://doi.org/10.18178/ijscer.6.3.182-185

RTA (2006) Bridge types in NSW-historical overviews. Roads and maritime services (formerly Roads and Traffic Authority). NSW State Government Agency

Teychenné DC, Franklin RE, Erntroy HC, Nicholls J, Marsh BK (1997) Design of normal concrete mixes, 2nd edn, UK

Agro-Industrial Discards and Invasive Weed-Based Lignocelluloses as Green Building Materials: A Pertinent Review

Nadeem Akhtar and Seema Patel

Abstract The detrimental impacts of conventional building materials on environmental health have raised global concerns. While the fast-depleting fossil-derived materials are adding to global warming, chemically-synthesized materials are polluting the air, water, and soil. To promote health and to meet sustainability target, the importance of eco-friendly green building materials is being realized. In this regard, the utilization of most abundant and bio-renewable lignocellulosic biomass appears promising. Biorefineries based on lignocellulosic waste materials have reached a good degree of maturity for the production of sustainable biofuels and other value-added products. However, green building materials reinforced with lignocellulose-derived products are at a nascent stage. Their wider acceptance would require robust technical development and the evaluation of product performance to hit the marketplace. Aiming for a clean future through smart technology, this chapter delineates the viability potential and hurdles in the path of using agro-industrial discards and invasive weed-based lignocelluloses in building materials. How rational planning and execution of these objectives might revolutionize urbanization with minimal threat to the environment has been highlighted.

Keywords Lignocelluloses · Biorefinery · Building materials · Green concrete
Sustainability

N. Akhtar (✉)
Department of Animal Biosciences, University of Guelph, Guelph, ON N1G 2W1, Canada
e-mail: nadeembiotech@gmail.com

S. Patel
Bioinformatics and Medical Informatics Research Center, San Diego State University,
5500 Campanile Dr, San Diego, CA 92182, USA

V. Achal and A. Mukherjee (eds.), *Ecological Wisdom Inspired Restoration Engineering*, EcoWISE, https://doi.org/10.1007/978-981-13-0149-0_7

1 Introduction

The detrimental impact of conventional building materials on environmental health, due to raised levels of greenhouse gases (GHGs) has fueled the search for alternative resources (Ma et al. 2016). According to the Environmental Protection Agency (EPA) Web site, buildings in the United States (USA) are major consumers of energy. They are accountable for 39% of total energy use, 12% of the total water consumption, 68% of total electricity consumption, and 38% of the carbon dioxide emissions. Asphalt pavement construction is an energy consuming method. Indoor environmental quality is compromised by volatile organic compounds (VOCs) such as benzene, toluene, ethylbenzene, m-, p-, and o-xylene (BTEX) emanating from adhesives, sealants, paints, thinners, etc., used in hardwood, plywood, and laminate floorings. Ozone reaction with building materials results in secondary emissions of aliphatic aldehydes, secondary organic aerosols, and other products. Inhalation of these gases leads to headaches, nausea, wheezing, and neurological symptoms (Institute of Medicine (USA) Roundtable on Environmental Health Sciences 2007). Major detrimental effects on the environmental health due to the usage of conventional building materials are shown in Fig. 1. So, big cities are transitioning toward the goal of climate neutrality. Leadership in Energy and Environmental Design (LEED) green building certification label has soared in popularity since its

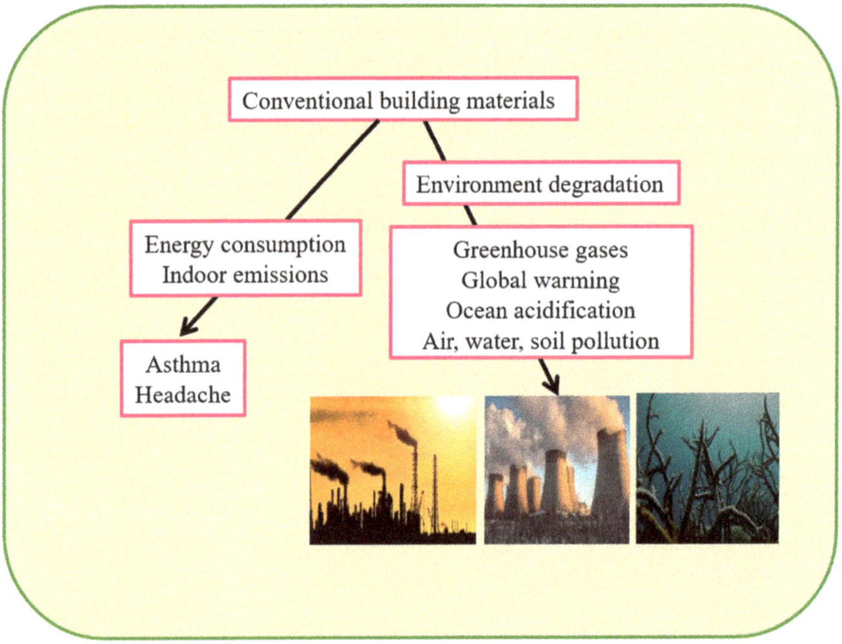

Fig. 1 Major detrimental effects on the environmental health due to the usage of conventional building materials

inception almost two decades back in 2000 (Scofield 2013). Architects are exploring the options for innovative, eco-friendly buildings. The interest in passive houses, which have become benchmark for low energy and efficiency, is soaring. Also, the green building materials minimize the chemical emission. However, the sustainability of the modern architecture in affordable price is still a big challenge for the construction industry across the globe. On the other hand, agro-industrial wastes are piling up. These scenarios have led to the concept of eco-friendly building materials. However, till date, only a limited effort has been made in the utilization of available renewable resources to produce building materials for market entrance.

Lignocellulosic residues, the most abundant, widely available renewable resources have gained attention in this area of research (Kumar and Sharma 2017). Lignocellulosic biomass is composed of cellulose, hemicellulose, lignin, and some amount of pectin (Akhtar et al. 2016). It has already found multiple uses such as the conversion into biofuels by pyrolysis or hydrolysis. Excellent thermo-physical characteristics of the lignocellulosic materials are considered to be ideal for low-carbon buildings (Cheng et al. 2014). Building a passive house is believed to be a sustainable concept for a comfortable living since it has a lesser impact on the environmental health. The importance of lignocellulosic materials as building blocks of such energy-efficient house is promising. The new trend of incorporating agro-wastes to produce construction materials is a key research area towards the eco-friendly construction projects. The traditional concrete is made from cement (12%), water (8%), and aggregates (80%), where the latter two is from natural sources. Cement production pollutes the environment by liberating NO_2, SO_2, PM (10), metals, polychlorinated dibenzo-p-dioxins, and dibenzofurans (Schuhmacher et al. 2004). The usage of cement can be reduced by the addition of alternative, environmentally-benign materials. Hence, it is a judicious move to utilize agro-wastes and contribute towards energy conservation and pollution mitigation.

A wide range of cellulosic waste material is being produced as a result of improved modern agricultural practices, which may pose a threat to the environment, if not recycled or disposed of properly. Currently, most of the wastes are subjected to landfilling, composting, or anaerobic digestion (Väisänen et al. 2016). It is proposed that the agro-waste natural fibres can be used as bio-composites. They are inexpensive and safer alternative to conventional petroleum-derived composites (Väisänen et al. 2016). The plants such as flax, jute, hemp, henequen, kenaf, pineapple leaf fiber, and sisal have been recognized as good sources of biofibres (Mohanty et al. 2002). Using these fibres with natural polymers such as cellulosic plastics; polylactides; starch plastics; soy-based plastics, polyhydroxyalkanoates (bacterial polyesters) are being assessed. The development of green bio-composites promises of multiple utilities, including in the construction sector. Oil palm, corn, bamboo, banana, cotton, soybean, and rice are some other plants with their stem, leaf, seed, fruit, stalk, grass, coir, bagasse, or husk being evaluated as reinforcement fibre source. These fibres are low weight, renewable, degradable, cheaper and possess low-abrasive property (Schaschke and Audic 2014).

Physical, mechanical, and chemical properties of different kind of lignocellulosic materials need to be accessed to judge their suitability for various applications in making building materials. These properties vary with growth conditions, chemical composition, and their extraction methods (Reddy and Yang 2009a, b). In particular, lignocellulosic aggregate and cellulosic fibers are a promising alternative of synthetic, glass, and asbestos fibres. The inhalation of asbestos fibres is responsible for mesothelioma, lung cancer, and pleural and interstitial abnormalities, among other pathologies (Goswami et al. 2013).

The cellulosic fibre and cement composites are primarily used for making partitions, flat ceilings, roofing tiles, and pre-manufactured components. The addition of 0.15% cellulose nanofibres to cement led to a 15 and 20% increment in the flexural and compressive strengths of cement paste (Jiao et al. 2016). The augmented mechanical strength is due to the high degree of hydration and dense microstructure of cement pastes after the fibre addition (Jiao et al. 2016). Similarly, hemp-lime composite showed low shrinkage and high thermal and acoustic insulating properties (Arizzi et al. 2015). Sisal fibre was used in composites, as it is a low-cost resource having high-density and high-specific strength and modulus, with least health hazard. Wheat straw, barley straw, and wood shavings were used as reinforcement fibre in plaster material, which showed lower thermal conductivity when straw fibre content was increased.

Physical, mechanical, and chemical properties of various agricultural wastes have been examined to assess their suitability for various applications. Such characterization will not only help in opening up a new avenue for these fibres but also to emphasize the importance of natural fibres from agricultural waste as future material. The benefits of using lignocellulosic biomass and their products in the building industry have been outlined in Fig. 2.

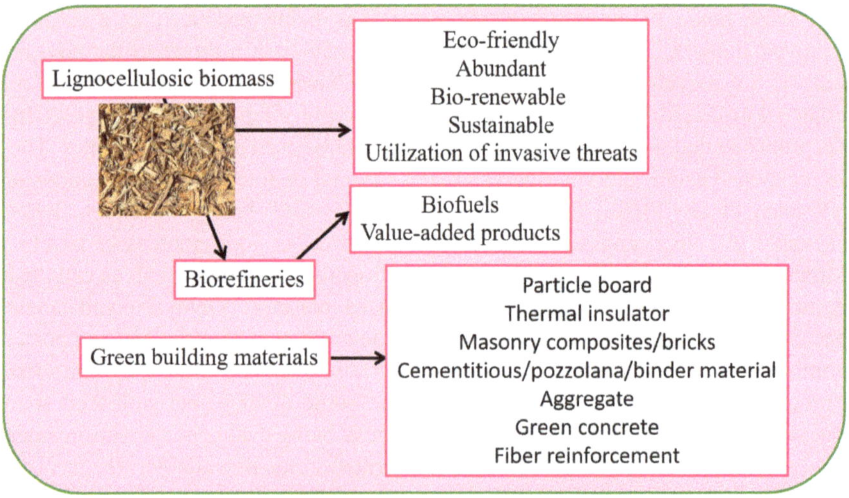

Fig. 2 Benefits of using lignocellulosic biomass and their products in building industry

2 Construction Materials from Lignocellulosic Waste

2.1 *Particle Board*

Particle board is the wood-based panel usually made from wood chips and shavings from the forest waste. The high quality of these materials is mainly due to better control of the homogeneity of the raw materials (Rodolfo de Melo et al. 2014). An environment-friendly binder-less cotton stalk fiberboard (BCSF) was made from cotton stalk without any addition of chemicals (Hua 2016). Fiberboard was made using non-synthetic binders after pre-treatment of the banana bunch using thermo-mechanical aqueous vapour steam explosion technique pre-treatment (Quintana et al. 2009). Particle boards were made from a mixture of the solid waste from tissue paper manufacturing and corn peel and bamboo with good physico mechanical properties (Dugmore et al. 2017).

2.2 *Thermal Insulator*

A thermal insulator is used to reduce the heat transfer between two objects at different temperatures. A new insulating material was developed from lignocellulosic material (jute, flax, hemp, straw) with comparable physical and mechanical properties, which could be used as an insulating material in building industry (Mengeloglu and Karakus 2008; Pandey et al. 2016). Reinforcement of the ratified bagasse fibres to the cementing material reduces the thermal conductivity of the composite, and yields a weaker specific heat.

2.3 *Masonry Composites/Bricks*

Paper mills waste (89–85 wt%) and cotton waste (1–5 wt%) were utilized for the manufacture of waste-create bricks (WCBs) with varying and fixed content of Portland cement (10 wt%) with matching IS 3495 (Part1–3): 1992 standard (Khatib 2016). The addition of un-burnt olive husk to clay bricks resulted in the lowering of thermal conductivity than that of conventional clay bricks (Alami 2009). In another study, the addition of 10 and 20 wt% of olive pomace bottom ash to bricks influenced bulk density and compressive strength, while fulfilling the standards of clay masonry units (Eliche-Quesada and Leite-Costa 2016).

2.4 Cementitious/Pozzolana/Binder Material

Sugar industry waste (bagasse ash) is an effective cement replacement in concrete and pozzolana (a type of volcanic ash used for mortar or for cement that sets under water) with the original ratio of 20% cement, which resulted in 50% reduction of the chloride diffusion, without affecting the properties of the hardened concrete (Amin 2011). Another study also found sugar cane bagasse ash to be an interesting source for preparing alkali-activated binders (Castaldelli et al. 2013). The mortar made of slag, palm oil fuel ash, and rice husk ash as an alkali-activated binder can be used as an alternative to cement (Karim et al. 2014). Ground ash has also been reported to be an effective replacement of the cement in the range of 20–40%. Rice husk ash when used in optimum quantity increases the mechanical properties of concrete, and it might act as a mineral admixture to replace cement.

2.5 Aggregate/Concrete

Structural concrete was made from a coarse aggregate from oil palm shell (OPS). The bulk density and compressive strength of the product was found to be 1850 kg/m^3 and 20–24 N/mm^2, respectively, which meet the strength requirement of lightweight structural concrete. The performance of the mortars and concretes using sugarcane bagasse ash showed comparable results as that of sand-prepared mortars and concretes. The inclusion of paper sludge increased the calorific value while manufacturing lightweight aggregates by sintering at high temperatures (Chen et al. 2016). Brewing industry waste was used with clay to generate lightweight aggregates. The resultant aggregates have good insulating properties and seemed suitable for use on green roofs (Farías et al. 2017).

Bamboo-reinforced lightweight concrete beams were developed and studied for their mechanical properties (Karthik et al. 2017). The heartwood of palmyrah palm has the potential to be exploited as reinforcement in lightly loaded slabs and beams (Baskaran et al. 2014).

3 Ecological Wisdom for Sustainability of Natural and Urban Ecosystem

Rapid industrialization and globalization as a result of population explosion resulted in the invasion of urban infrastructure in agricultural lands and wilderness. To provide employment and shelter to the migrant populations in urban areas, construction technologies use too much material and energy and produces huge quantities of GHGs. The establishment of sustainable cities must be taken into consideration since the percentage of global urban dwellers has risen sharply and is expected to rise further. It is expected to hit 5 billion by 2030 (Lutz and Samir 2010). The megacities require innovative infrastructure to handle this pressure (Achal et al. 2016). Sustainability refers to the creation and maintenance of a productive harmony state that permits the coexistence of humans and nature, fulfilling the socioeconomic and other requirements of the present and future generations (Burger et al. 2015). The management of the concept requires both acceptance of a changing environment and the recognition that appropriate manipulation of the environment to do "real and permanent good" (Wang and Xiang 2016). The ecological wisdom has a profound effect in influencing the sustainability of holistic human and natural ecosystems. To limit the global CO_2 emission, recycling of agro-industrial waste and the production of green building materials can mitigate the detrimental effect on the environment.

4 Issues to Surmount in Lignocellulose-Based Building Materials

A major limitation in using green building materials is that their inclusion in construction is too expensive to be considered economically feasible. Industries are continuously hunting for circular-economy approaches to design, use, and recycle durable products in a cost-effective manner.

The hemp-lime composite is highly hygroscopic, which can affect long-term durability of the building (Arizzi et al. 2015). Such negative aspects ought to be evaluated before using the material. The use of recycled green materials is suspected to be a fire hazard. It is a matter of concern in the certification of the green housing designs.

5 Future Prospective in Lignocellulose-Based Building Materials

Limited resources are available for the developing nations due to the population growth, industrialization, and globalization. Globally, the largest consumer of the materials is the construction sectors and buildings are the sector consuming the largest fraction of energy worldwide. In total, the infrastructure and building construction accounts for approximately 60% of the raw material extracted from the Earth. The buildings made from these materials consume a multitude of products and biological nutrients, and hence, they impose a great threat on water and air quality, energy cycles, natural flora and fauna, and socioeconomic factors.

The use of ash as a filler material in cement and concrete depends on the boiler temperatures and the physical characteristics of the ash (Clark et al. 2017). So, these parameters need to characterized if the ash is intended to be used in building materials.

The inclusion of invasive alien plant species like *Eichornia cressipes*, *Parthenium hysterophorus*, *Lantana camara* as cement constituent can be explored (Patel 2011a, b, 2012). As these plants are weeds and abundant, their incorporation in building material is a clean, sustainable solution.

The dry stalks of *Agave salmiana*, the hardy century plant, are used for house construction in some regions of Mexico (Rangel-Landa et al. 2016). In the deserts of South West, smart homes with 'heating and cooling systems' are being made using straws, among other sustainable materials. So, the green buildings have to adopt and adhere to some codes, before they earn the certificates of safety. Optimization is needed for cost-effective, viable transition into green building technology. A life cycle assessment (LCA) determines the "greenness" of building materials and products. LCA of the proposed green materials can determine their usability in construction.

6 Conclusion

The major concern for the environment is piling up of the agro-industrial waste and less-strategic management and recycling. Various construction building materials showed promising physico-mechanical properties that could be of help in their low-cost production. Efforts have also been made to significantly reduce the energy consumption and the detrimental impact on the environmental health as a result of cement industry as it contributes approximately to 10% of the global carbon dioxide (CO_2) emissions. Green concrete or eco-friendly building materials could substitute and the traditional building material usage to a greater extent. The main highlight of a sustainable building is intended toward beneficial impact on their residents as well as environmental health. However, the designing and construction approach of the sustainable buildings using renewable resources needs full-proof planning throughout the construction as well as operation phase. The utilization of green materials ushers in an exciting trend to the construction sector and promises to reduce the burden of pollution. This chapter summarizes and proposes key points in this regard.

References

Achal V, Mukherjee A, Zhang Q (2016) Unearthing ecological wisdom from natural habitats and its ramifications on development of biocement and sustainable cities. Landsc Urban Plan 155:61–68. https://doi.org/10.1016/J.LANDURBPLAN.2016.04.013

Akhtar N, Gupta K, Goyal D, Goyal A (2016) Recent advances in pretreatment technologies for efficient hydrolysis of lignocellulosic biomass. Environ Prog Sustain Energy 35:489–511. https://doi.org/10.1002/ep.12257

Alami AH (2009) Experiments on olive husk-addition to masonry clay bricks on their mechanical properties, and their application and manufacturability as an insulating material. Adv Mater Res 83–86:874–880. https://doi.org/10.4028/www.scientific.net/AMR.83-86.874

Amin N (2011) Use of bagasse ash in concrete and its impact on the strength and chloride resistivity. J Mater Civ Eng 23:717–720. https://doi.org/10.1061/(ASCE)MT.1943-5533.0000227

Arizzi A, Brümmer M, Martín-Sanchez I et al (2015) The influence of the type of lime on the hygric behaviour and bio-receptivity of hemp lime composites used for rendering applications in sustainable new construction and repair works. PLoS ONE 10:e0125520. https://doi.org/10.1371/journal.pone.0125520

Baskaran K, Mallikarachchi HE, Jayasekara MJPLM, Madushanka GAT (2014) Feasibility of using palmyrah strips as reinforcing material in cost effective houses. J Constr Eng 2014:1–9. https://doi.org/10.1155/2014/589646

Burger J, Gochfeld M, Clarke J et al (2015) Environmental assessment for sustainability and resiliency for ecological and human health

Castaldelli VN, Akasaki JL, Melges JLP et al (2013) Use of slag/sugar cane bagasse ash (SCBA) blends in the production of alkali-activated materials. Mater (Basel, Switzerland) 6:3108–3127. https://doi.org/10.3390/ma6083108

Chen H-J, Hsueh Y-C, Peng C-F, Tang C-W (2016) Paper sludge reuse in lightweight aggregates manufacturing. Mater (Basel) 9:876. https://doi.org/10.3390/ma9110876

Cheng Z, Xu Z, Zhang L, Wang X (2014) Thermophysical properties of lignocellulose: a cell-scale study down to 41 K. PLoS ONE 9:e114821. https://doi.org/10.1371/journal.pone.0114821

Clark MW, Despland LM, Lake NJ et al (2017) High-efficiency cogeneration boiler bagasse-ash geochemistry and mineralogical change effects on the potential reuse in synthetic zeolites, geopolymers, cements, mortars, and concretes. Heliyon 3:e00294. https://doi.org/10.1016/j.heliyon.2017.e00294

Dugmore TIJ, Clark JH, Bustamante J et al (2017) Valorisation of biowastes for the production of green materials using chemical methods. Top Curr Chem 375:46. https://doi.org/10.1007/s41061-017-0133-8

Eliche-Quesada D, Leite-Costa J (2016) Use of bottom ash from olive pomace combustion in the production of eco-friendly fired clay bricks. Waste Manag 48:323–333. https://doi.org/10.1016/j.wasman.2015.11.042

Farías RD, Martínez García C, Cotes Palomino T, Martínez Arellano M (2017) Effects of wastes from the brewing industry in lightweight aggregates manufactured with clay for green roofs. Mater (Basel, Switzerland). https://doi.org/10.3390/ma10050527

Goswami E, Craven V, Dahlstrom DL et al (2013) Domestic asbestos exposure: a review of epidemiologic and exposure data. Int J Environ Res Public Health 10:5629–5670. https://doi.org/10.3390/ijerph10115629

Hua GB (ed) (2016) Smart Cities as a solution for reducing urban waste and pollution. IGI Global, USA

Institute of Medicine (US) Roundtable on Environmental Health Sciences R and M (2007) The health aspects of green buildings

Jiao L, Su M, Chen L et al (2016) Natural cellulose nanofibers as sustainable enhancers in construction cement. PLoS ONE 11:e0168422. https://doi.org/10.1371/journal.pone.0168422

Karim MR, Hossain MM, Khan MNN, et al (2014) On the utilization of pozzolanic wastes as an alternative resource of cement. Mater (Basel, Switzerland) 7:7809–7827. https://doi.org/10.3390/ma7127809

Karthik S, Rao RM, Awoyera PO (2017) Strength properties of bamboo and steel reinforced concrete containing manufactured sand and mineral admixtures. J King Saud Univ Eng Sci 29 (4):400–406.https://doi.org/10.1016/j.jksues.2016.12.003

Khatib JM (2016) Sustainability of construction materials. Woodhead Publishing

Kumar AK, Sharma S (2017) Recent updates on different methods of pretreatment of lignocellulosic feedstocks: a review. Bioresour Bioprocess 4:7. https://doi.org/10.1186/s40643-017-0137-9

Lutz W, Samir KC (2010) Dimensions of global population projections: what do we know about future population trends and structures? Philos Trans R Soc Lond B Biol Sci 365:2779–2791. https://doi.org/10.1098/rstb.2010.0133

Ma F, Sha A, Lin R et al (2016) Greenhouse gas emissions from asphalt pavement construction: a case study in China. Int J Environ Res Public Health. https://doi.org/10.3390/ijerph13030351

Mengeloglu F, Karakus K (2008) Thermal degradation, mechanical properties and morphology of wheat straw flour filled recycled thermoplastic composites. Sens (Basel) 8:500–519. https://doi.org/10.3390/s8010500

Mohanty AK, Misra M, Drzal LT (2002) Sustainable bio-composites from renewable resources: opportunities and challenges in the green materials world. J Polym Environ 10:19–26. https://doi.org/10.1023/A:1021013921916

Pandey P, Bajwa D, Ulven C, Bajwa S (2016) Influence of hybridizing flax and hemp-agave fibers with glass fiber as reinforcement in a polyurethane composite. Mater (Basel, Switzerland). https://doi.org/10.3390/ma9050390

Patel S (2011a) Harmful and beneficial aspects of Parthenium hysterophorus: an update. 3. Biotech 1:1–9. https://doi.org/10.1007/s13205-011-0007-7

Patel S (2011b) A weed with multiple utility: *Lantana camara*. Rev Environ Sci Bio/Technology 10:341–351. https://doi.org/10.1007/s11157-011-9254-7

Patel S (2012) Threats, management and envisaged utilizations of aquatic weed *Eichhornia crassipes*: an overview. Rev Environ Sci Bio/Technology 11:249–259. https://doi.org/10.1007/s11157-012-9289-4

Quintana G, Velasquez J, Betancourt S, Ganan P (2009) Binderless fiberboard from steam exploded banana bunch. Ind Crops Prod 29:60–66. https://doi.org/10.1016/J.INDCROP.2008.04.007

Rangel-Landa S, Casas A, Rivera-Lozoya E et al (2016) Ixcatec ethnoecology: plant management and biocultural heritage in Oaxaca. Mexico. J Ethnobiol Ethnomed 12:30. https://doi.org/10.1186/s13002-016-0101-3

Reddy N, Yang Y (2009a) Natural cellulose fibers from soybean straw. Bioresour Technol 100:3593–3598. https://doi.org/10.1016/j.biortech.2008.09.063

Reddy N, Yang Y (2009b) Properties and potential applications of natural cellulose fibers from the bark of cotton stalks. Bioresour Technol 100:3563–3569. https://doi.org/10.1016/j.biortech.2009.02.047

Rodolfo de Melo R, Stangerlin DM, Santana RRC, Pedrosa TD (2016) Physical and Mechanical Properties of Particleboard Manufactured from Wood, Bamboo and Rice Husk. Mat. Res. 17 (3):682–686.http://dx.doi.org/10.1590/S1516-14392014005000052

Scofield JH (2013) Efficacy of LEED-certification in reducing energy consumption and greenhouse gas emission for large New York City office buildings. Energy Build 67:517–524. https://doi.org/10.1016/J.ENBUILD.2013.08.032

Schaschke C, Audic J-L (2014) Editorial: biodegradable materials. Int J Mol Sci 15:21468–21475. https://doi.org/10.3390/ijms151121468

Schuhmacher M, Domingo JL, Garreta J (2004) Pollutants emitted by a cement plant: health risks for the population living in the neighborhood. Environ Res 95:198–206. https://doi.org/10.1016/j.envres.2003.08.011

Väisänen T, Haapala A, Lappalainen R, Tomppo L (2016) Utilization of agricultural and forest industry waste and residues in natural fiber-polymer composites: a review. Waste Manag 54:62–73. https://doi.org/10.1016/j.wasman.2016.04.037

Wang X, Xiang W-N (2016) Ecological wisdom for urban sustainability: doing real and permanent good in ecological practice. Landsc Urban Plan 155:1–2. https://doi.org/10.1016/J.LANDURBPLAN.2016.08.012

Integration of Ecosystem Services in the Structure of the City is Essential for Urban Sustainability

Eva Stache, Henk Jonkers and Marc Ottelé

Abstract Examining the actual major environmental and social problems of the modern city, e.g., pollution, difficulties in food and water supply, poverty or homelessness, this study argues that insights from the field of ecology could offer structural solutions. In particular, specific ecosystem services could be used to fight/solve specific urban problems. Today, on a large scale, only a few different ecosystem services from outside the urban area are used as ecosystems, as biotopes are insufficiently available inside the city boundary. Their physical absence obstructs the use of their benefits and leaves an important potential of urban ecological space unused. Most vegetation was banned from cities during urban history, what may have been the fundamental cause of several major urban problems emerging today. Therefore, the solving potential was analyzed of 20 ecological services if consistently located inside the urban boundary. According to the results, respectively, 14 and 7 ecosystem services can be linked as solutions to 10 environmental and 8 social problems eminent in contemporary cities. This study, therefore, concludes that structural integrating ecosystem services in the built-up urban space: (1) could solve major urban environmental and social problems; (2) improve urban sustainability; (3) revitalize degraded urban areas.

Keywords Urban problems · Urban ecological services · Urban nature
Sustainability of cities

1 Introduction

> Linking biophysical aspects of ecosystems with human benefits through the notion of ecosystem services is essential to assess the trade-offs (ecological, socio-cultural, economic and monetary) involved in the loss of ecosystems and biodiversity in a clear and consistent manner. The Economics of Ecosystems and Biodiversity (TEEB 2010)

E. Stache (✉) · H. Jonkers · M. Ottelé
TU Delft, CiTG (Faculty of Civil Engineering and Geosciences), Materials and Environment, Stevinweg 1, 2628 CN Delft, The Netherlands
e-mail: E.Stache@tudelft.nl

V. Achal and A. Mukherjee (eds.), *Ecological Wisdom Inspired Restoration Engineering*, EcoWISE, https://doi.org/10.1007/978-981-13-0149-0_8

A significant step in the development of the ecology was the recognition of its connection with economical processes. The economical approach brought understanding of the cost-effective benefits for modern society offered by ecosystem services. The interdisciplinary science named ‘economical ecology’ demonstrated that vegetation plays a significant role in the creation of human well-being (Costanza et al. 1997).

The dependence of human well-being on ecosystem services was further investigated by two major international researches: the Millennium Ecological Assessment and the TEEB Studies (MA 2005[1]; TEEB 2010[2]). Both investigations were initiated and financed by an international cooperation between European, Asian, and American countries. Results of both studies clearly pointed out the necessity of using ecosystem services to solve environment related problems also in urban areas.

According to the results of these studies, ecosystem services should largely be implemented in the city to reduce the negative effects of urban activities on the environment. However, the application of vegetation in urban areas has improved little in recent years. Until now, only a minor amount of ecological services is used inside the city as green is not yet considered a functional element in urban developments (Gómez-Baggethun et al. 2013; Boyd and Banzhaf 2007; State of the world’s cities 2012–2013).

Environmental problems, including climate change, are essentially caused by pollution as a result of human activities. Pollution is a product of local actions, which takes place predominantly in urban areas. Recent research shows that 27 megacities are responsible for the total use of 9% of electricity, 10% of gasoline (sources of massive pollutant emissions including CO_2), and 13% of solid waste production globally. These statistics show the prevalent role that megacities/cities and their local activities have today on the global environmental scene (Kennedy et al. 2015) (Fig. 1).

Linking the disciplines of urbanism and ecology, urban vegetation can be analyzed as part of a physical space in the city. The economical approach of the physical urban space can be explicitly linked to the economical approach of ecological services. Consequently, ecosystem services can be integrated into the urban economic cycle.

In the urban economic process, development represents an answer or solution for a problem. Urban problems emerge in urban spaces, and urban developments are

[1]The Millennium Ecosystem Assessment was called for by United Nations Secretary-General Kofi Annan in 2000 in his report to the UN General Assembly. More than 2,000 authors and reviewers worldwide contributed to its realization.

[2]In March 2007 agreed at Potsdam the G8 and the five NIC countries to do a study on the economic importance of biodiversity and ecosystem services (abbreviated as TEEB). The purpose of the study was ‘to sharpen awareness of the value of biodiversity and ecosystem services and to facilitate effective policy, as well as business and engaged citizen responses.’ In short, bring green benefits in the spotlight and offer prospects for action.

Fig. 1 These two pictures illustrate the minor amount of ecological services available in urban areas. The 9 m^2 green per capita as advised by the WHO is visibly not available: industrial quarter in Sidney, Australia, and housing quarters in Beijing, China (*Source* Google Earth)

meant to solve them 'in situ.' Linking ecological services to urban problems also connects them to the physical urban space and makes them part of the urban developments.

Therefore, to define vegetation as a functional element of the urban structure, an integrated approach from urbanism to ecology is required. A historic literature survey may reveal the historical reasons for the structural absence of functional green in the modern city. Identifying urban problems which could potentially be solved by ecosystem services is the second step of this interdisciplinary research. After defining the urban ecological services, a comparison with the urban problems establishes the possible correspondences.

1.1 Research Scope

This interdisciplinary study examines whether current major urban problems can be linked to the structural absence of coherent ecosystem services inside the city boundary.

The study assumes that the present urban problems are caused—partly or totally—by a structural 'system failure' of the city: the absence of sufficient urban ecosystem services inside the city boundary. Its aim is to establish the causes of this absence and its correlation with actual urban problems. The study also wishes to formulate the prerequisites for the (re-) integration of ecological services by implementing vegetation in the urban texture.

The sub-questions this research investigates are:

- What is the historical development of urban greenery?
- What are the major problems cities of today are struggling with?
- Which urban ecological services can be identified as essential/functional to solve urban problems?
- What is the applicability of ecological services in the urban space?

2 Background

2.1 *Historical Development of the City—A Short Literature Review of Urbanization*

The historical reason for the limited use of urban ecological services is that vegetation was excluded from the city during the development of urban settlements and along their urbanization.[3] The sedentary settlements in prehistoric times or cities later on emerged to enhance the safety and quality of human life and offer an appropriate background to its activities (e.g., to exchange products, organizational or cultural activities). By building a road or a house, nature is eliminated from those specific locations. Settlements, towns, and cities were in fact built *to protect* humanity *from nature* and not to integrate with it (Azar Gat 2006; Naerebout and Singor 1995; Landes 1998; Blockmans and Hoppenbrouwer 2009).

The industrialization in the eighteenth century was in this context an answer to the fluctuating results of agricultural activities that depended on the unpredictability of the climate. For the rural population without perspective, the industrial city offered better living standards, economic success, and work opportunities. Dwellings for the new working class and the emerging industrial buildings exiled nature beyond the city boundary. At the same time, the production of environmental pollution became a permanent side effect.

As a response to the serious pollution problems experienced during the early industrial age, architects developed a new concept for the urban structure. They described 'The functional city' during the congresses of CIAM.[4] In the "Charter d'Athénes" in 1933, the new urban functions were defined as: **living, working, traffic**, and **recreation**. According to the charter, these functions should be built physically separated (segregated) in order to prevent that negative effects (pollution) were released from one to the other.

The segregation of urban functions has been the usual practice until nowadays. It resulted in an unprecedented growth of mono-cultural urban spaces: residential areas, industrial districts, shopping malls, shopping quarters, and so on. Among these, recreation is the only nature-related activity.

The separated quarters are connected by roads. Direct and short transportation, in order to lower costs, is not indispensable anymore as carrying goods is cheap even across very long distances.

Despite the uninterrupted urbanization, the functions cities fulfill today are basically the same as in ancient times: security, dwellings, social contacts, work opportunities. In the groups of security, social contacts, and work opportunities, the following functions are also included: organization, education, health care, sport,

---

[3]http://www.nlm.nih.gov/cgi/mesh/2014/MB_cgi?term=Urbanization: Urbanization is the process whereby a society changes from a rural to an urban way of life. It refers also to the gradual increase in the proportion of people living in urban areas.

[4]Congrès International d'Architecture Moderne 1928–1959.

Fig. 2 These two pictures illustrate the discrepancy between the built area and the available vegetation on the same surface. Vegetation is not visible: industrial quarter in New York in the USA and dwelling quarter in Teheran, Iran (*Source* Google Earth)

recreation, and culture. All functions are facilitated by the services of transportation, food/water supply, and waste management. These facilitating services are an indispensable requirement for the survival of the city (Fig. 2).

Nature is present in urban areas in a limited amount. There is a substantial discrepancy between the scale of available urban nature and the scale of built urban surfaces (see Fig. 1). As a result, the volume and diversity of urban green today are low and consequently not sufficient to provision and regulate the socio-environmental needs of a modern city (Pugh et al. 2013; Lenzholzer 2013).

2.2 Environmental and Social Problems in Modern Cities

Today, more than 54% of the world population lives in cities. Urban areas occupy around 4% of the global land. These percentages are increasing every day. Contemporary cities have grown far beyond their original borders. Urban agglomerations are placed under great pressure by the continuing growth of worlds population and the massive move to urban areas of rural natives.

The expanding cities are confronted with exacerbating major problems. Difficulties in facilitating basic needs like water, food and energy supply, waste management or social problems like slum districts, deserted industrial areas, poverty, homelessness, or decreasing work opportunity are aggravated by the growing amount of waste, pollution and loss of biodiversity (Kennedy et al. 2015; Pilon 2014; Davies et al. 2008; Ref[5]).

[5](World Urbanization Prospects 2014 UN, Yedla 2003, EEA–European Environment Agency report no. 10/2012, Climate Change 2014 Synthesis Report–IPCC, Geng et al. 2011, State of the world's cities 2012–2013, UN Habitat, http://www.citypopulation.de/world/Agglomerations.html).

2.3 Ecosystem Services[6]

The surface of nature, used by a city today to provide the products necessary for the surviving and well-being of its inhabitants (such as at least food, water, or raw materials) is a multiple of its own built surface and is located therefor outside the city boundary. The providing ecosystems are spread over different continents, and their exploitation causes massive losses of biodiversity worldwide. Cheap and widely available transportation means make such a system possible, resulting in an uncontrolled inflation of energy use for the purpose of transportation and industrial production. However, cheap transport and production outside the city are not the only reasons for this development. **Ecosystem services are used from outside the urban area because they are not available inside the city boundary** (Costanza et al. 1997; Gómez-Baggethun et al. 2013; Boyd and Banzhaf 2007).

Most provisioning ecological services, e.g., the production of food and raw materials, used by cities today are coming from outside the urban area, as established by the Millennium Ecological Assessment.[7] Provisioning services originated from inside the city boundary are used only to a limited extent. Regulating services such as urban temperature regulation, noise reduction or air purification and habitat services from inside the city boundary are even less available. In contrast, cultural services are most used from inside the city, though not structurally (CIAM 1933; MA 2005; TEEB 2010).

The existing urban ecosystems, i.e., vegetation located inside the city boundary, are placed incidental and interrupted, mostly on streets, squares, city parks or gardens, and rarely on buildings. They carry out their processes on a local climatologic layer, but may have significant effects regionally and globally. The sporadic and discontinuous presence of ecosystems in the urban space affects their functioning and services, facilitating only very low performances (Costanza et al. 1997; Lenzholzer 2013; Fisher et al. 2007; MA 2005; TEEB 2010).

---

[6]Costanza's Definition of Ecosystem Services: "Ecosystem Goods (Such as Food) and Services (Such as Waste Assimilation) Represent the Benefits Human Populations Derive, Directly or Indirectly, from Ecosystem Functions" (the Value of the World's Ecosystem Services and Natural Capital—Costanza et al. 1997, p. 253).

[7]The classification by the Millennium Ecological Assessment (MA) in 2005 and later by the TEEB studies (The Economics of Ecosystems and Biodiversity) in 2010 ranked the ecological services according to the effects on human well-being: Provisioning services (raw materials, food, and water supply), regulating services (urban temperature regulation, noise reduction, air purification, moderation of climate extremes, runoff mitigation, waste treatment, pollination, pest regulation and seed dispersal, global climate regulation), cultural services (recreation, aesthetic benefits, cognitive development, place values, and social cohesion), habitat services (habitat for biodiversity), and ecosystem disservices.

3 Methods

The major urban problems and the widely accepted ecological services were identified in this study by reviews of literature, statistics, and reports. A list of urban problems and one of ecological services was composed. Subsequently, problems were linked to the possible solutions ecological services can provide.

The actual major problems of urban areas were identified:

- A literature review was conducted on the historical evolution of the cities in order to understand the development of vegetation in urban areas and the reasons of its present status.
- Analyses of official reports and statistics were conducted to identify the problems cities are struggling with today.
- Scientific researches were investigated to identify the actual causes of the problems.
- Seven megacities were examined as examples.

The ecological services offered by urban nature were determined:

- The ecological services and the natural functions on which they are based were almost simultaneously described by Robert Costanza and Gretchen Daily in 1997. For this research, the list of Costanza is chosen as being the most complete one (Daily et al. 1997; Costanza et al. 1997).
- Recent studies were analyzed in order to investigate information on recently recognized ecological services.
- The working of each urban ecosystem service was analyzed and shortly described.

The following reasoning/interpretation was used: as the presence of an ecological service is considered to have a positive effect, its absence will consequently be a negative one. The coincidence between the presence of a specific urban problem and the absence or scarce presence of the related specific ecological service may indicate a causal relation between the problem and the analogue missing ecosystem service.

4 Results

4.1 Selection of Urban Problems

See Table 1.

The most serious urban problems were identified from official reports and statistics.

Reports released by the *Intergovernmental Panel on Climate Change* (IPCC), the UN Habitat[8] and the *Netherlands Environmental Assessment Agency*

[8]State of the world's cities 2012–2013.

Table 1 Results of the literature review carried out to identify the main urban problems at the present time

		fields of social problems																fields of environmental problems									
		powerty	homelessness	violence	high living costs	urban trafic	lack of diversity	energy	urban safety	education	noise	health care	work opportunity	social services	housing	economical stability		air pollution	water supply	mounting waste	drought or flood	extreme weather	food shortage	desertification	loss of biodiversity	empovering soil	heat island
REPORTS AND STATISTICS																											
1	UN Habitat 2012-2013	1	1	1	1	1			1	1	1	1	1		1			1	1	1				1			1
2	PBL, 2014																	1									
3	EEA, 2012					1												1									
4	UN WUP, 2014	1				1							1	1	1							1		1		1	
5	Quinn, 2014																		1								
6	IPCC, 2014					1						1		1	1	1			1		1		1		1		1
7	UNEP, 2004							1				1						1	1	1			1		1		
totals reports and statistics		2	1	1	1	4	0	1	1	1	1	3	2	2	3	1		4	4	2	1	1	2	2	2	1	2
SCIENTIFIC STUDIES																											
8	Pilon, 2013	1	1	1	1	1		1	1	1		1	1	1	1	1		1	1	1							
9	Wendt, 2003-J.Jacobs, 1961	1	1	1			1		1					1													
10	Y. Geng, 2011					1												1									
11	M. Georgescu, 2014							1							1			1			1						1
12	Hooper, 2012																					1			1		
13	KNMI, 2015					1												1	1	1		1	1				1
totals scientific studies		2	2	2	1	3	1	2	2	1	0	1	1	2	2	1		4	2	2	1	2	1	0	1	0	2
SPECIFIC CITIES																											
14	Teheran, Iran			1		1		1	1				1	1				1	1	1	1			1	1		
15	New York USA		1	1										1				1									
16	Tokyo, Japan																	1	1					1	1		
17	Rio de Janeiro, Brasil	1	1	1	1	1			1	1					1			1	1						1		
18	Sidney, Australia	1			1	1	1								1			1	1	1					1		
19	Paris, France					1								1		1		1							1		
20	Mumbay, India	1	1															1		1							
21	27 Megacities, 2014	1			1			1	1			1		1		1		1		1			1		1		
totals specific cities		4	3	3	3	4	1	2	3	1	0	1	1	4	2	2		8	4	4	1	0	1	2	6	0	0
TOTALS		8	6	6	7	11	2	5	6	3	1	5	4	8	7	4		16	10	8	3	3	4	4	9	1	4

(PBL) documented the increasing risks caused by environmental unbalances in urban areas. Scientific studies reported the socio-environmental problems of the city. As recent research shows that megacities are today's leading pollution producers, a review was done for one megacity on each continent: Tokyo, New York, Mumbai, Rio de Janeiro, Paris, Sidney, and Teheran (Kennedy et al. 2015; Wendt 2009).

A list of fifteen social urban problems and ten environmental urban problems was identified.

The selected urban problems are those which appear more than three times in the analyzed reports and studies. There is an extensive amount of literature of different kinds available describing urban problems in the modern city today, and this number is exponentially increasing every day. Therefore, this research did not intend to find all specific problems but the most generally experienced ones.

The identified urban problems can be divided into two groups: problems having environmental roots and problems originated from social unbalances. There is strong indication that social and political difficulties may also be linked to environmental shortcomings, e.g., increasing health problems and related growing healthcare costs can also be caused by local air pollution, urban heat islands, or lack of an adequate waste management. As the approach of this study is an environmental one, the inquiry only concentrates on ecological issues including through the related social problems.

Air pollution including greenhouse gasses (mentioned 16x), deficiency of water supply (mentioned 10x), and mounting waste (mentioned 8x) is found as being the most stringent environmental urban problems. Loss of biodiversity (mentioned 9x) and extreme and dangerous weather (mentioned 3x) are noted in the official statements about environmental problems and climate change, however, rarely as an urban problem. Inadequate transit or public transportation (mentioned 11x) is found to be the biggest social problem, next to poverty related matters (mentioned 8x) the lack of social services (mentioned 8x) or proper housing (mentioned 7x).

4.2 *Selection of Ecological Services*

Costanza defined ecological services by describing natural processes in light of the economic benefits they offer to humans. Also, other ecological economists defined ecosystem services using the concepts of economy. Their sequences assume a theoretical situation, in which the location, quantity, type, quality, and morphology of the ecosystem are not widely discussed (Costanza et al. 1997).

Therefore, the starting point of this research is formed by the location of ecological services inside the urban texture and their position among urban functions. As such, the **urban ecological services** are the nature's benefits used by a city for its provisioning, regulating, and cultural needs. They can be located inside or outside the city boundary. In both cases, the city enjoys the benefits.

The locations of vegetation inside urban areas are the physical places where ecosystems can develop and ecological services can be provided. Different ecological services can be offered by many different types of vegetation in various morphological structures. The morphology of urban greenery inside the city depends mainly on the morphology of its built surroundings.

The following ecosystem services may play an important role in provisioning and regulating processes and cultural or recreational needs:

1. Gas regulation	Regulation of atmospheric chemical composition.
2. Climate regulation	Regulation of global temperature, precipitation, and other biologically mediated climatic processes at global or local levels.
3. Disturbance regulation	Capacitance, damping, and integrity of ecosystem response to environmental fluctuations.
4. Water regulation	Regulation of hydrological flows.
5. Water supply	Storage and retention of water.

6. Erosion control and sediment retention	Retention of soil within an ecosystem.
7. Soil formation	Soil formation processes.
8. Nutrient cycling	Storage, internal cycling, processing, and acquisition of nutrients.
9. Waste treatment	Recovery of mobile nutrients and removal or breakdown of excess or xenic nutrients and compounds.
10. Pollination	Movement of floral gametes.
11. Biological control	Trophic-dynamic regulations of populations.
12. Refugia	Habitat for resident and transient populations.
13. Food production	That portion of gross primary production extractable as food.
14. Raw materials	That portion of gross primary production extractable as raw materials.
15. Genetic resources	Unique biological materials and products.
16. Recreation	Providing opportunities for recreational activities.
17. Cultural	Providing opportunities for commercial and non-commercial uses.

The process of defining specific ecological services is still ongoing. According to the results of recent research, there are about three more ecological services that could be added to the list:

18. Health improvement	Providing mental and physical health effects (according to recent research, (Alcock et al. 2013).
19. Noise reduction	Vegetation has a damping/mitigating effect on urban noise.
20. Thermal insulation	Wall and roof vegetation influences the thermal behavior of building materials.

4.3 Comparison of Ecosystem Services and Urban Problems

Ten urban environmental and eight social problems were compared with twenty ecological services. The effects of the ecological services were compared with the causes of the problems, in order to find possible functional relationships. The quantitative and qualitative characteristics were also compared.

The following correspondences were found (Tables 2 and 3):

Table 2 Possible use of ecological services of urban nature to solve urban environmental problems

Environmental problems		Ecological services
Air pollution	↔	Gas regulation [R]
Water shortage	↔	Water supply, water regulation [P, R]
Loss of biodiversity	↔	Biological control, pollination, nutrient cycling, refugia [H]
Urban waste and pollution	↔	Waste treatment [R]
Extreme weather	↔	Disturbance regulation [R]
Food shortage	↔	Food production [P]
Heat island	↔	Climate regulation, insulation [R]
Empowering soil	↔	Erosion control, soil formation [R]
Desertification	↔	Soil formation, water regulation [R]
Noise pollution	↔	Noise reduction [R]

The following abbreviations were used: R for regulatory services; P for provisioning services; HB for habitat; C for cultural services. These categories correspond with the categorization of the MA 2005 and the TEEB Studies. See page 4 note nr. 7

Table 3 Possible use of ecological services of urban nature to solve urban social problems

Social problems		Ecological services
Poverty and hunger	↔	Food production [P]
Work opportunities, unemployment	↔	Raw materials [P]
Rising costs for health care	↔	Health [R], genetic resources [P]
Unbalanced social life	↔	Recreation, cultural [C]
Violence, crime, and corruption	↔	Cultural [C]
Education, spiritual development	↔	Recreation, cultural [C]
Scientific development	↔	Cultural, genetic resources [C]
Lack of diversity, monotony	↔	Cultural [C]

The following abbreviations were used: R for regulatory services; P for provisioning services; HB for habitat; C for cultural services. These categories correspond with the categorization of the MA 2005 and the TEEB Studies. See page 4 note nr. 7

4.4 *Description of the Urban Ecological Services as Solutions for Place-Related Urban Problems*

4.4.1 *Gas regulation* is a regulating service of the atmospheric chemical composition, which influences locally produced air pollution directly. Its effects are (1) local; (2) regional; (3) global[9]:

---

[9](1) Reduction of air pollution on street, neighborhood, and regional level by PM2,5 + PM10 + NOx; (2) Reduction of air pollution on global level by CO_2; (3) reduction of

The mechanism is photosynthesis taking in CO_2 or adsorbing PM2,5 + PM10 + NOx by stomata and taking PM10 in by deposition on leaves (Ottelé 2011; Pugh et al. 2013; CROW 2012).

According to recent research, green facades are able to reduce up to 60% of the particulate matter and 40% of the NOx emitted by the local road traffic. A grown-up beech tree emits 1.71 kg of O_2 using 2.4 kg of CO_2 and 25.5 kJ thermal energy every hour.

4.4.2 *Water supply and water regulation* are regulating ecological services which counteract the dehydration of drinking water sources and dehydration of hydrological flows.[10] The urban water regulating service has effects on local, regional, and on global level. Its mechanism is storage and retention of water by vegetation.

The ecosystem process is the water–soil–plant continuum, where the uptake rate depends on the transpiration rate and on the extent of the root system. This process is known as the water cycle (Kirkham 2004). The retention rate realized by green roofs may reach 50–75% in European countries and 65% in tropical countries. A normal adult beech tree absorbs 96 kg of H_2O every hour (Köhler et al. 2001).

4.4.3 *Biological control, pollination, nutrient cycling, and creating refugia* are responses to the loss of biodiversity. As a result of the growing lack of refugia and habitat in today's urban built surroundings, urban biodiversity is rapidly decreasing (Hooper et al. 2012; Jagers op Akkerhuis 2012; Mant et al. 2014).

As biodiversity is an insurance for the survival of the ecosystem, every loss is threatening the balance and its surviving capacity. The more participants there are in an ecosystem the better is its capacity of crisis interception. Biodiversity is a strong supporting service. Its effects are local, regional, and global (Rees 1996; Jagers op Akkerhuis 2012; Mant et al. 2014; Yachi and Loreau 1998).

4.4.4 *Waste treatment* is a regulating service of CO_2 and energy release in the atmosphere from biomass, filtering of water, or recycling of nutrients to close nutrients loop.

The mechanisms are energy release from biomass by combustion and water filtering by macrophytes or helophytes filters. It uses the work of aerobe and anaerobe bacteria's and microorganisms in symbiosis. (Emis–VITO, http://emis.vito.be/).

4.4.5 *Disturbance regulation* is a regulating service assuring capacitance, damping, and integrity of ecosystem responses to environmental fluctuations. In the city, it consists of wind regulating effects and diminishing the extremes of heat, storm, flooding, and drought. Its effects are local and regional (Lenzholzer 2013).

---

greenhouse effect on global level = reduction of global warming (CBS 2009; TNO-LOTOS-EUROS 2013).

[10]http://reports.weforum.org/global-risks-2015/part-1-global-risks-2015/introduction/;WHO 1000m3/jaar/pp).

4.4.6 *Food production and raw materials* are providing services in the form of gross primary production, extractable as food, biomass, or raw materials. Most food resources and raw materials have their origin outside the city. Food production is, however, also possible inside the city boundary (Costanza et al. 1997).

4.4.7 *Climate regulation* in urban area means a strategy to control global and local temperature, precipitation, and other biologically mediated climatic processes. Urban areas present raised temperatures compared with the surroundings being able to affect health and daily functioning.

The urban heat island effect can swell from local to regional as distinct metropolitan complexes merge into a network of cities. The mechanism of urban climate regulation is that greenery in the city prevents stone-like materials and the air from heating up. The effects of this ecological service are local although they may have regional influences (Georgescu et al. 2013).[11]

4.4.8 *Erosion control and soil formation* are regulating ecological responses on desertification. Desertification is the persistent degradation of dryland ecosystems by human activities or variations in climate (MA 2005). Being a result of deforestation or other ecosystem damages, desertification causes scarcity of food and water, affects climate regulation, hinders recreation, and obstructs soil conservation.

By the effect of water–soil–plant continuum, the soil will be retained within an ecosystem (Costanza et al. 1997). This problem is occurring only in some specific cities or in parts of cities.

4.4.9 *Noise reduction* is a regulating service based on the capacity of greenery to damp sound waves (Ottelé 2011). Its effects are local.

4.4.10 *Health improvement*, as demonstrated by recent research, means a positive effect on the mental and physical health of human beings (MA 2005). As health damages are directly translated in the costs of the needed care, this ecological service has an important effect also on the local and regional economy.

4.4.11 *Genetic resources* are providing opportunities for scientific uses as sources of unique biological materials and products (Costanza et al. 1997).

4.4.12 *Recreation and cultural services* mean providing green opportunities for recreational activities and non-commercial uses. It is the most present type of vegetation inside the urban boundary, and it has local and regional effects (Costanza et al. 1997).

[11]Therefor are the following ecosystem processes needed:—evapotranspiration (20–40% Ottelé 2011);—low albedo (gras 15–20%, leaf 5–30% Ottelé 2011);—passed through radiation (leaf 5–30% Ottelé 2011);—high emissivity (gras 90–95%, Lenzholzer 2013, leaf 90–95% Ottelé 2011); —low thermal conductivity (Lenzholzer 2013);—manipulation of turbulencies by green mass (Lenzholzer 2013).

5 Discussion

5.1 Historical Development of Urban Greenery and the Relationship with Actual Urban Problems

Buildings and roads were and are the two necessary physical and spatial ingredients of a city or town. Being built, they replaced the nature which was present on the very same place. By substituting vegetation through buildings or roads, also its ecological services are lost on those specific places.

Biodiversity supports almost all other ecological services, and its loss has an exacerbating effect on the whole ecosystem. In urban areas, vegetation presence is already limited and the city depends substantially on land and green from outside its boundary. Therefore, losing biodiversity exterior to the city affects also its interior. (Climate Change 2014; Synthesis Report–IPCC; Omer et al. 2013).

The extreme urbanization from recent decades produced pollution, water shortage, augmenting urban waste, and emission of air pollution including greenhouse gasses. These effects caused serious damages to the environment and contributed also to the development of diseases, poverty, and unbalanced social life (Fig. 3).

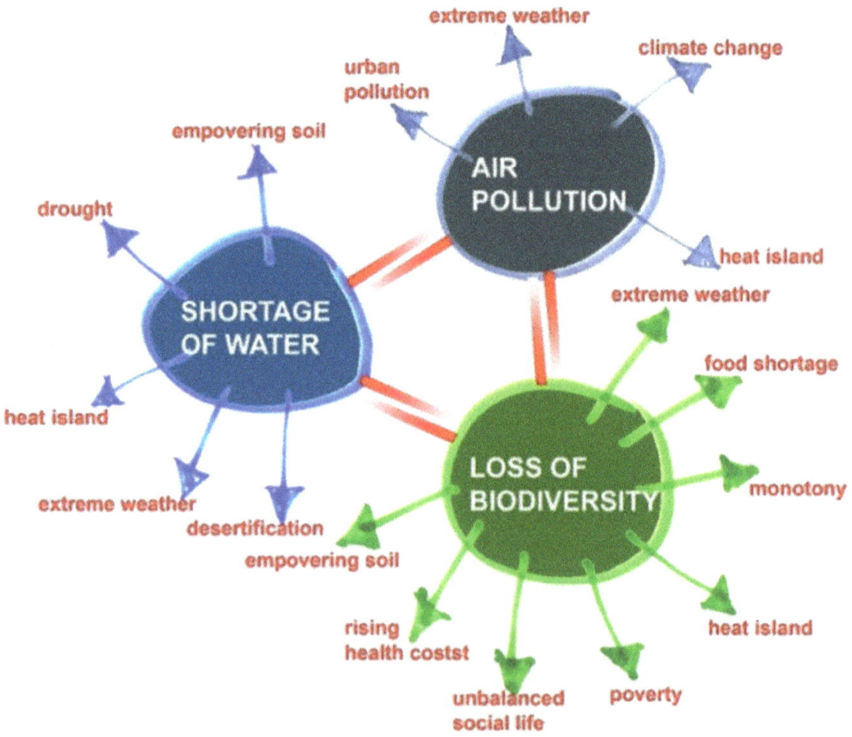

Fig. 3 Social and ecological urban problems are interrelated

The results are higher costs, loss of human well-being, and social control (Geng et al. 2011; EEA–European Environment Agency report no. 10/2012; David Thorpe 2014; Okata and Akito Murayama 2015; Yongjian et al. 2005; Climate Change 2014; Synthesis Report–IPCC).

Social and ecological urban problems are interrelated. Environmental shortcomings are a product of the overwhelming urbanization and its poverty, social problems, ill-functioning urban traffic, or inadequate governance. The other way around, although social problems originate from many different roots, there is observed a definite causal relationship with environmental issues.

5.2 Urban Ecological Services and Their Applicability to Solve Specific Actual Urban Problems

5.2.1 Essentially, all ecological services may be able to solve the problems in urban areas if nature would be sufficiently available in the city. Therefore, the amount and spatial distribution of greenery a city should possess in order to provide an adequate amount of ecological services need to be defined based on calculations. Calculations should reflect the local specific requirements, i.e., how much of a specific ecosystem function is required to solve a specific socio-environmental problem.[12]

Cities show significant discrepancies regarding the surface of greenery per capita[13] as there are today no general rules based on calculations for the expected benefits. To increase the performance of ecological services in the city, new physical places need to be created or allocated for vegetation inside the city boundary. Finding new places in the existing urban texture supposes a creative decision-making process and a different design approach (Fig. 4).

5.2.2 New processes bring new conflicts during the implementation that have to be recognized at an early stage. One such conflict could be related to the ownership of the ecosystem. Wherever vegetation is located in a town, there is an owner of the physical place. The ownership is related to administrative and juridical rules. These

[12]The World Health Organization (WHO) advises 9 m^2 green per capita. As shown in the comprehensive research of Dobbs (2014) about three urban vegetation services—recreation, habitat provision, and carbon storage—out of 100 investigated cities only 26 meet the recommendation of the WHO and 12 had more than 20 m^2 green per inhabitant, amounts depending also on the grade of development of the inhabitants and the political system (Dobbs et al. 2014).

[13]In the Netherlands, the minimum (horizontal measured) surface of vegetation per inhabitant is stated at 75 m^2, which cities almost never achieve (Alterra 584, 2002; Nota Ruimte 2006, Rijskoverheid).

In Teheran (Iran) in the central area, there is less than 0.025 m^2 green per inhabitant; in other districts, there is an increased array of 0.35 m^2. The advised amount is in the region of 7–8 m^2 green per inhabitant. This amount is in Teheran never attained (Faryadi et al. 2008).

Fig. 4 These two pictures illustrate the discrepancy between the built area and the available vegetation on the same surface. The 9 m^2 green per capita as advised by the WHO is visibly not available: dwelling quarters in Tokyo in Japan and Mumbai in India (*Source* Google Earth)

rules do not include indications about the ownership of the benefits offered by the ecosystems.

Based on the above argumentation, a conflict can be identified between the ownership of the ecosystems in the city and the beneficiaries of their services. The possessor of a property containing a whole or partial ecosystem or biotope is free to decide about the vegetation which forms it. Also, maintenance and the costs of the greenery are the responsibility of the owner. At the same time, the benefits are available on a local level for the owners of other lots as well,[14] and on a mezzo[15] or macro level[16] also for the region (Bonthoux et al. 2014; Cameron et al. 2012; Engelsing 2011) (Fig. 5).

As ecosystems are mainly divided over more allotments and owners, a consequent and congruent strategy is needed for their maintenance or development. At this moment, there is minor attention for the development, continuity, or maintenance of the urban vegetation, which results in fragmented ecosystems. As they include the water cycle as well, this problem is not only a 'green' but also a 'blue' one, (Bonthoux et al. 2014; Köhler et al. 2001) (Fig. 6).

The disconnected ecosystems in urban regions need to be reconnected. A common strategy adopted by the different owners is needed to assure an effective all-embracing planning for the total green/blue framework in the city. Only an integrated approach of the green and blue infrastructure, supported by all involved stakeholders can assure a healthy urban ecosystem (Cameron et al. 2012; Engelsing 2011; Bonthoux et al. 2014).

[14]E.g., adsorbing of particulate matter.

[15]E.g., like reducing the heat island effect.

[16]E.g., CO_2 reduction.

Fig. 5 Ecosystem located over a space with multiple owners

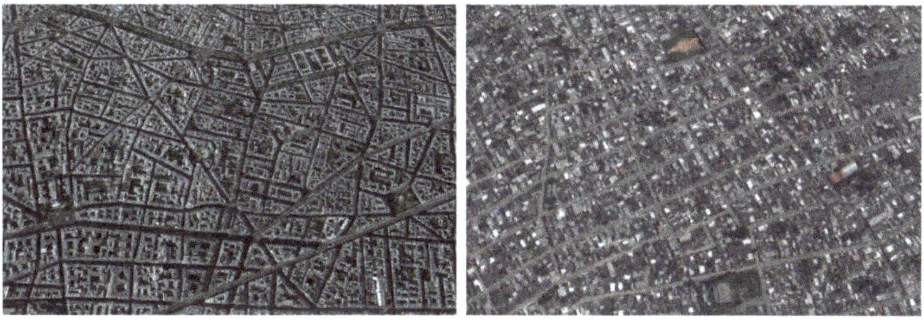

Fig. 6 These two pictures illustrate that the benefits of vegetation in the city are available not only for their owners but for the owners of other lots as well: dwelling quarters and local parks in Paris in France and Rio de Janeiro in Brasil (*Source* Google Earth)

6 Conclusions

As this paper has shown, the absence of ecosystems in urban areas is due to the historical development of the cities. Confronted with the unprecedented urbanization, urban development fails to solve increasing environmental and social problems. There is a complex interaction between urban social-environmental problems and the absence of services offered by nature.

This research concludes that the presence of sufficient and adequate nature in the city may give an answer to most of the current environmental and social urban problems. Although the benefits provided by nature do not solve all the problems immediately or completely, they represent a substantial and decisive part of the solution.

The remarkable coincidence of urban problems and the solutions provided by nature tell us that they might be causally linked. Urbanization reached a point where not taking nature into account and not using its ecological services generate major problems in the city. The current urban difficulties cannot be solved anymore by only technical means.

The descriptions of the natural services in this paper show that there is a gap between the scale of built urban surfaces and the amount of vegetation available. To intensify urban ecosystem services, the structural integration of vegetation inside the city space is needed. Its sporadic and discontinuous presence needs to be transformed in a permanent and continuous one in order to increase the performance of the provided services. An increased performance of ecological services in the city will improve urban sustainability and revitalize degraded urban areas.

To incorporate vegetation in the complex system of urban functions/services requires a considerable change in the decision making, management, and design practice of urban spaces. The functions and services of the city need to be rethought. The ecological services have to be structurally included in the services provided by the city, and physical places for vegetation have to be allocated. The conflicts of ownership should be solved, and the continuity of ecosystems should be ensured. It is the task of further investigation to establish and calculate the right place, required quantity, quality, and morphology of the needed urban vegetation.

"Biodiversity is our passport to survival. Let's make place for it"—em. prof. dr. ir. Michiel Haas

Acknowledgements I would like to thank to em. prof. dr. ir. Michiel Haas for his kind support and help.

References

Alcock et al (2013) Longitudinal Effects on Mental Health of Moving to Greener and Less Green Urban Areas. Environ Sci Technol 2014

Azar Gat (2006) War in Human Civilization. Oxford University Press, New York

Bezemer V Groene meters, Alterra Rapport 584/V/2002

Blockmans W, Hoppenbrouwer P (2009) Eeuwen des Onderscheids, Een geschiedenis van middeleeuws Europa, Uitgeverij Bert Bakker Amsterdam

Bonthoux S et al (2014) How can wastelands promote biodiversity in cities? A review. Elsevier

Boyd J, Banzhaf S (2007) What are ecosystem services? The need for standardized environmental accounting units, Energy and Natural Resources Division, Resources for the Future, 1616 P St., NW Washington, United States

Cameron et al (2012) The domestic garden: its contribution to urban green infrastructure. Elsevier

CBS (2009) http://www.cbs.nl/nl-NL/menu/themas/natuur-milieu/publicaties/compendium/milieu2/lucht/default.htm

CIAM (1933)

Climate Change (2014) Synthesis Report–IPCC

Costanza et al (1997) The value of the world's ecosystem services and natural capital. Nature 387

CROW (2012) Beplanting en Luchtkwaliteit. CROW. ISBN 978-906628-600-9 (51–58)

Daily GC et al (1997) Ecosystem services: benefits supplied to human societies by natural ecosystems. Ecological Society of America Number 2. Spring

David Thorpe (2014) Sustainable cities collective, what can other cities learn from Paris' response to urban smog?

Davies et al (2008) A national scale inventory of resource provision for biodiversity within domestic gardens. Department of Animal and Plant Sciences, University of Sheffield

Dobbs et al (2014) Global drivers and tradeoffs of three urban vegetation ecosystem services. Plos, UK. https://doi.org/10.1371/journal.pone.0113000

EEA–European Environment Agency report no. 10/2012

Emis–VITO http://emis.vito.be/techniekfiche/kleinschalige-biologische-zuivering

Engelsing (2011) Groningen, April 2012. De groene megastad. http://2012.realissues.eu/?p=70
Faryadi et al (2008) Interconnections of Urban Green Spaces and Environmental Quality, Sh.* Department of Environmental Planning, Graduate Faculty of Environment, University of Tehran, P. O. Box 14155-6135, Tehran, Iran 2008
Fisher et al (2007) Defining and classifying ecosystem services for decision making, p. 4, CSERGE Working Paper EDM, No. 07-04
Geng Y et al (2011) Energy Procedia 5 (2011) 370–376. IACEED 2010 Energy Use and CO_2 Emission Inventories in the Four Municipalities of China a School of Earth Sciences and Resources, China University of Geosciences, Beijing, China, Elsevier, Science Direct 2011
Geography Human Environments-Urban Change-A Case Study of Rio de Janeiro-Learning and Teaching Scotland gratefully acknowledges this contribution to the National Qualifications support programme for Geography 2007
Georgescu et al (2013) Urban adaptation can roll back warming of emerging megapolitan regions. PNAS, Feb 2014
Gómez-Baggethun et al (2013) The history of ecosystem services in economic theory and practice: from early notions to markets and payment schemes. Ecological Economics 2009 Elsevier
Hooper DU et al (2012) A global synthesis reveals biodiversity loss as a major driver of ecosystem change. Nature 486:105–108. https://doi.org/10.1038/nature11118
http://reports.weforum.org/global-risks-2015/part-1-global-risks-2015/introduction/
Jagers op Akkerhuis (2012) De Soortenstorm: minder soorten, meer biodiversiteit, KNNV Uitgeverij, Zeist, WUR
Kennedy et al (2015) Energy and material flows of megacities. PNAS
Kirkham MB (2004) Principles in soil-plantwater relations, Manhattan, Kansas April 6, 2004. Kansas State University, Elsevier
Köhler M, Schmidt M, Grimme FW, Laar M, Gusmão F (2001) Technology resource management & development. Sci Contrib Sustain Dev 2
Landes DS (1998) Arm en Rijk, Uitgeverij Het Spectrum B.V
Lenzholzer (2013) Het weer in de stad, nai 010 Uitgevers
MA (2005) The millennium ecosystem assessment called by United Nations Secretary-General Kofi Annan in 2000 in his report to the UN General Assembly
Mant R et al (2014) Using criteria to strengthen biodiversity considerations: Recommendations for the Inter-national Climate Initiative (IKI), UNEP World Conservation Monitoring Centre, Cambridge, UK
Naerebout FG, Singor HW (1995) De Oudheid, Ambo Amsterdam, 1995, 2001, 2008, 2009
New York http://www.demographia.com/db-nyc-landuse.htm
Okata and Akito Murayama (2015) Tokyo's Urban Growth, Urban Form and Sustainability, Junichiro Okata and Akito Murayama
Omer et al (2013) Values, ethics and built environment, Urbanization: problems and solutions 2013 Academia.edu
Ottelé M (2011) The green building Envelope, Vertical Greening, Delft
Pilon (2014) André F. Pilon environmental sustainability, Urban problems and quality of life. Urban Gateway for the International Urban Development Community, Jan 2013
Pugh et al (2013) Effectiveness of green infrastructure for improvement of air quality in urban street canyons. Lancaster Environment Centre, Lancaster University, Lancaster
Rees (1996) Revisiting carrying capacity: area-based indicators of sustainability. The University of British Columbia. In: Population and environment: a journal of interdisciplinary studies, vol 17 (3)
State of the world's cities 2012–2013
TEEB 2007 (2010) Integrating the ecological and economic dimensions in biodiversity and ecosystem service valuation. Coordinating Lead Author: Rudolf de Groot, Lead Authors: Brendan Fisher, Mike Christie. Contributing Authors: James Aronson, Leon Braat, John Gowdy, Roy Haines-Young, Edward Maltby, Aude Neuville, Stephen Polasky, Rosimeiry Portela, Irene Ring, Mar 2010

TNO-LOTOS-EUROS, Kranenburg R et al (2013) Copernicus Publications on behalf of the European Geosciences Union

Trends in Global CO_2 Emissions 2013, PBL 2013 report

UN Habitat http://www.citypopulation.de/world/Agglomerations.html

Wendt M (2009) The importance of death and life of great American cities (1961) by Jane Jacobs to the Profession of Urban Planning, New Visions for Public Affairs

World Urbanization Prospects 2014 UN

Yachi S, Loreau M (1998) Biodiversity and ecosystem productivity in a fluctuating environment: the insurance hypothesis Laboratoire d'Ecologie, Unité Mixte de Recherche 7625, Ecole Normale Supérieure, 46, rue d'Ulm, F-75230 Paris Cedex 05, France Stanford University, Stanford, CA, 30 Nov 1998

Yedla S (2003) Urban environmental evolution: the case of Mumbai. Dr. Indira Gandhi Institute of Development Research, Mumbai, India. Commissioned Report Prepared for Institute for Global Environmental Strategy (IGES), Japan

Yongjian et al (2005) For Greenpeace = Yellow River at Risk An Assessment of the Impacts of Climate Change on the Yellow River Source Region, Greenpeace

Integrated Blue and Green Corridor Restoration in Strasbourg: Green Toads, Citizens, and Long-Term Issues

Paul Bois, Jean-Nicolas Beisel, Carine Heitz, Léa Katinka, Julien Laurent, Marjorie Pierrette, Milena Walaszek and Adrien Wanko

Abstract The Ostwaldergraben is an urban stream located in Strasbourg (northeast of France). Mostly fed by groundwater, it was enlarged some forty years ago, which led to a radical alteration of the flow dynamics and a strong siltation. According to the European Water Framework, the stream displayed a bad status with sediments polluted by discharges of former tanneries. Hence, a project of restoration—both of the stream and the adjacent wasteland—was launched by the City of Strasbourg in 2010 to solve these issues of environmental degradation in accordance with the European regulation. The stream bed was redesigned to energize the flows and to create meanders and vegetated benches. To improve the connectivity between two adjacent wetlands, new habitats and a network of ponds have been created. A hybrid type of stormwater treatment system—a pond followed by a constructed wetland—was implemented to complete the restoration project. In this chapter, we propose to study this project from its construction to its current development, through the lens of ecological engineering and a perspective on long-term issues. We aim at illustrating the facts that nature-based solution management can differ from technological management and that the ecosystem services provided by a nature-based solution result from trade-offs, which requires a global analysis of such restoration project. To reach this goal, the project will be studied from ecological, engineering, and sociological perspectives. Our study shows that the restored socio-ecosystem works on a rustic basis and provides several ecosystem services: supporting services (habitat for

P. Bois (✉) · J.-N. Beisel · J. Laurent · M. Walaszek · A. Wanko
ICube, UMR 7357 ENGEES/CNRS/Unistra, 2 Rue Boussingault,
67000 Strasbourg, France
e-mail: p.bois@unistra.fr

P. Bois · J. Laurent · M. Walaszek · A. Wanko
LTSER France, Zone Atelier Environnementale Urbaine, 3 Rue de L'Argonne,
67000 Strasbourg, France

C. Heitz · M. Pierrette
Irstea, UMR GESTE, ENGEES, 1 Quai Koch, BP61039, 67070 Strasbourg, France

L. Katinka
ENGEES, UMR GESTE, 1 Quai Koch, BP61039, 67070 Strasbourg, France

V. Achal and A. Mukherjee (eds.), *Ecological Wisdom Inspired Restoration Engineering*, EcoWISE, https://doi.org/10.1007/978-981-13-0149-0_9

amphibians), regulating services (water quality enhancement), and cultural services (urban landscape greening).

Keywords Stormwater treatment · Ecosystem services · Green toads
Urban landscape · Restoration

1 Introduction

The main target of the Ostwaldergraben restoration's project (2010–2015) was to create an ecological corridor in an urban area by restoring a length of about 600 m of the Ostwaldergraben stream and its floodplain. The main ambition was then to rebuild the stream's function as a blue and green corridor for a target species (*Bufotes viridis*, commonly referred to as the green toad) between two upstream and downstream sites that were previously restored. In 2008, the upstream zone originally composed of an agricultural area was turned into three ponds located in a compensatory afforestation north of a large one called 'Etang du Bohrie' (Fig. 1). This area is surrounded by a mosaic of different habitats. In 2009, on the downstream area, a wetland was restored around the Ill River at the confluence with the Ostwaldergraben. At this location, the restoration consisted in the creation of various environments (wet meadows, woodlands, reed beds) and a flooded island enclosed by arms with various depths. In less than one year, the site was colonized by at least 12 regionally declining species, of which 7 are listed as endangered (birds, reptiles, and insects such as *Odonata* and *Lepidoptera*). But the stream stretch that connects these two wetlands was preventing these species from spreading or freely moving, especially the green toad. The green toad is a patrimonial species, an important and emblematic element of the regional fauna. The species easily colonizes early successional habitats within its area of occurrence as long as the vegetation is not too developed. This species depends on two major types of habitat: ponds as suitable habitats for breeding and larval development and terrestrial pioneer habitats for juveniles and adults (foraging, hibernating, and/or traveling). Several indications of the presence of *B. viridis* were observed upstream (recurrent use of breeding sites) and seldom downstream the restored area along the Ostwaldergraben in a large pond called 'Etang Gerig.'

In that context, the main aims were (i) to provide *B. viridis* with a corridor to reach and leave reproduction sites, allowing exchanges between two close sub-populations; (ii) to reinforce habitats available to complete its life cycle. Organisms that require two different habitat types to fulfill their life cycles, such as pond-breeding amphibians, are especially vulnerable to habitat loss and degradation (Becker et al. 2007). For them, landscape complementation can be defined as the process by which the proximity of two critical habitat patches of different types essential for a major ontogenetic niche shift complements occupancy, abundance, or persistence in each patch (Dunning et al. 1992). The project was designed to both allow displacement of individuals along the river and increase the landscape complementation in an urban context for the green toad.

Fig. 1 Location of the restored section (white double arrow) between two adjacent sites previously restored for environmental purposes (view extracted from Google earth). The flow direction is from west to east

2 Issues and Potential of the Socio-Ecosystem at Stake

2.1 *What Alterations Have Motivated the Restoration Program?*

The Ostwaldergraben is an urban stream mostly fed by groundwater. Forty years ago, it was enlarged which led to a radical alteration of the flow dynamics and a strong siltation. The environmental characteristics change along the course of the stream from the upstream part ('Etang du Bohrie') to the downstream one. Before the confluence with the Ill, the stream flows under a bridge dedicated to the trams' and cars' traffic. The wet bed occupied the entire width of the bridge deck, meaning there were no banks at this point. The passage of terrestrial fauna was impossible there, while the bank vegetation continuity was interrupted and fish passage restricted because of a very low water depth (Fig. 2a). An earthen bund contained the river on the left bank; it abruptly severed the aquatic compartment from the terrestrial one.

In terms of nuisances, pylons of power lines are placed overhang from the stream on the left bank. A regulation obliges the landowners to manage the vegetation that develops under and around the pylon within a radius of 5 m. Since these lines cannot be moved, regular vegetation management operations (cutting of trees and shrubs, mowing) are scheduled.

Before the restoration program, a monitoring highlighted that the stream displayed a bad status, notably because of polluted sediments due to the former tanneries. The district's stormwater was directly discharged into the watercourse without any prior treatment, and the ducts were overhanging the natural environment.

In this context, the challenge was (i) to revitalize the stream in an environment with fine sediments of very poor quality, (ii) to allow the creation of a mosaic of environments favorable to the movements and reproduction of the green toad, (iii) to decrease the negative effects of the passage under the bridge in terms of longitudinal continuity.

2.2 *First Project, Public Inquiry, and Social Adjustments*

From the aforementioned alterations, the Ostwaldergraben project was first imagined as a three-pronged approach:

1. Stream restoration and creation of an ecological continuum allowing animal species to come back, among them the green toad (*B. viridis*);
2. Stormwater depollution performed by treatment systems (we will explain in the next part why this was included);
3. Creation of a bike lane along the restored stream.

Closing the whole area to the public was also an option considered in the project. A public inquiry was carried out in 2011 as a prior step to the whole project. A public inquiry is a regulatory and mandatory step (French regulation) that takes place prior to any land use planning project. It is meant as a democratic consultation tool, where any citizen or environmental/local associations can freely express their view on a given land use project. Three environmental associations answered the inquiry and insisted mainly on the creation of the bike lane, perceived as in total contradiction with the species comeback. Public meetings were held in 2012 in Strasbourg and Ostwald (the two municipalities at the border of the Ostwaldergraben). The organization of public meetings is the second mandatory requirement in the process of public consultation in a land use planning project. Following the public inquiry and the public meetings, the local authorities gave up

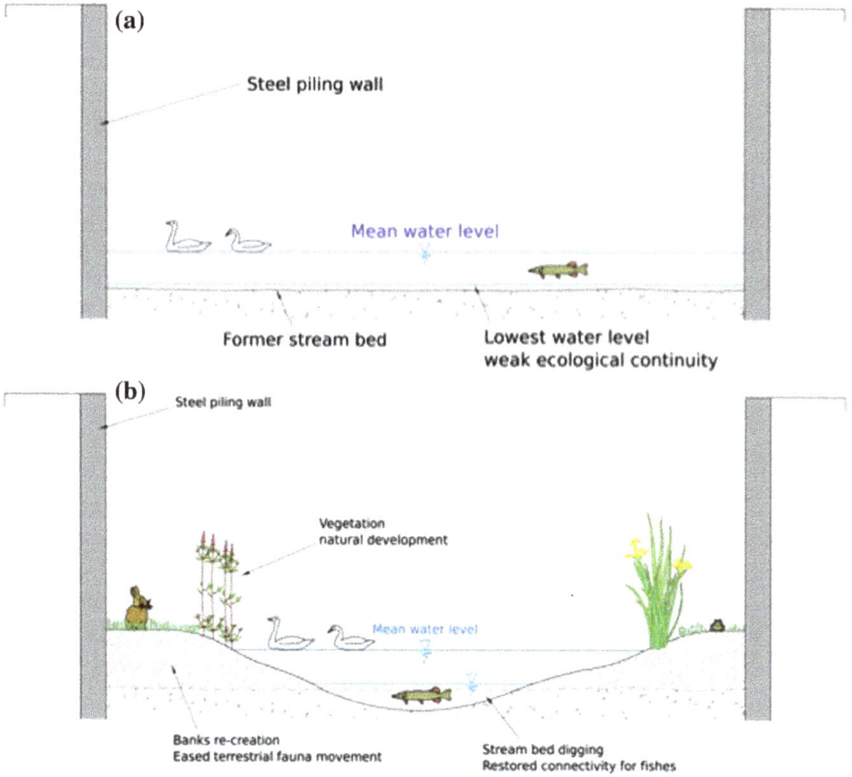

Fig. 2 **a** Former situation. The space under the bridge breaks the terrestrial continuity (banks severed) and the aquatic continuity (low water depth). (Modified from a document produced by the engineering consultant SINBIO), **b** Restored situation. The space under the bridge was modified to tighten the width of the streambed, revitalize the flows, and avoid siltation. The tightening of the minor bed and digging of the middle part of the channel make it possible to obtain water height and flow velocities compatible with fish continuity. The naturally vegetated side banks allow the passage of the terrestrial fauna. (Modified from a document produced by the engineering consultant SINBIO)

on the bike lane project and concluded also that the comeback of given species (e.g., mosquitoes, amphibians, etc.) subsequent to the stream restoration was an issue for some citizens. The stream flow was very low at that time, so mosquitoes were largely able to nest in even before restoration. The report also highlighted that the residents preferred closing the site to avoid burglary or potential nuisances.

The project was eventually carried out in two phases to restore the site (stream and flood channel) over more than 600 m (2015) and setup stormwater treatment systems (2012).

2.3 The 'Dirty' Stormwater Problem and the Way It Was Solved

The restoration of a stream relies on its morphologic features, but also on the various water inputs, mainly stormwater in this case. Two main effects are at stake: (i) banks physical alteration due to water discharge and (ii) water pollution through contaminant transports by stormwater.

The first effect is caused by stormwater collection in separate networks and direct discharge into streams; the resulting peak flows during storm events create a physical effect of digging on the riverbanks. This at least disturbs the morphology of the stream, if not reshaping it and may generate suspended solids resuspension in the stream; the consequences unveil very quickly. Additionally, in the case of the Ostwaldergraben stream, the outlet pipes of the separate network were positioned overhang and discharged directly into the water bodies, which increases the physical impact of runoff.

The second effect results from a quite complex chain of processes. When it rains, rainwater loads with airborne pollutants, such as heavy metals, hydrocarbons, pesticides, and gaseous species (Azimi et al. 2005; Scheyer et al. 2007; Fenger 1999). As it reaches the roofs, roads, gardens and if there is enough rain to start runoff, it will collect other compounds either by transport or dissolution. In the case of urban systems, these main compounds are (Barbosa et al. 2012): solid particles from dust, traffic, and animal feces (becoming suspended solids when carried in the water), heavy metals coming mainly from gutter, road and car material, PAHs from traffic, pesticides from gardening activities, bacteria from animal feces and miscellaneous compounds from light-headed point discharges. In the case of the Ostwaldergraben stream, the typical pollutant concentrations for eight runoff events analyzed in 2013 are listed in Table 1. As can be seen from the table, these are low but significant levels, and of course highly variable values. Urban streams are thus directly contaminated by polluted stormwater in case of untreated discharge. Eventually, even at these low levels of contamination, stormwater was shown to display ecotoxicological effects on aquatic ecosystems (Gosset et al. 2017; Chong et al. 2013); stormwater treatment is mandatory to keep the stream in a good state once restored.

To alleviate this environmental degradation and help maintaining the stream at a good status, many options can be chosen, from the most classically engineered ones to ecologically engineered ones, sometimes called nature-based solutions (NBS) (Erickson et al. 2013). Classically, engineered systems are generally characterized by a large environmental footprint due to the use of exogenous material and of their complex structures arranged with the intensive use of machines powered with fossil fuels. Eventually, the recycling of the materials at the end of their lifecycle is not neutral. Although the environmental footprint of NBS exists as well, it is greatly reduced compared with classically engineered systems (O'Sullivan et al. 2015). In the case of constructed treatment wetlands (CTWs), a large part of the ecological footprint is due to the fact that the basin

Table 1 Contamination levels in the Ostwaldergraben stormwater for eight runoff events in 2013 (adapted from Schmitt et al. 2015). COD = Chemical Oxygen Demand, TSS = Total Suspended Solids, TN = Total Nitrogen, TP = Total Phosphorus

	COD (mg_{O2}/L)	TSS (mg/L)	TN (mg_N/L)	TP (mg_P/L)	Cu (μg/L)	Cr (μg/L)	Pb (μg/L)	Zn (μg/L)
Minimum concentration	25	5	0.8	0	9.6	1.1	1.7	176
Maximum concentration	400	110	11	1.2	42.4	10.6	94.4	640
Number of detections	5	7	8	6	7	5	7	3

must be filled with sand of a given particle size, that is sometimes sampled and transported from afar.

Apart from their ability to clean stormwater and to protect the stream (feature of any stormwater treatment system), the reduced ecological footprint of NBS and the will to create green/blue connections with the to-be-restored stream were the main reasons these systems were set up along the urban stream. More specifically, constructed wetlands were set up. Among the ecosystem services provided by this type of system (Moore and Hunt 2012), regulating services were guiding the construction of a hybrid system composed of a pond and a constructed treatment wetland (CTW). The potential peak flows are buffered by the succession of a pond and a CTW, while the pollutant load is reduced by the combination of settling, photodegradation, physical filtration, biodegradation, and sorption phenomena happening in the pond-CTW system. This system is typical of an ecological engineering approach, as defined by Mitsch (2012): ‘the design of sustainable ecosystems that integrate human society with its natural environment for the benefit of both.’ It is (among others) characterized by reduced engineering during the construction phase, reduced compulsory maintenance and energy costs and use of natural properties from a complex ecosystem to provide ecosystem services—in this case, regulating services.

3 Tackling the Issues: The Restoration of the Ostwaldergraben Socio-Ecosystem

The implementation of this restoration program, the technical choices, and the construction resulted from the work of a consortium with an owner (City of Strasbourg), a contractor (the engineering consultant SINBIO), an engineering school (Ecole Nationale du Génie de l’Eau et de l’Environnement de Strasbourg), a research laboratory (ICUBE lab.), a water agency (AERM—public establishment of the Ministry for Sustainable Development), a NGO (BUFO—study of Amphibians and Reptiles of Alsace), and citizens. This co-construction exercise involved a participatory approach that marked the different stages and accomplishments that we describe here.

3.1 Technical Itinerary to Achieve the Restoration Objectives

In order to stimulate the flow and limit the siltation, the bed width was reduced to 2 m. Gently sloping banks likely to be naturally vegetated by helophytes were created as well as windings and meanderings (Fig. 3). Shrub willow cuttings were planted.

The first operations to improve the diversity of the habitats were devoted to the creation of a pond network in the upstream sector where the sediments were not polluted. These ponds have three main functions: They serve as potential breeding habitat for amphibians, they contribute to the habitat diversity available to wildlife, and they collect the water discharging from the stormwater treatment system (see next part). The excavation/embankment operations were conducted to erase the bund; this reconnected the river and its major bed, while removing an awkward feature in this landscape context.

Management operations (mowing) of the vegetation under and around electric pylons must be conducted regularly. These interventions allowed artificially rejuvenating the environment on the left bank where breeding ponds were created, the right bank being left in natural evolution. The fauna and flora thus have a mosaic of aquatic/terrestrial habitats at different stages of plant succession. Mowed habitats are banks of alluvium and sand adequate to the displacements of many species.

Fig. 3 Restored stream. A dead arm can be seen on the left side of the picture, and a meander on the right (*Source* ICube Laboratory)

The biological quality of the Ostwaldergraben was evaluated with biological indicators based on aquatic macroinvertebrates. The biotic index applied highlighted the bad status of the stream stretch. Deposits made during the period of the tanneries' activity resulted in sediment contamination with highly toxic metal trace elements, such as chromium (Table 2). These observations led to remove the bund and dispose of this part of the polluted materials in a nearby dump (the one from the old tanneries). The remaining contaminated sediments corresponded to finer fractions, with sludge dredged from the stream bed. They were confined on created mudflats in a part of the old bed (three sections short-circuited by the re-meandering). The objective was to confine all the polluted muds extracted from the reduced minor bed and the ponds created on the right bank. The mudflats were covered with a geomembrane and a geotextile to avoid contact with the air in case of a drop in the water table. The containment of polluted sediments enabled the project to stay economically realistic compared with an export-and-treatment option. Additionally, it avoids for humans as for animals any possibility of contact or ingestion of the pollutants when visiting the area.

Vegetated embankments were created under the bridge to ease the passage of wildlife under the structure and restore the ecological corridor function for terrestrial species (Fig. 2b). The flow was tightened in a dug channel to promote aquatic continuity, in particular for fish. This panel of technical options was chosen thanks to the property right of the site by the owner builder, the City of Strasbourg.

Table 2 Sediment quality of the stream observed at two locations in 2006. All measured organic compounds are polycyclic aromatic hydrocarbon (PAH). The limit values provided correspond to French guideline values for soil and water pollution

	Units	Location 1	Location 2	Limit value
Dry matter	%	23.5	41.0	/
Cd	mg/kg DM	5	5	<10
Cr	mg/kg DM	**1,600**	**5,700**	<65
Cu	mg/kg DM	**150**	53	<95
Hg	mg/kg DM	1	0.23	<3.5
Ni	mg/kg DM	34	15	<70
Pb	mg/kg DM	190	70	<200
Zn	mg/kg DM	1,100	190	<4,500
Benzo(b)fluoranthene(3,4)	µg/kg DM	1,200	400	/
Benzo(k)fluoranthene(11,12)	µg/kg DM	550	190	/
Benzo(g,h,i)perylene(1,12)	µg/kg DM	700	280	/
Indéno (1,2,3-c,d) pyrène	µg/kg DM	1,300	390	/
Fluoranthene	µg/kg DM	2,200	640	/
Benzo(a)pyrene(3,4)	µg/kg DM	1,100	410	/
Sum of the six PAH	µg/kg DM	7,050	2,310	/

3.2 Setting up the Nature-Based Solutions to Treat Stormwater

Three urban watersheds are discharging runoff water into the stream on the study site, so three nature-based systems were set up; each one collecting and treating the water from one watershed. The sizes of the watersheds are pretty alike (ca. 2 ha) but in order to test the effect of different configurations on treatment efficiency, the constructed wetland varied in size. The choice of the CTW being the variable was somewhat arbitrary, as the pond might have been another variable of choice for different setups. Thus on the location of the initial wasteland were eventually built these treatment systems (Fig. 4), made of (from upstream to downstream):

- An artificial pond followed by a vertical subsurface flow wetland (#1 and #3, with differences in the size of the porous media that were used) followed by a discharge pond;
- An artificial pond followed by a horizontal subsurface flow wetland (#2) followed by a discharge pond.

Of the three systems, only two (#1 and #3) have been studied so far, so we will not develop further on #2. The main geometric, hydraulic, and upstream watershed characteristics of the systems are summarized in Table 3. For each watershed, the {pond + CTW} combination area is around 1–2% of the watershed active area.

As the groundwater table is close to the surface on the premises, the whole system was conceived to be impervious, in order to prevent infiltration of untreated water. As the available soil was not watertight, the main question that arose was how to ensure such imperviousness? The choice was finally made to coat the bottom of the pond and the wetland with 30 cm clay to achieve this at a reasonable cost; luckily this was the most environmental-friendly solution (other solutions were geomembrane or concrete).

The pond is fed through a concrete duct that is the outlet of the separate network collecting runoff water from the urban watershed (Fig. 5). The water flows from the pond to the CTW through a floating weir, whose triggering depends on the initial pond water level and rain events characteristics (intensity, duration, dry period, water level, and return period) and that works only for large enough rain events. The constructed treatment wetland is fed through several PVC pipes reaching different parts of the system to try and feed the system homogeneously. After vertical flow in the CTW, the stormwater is discharged into the last artificial ponds that are hydraulically connected to the stream. Moreover, the first pond is equipped with overflows—made of concrete pipes, which increases the ecological footprint of the system—that discharge directly into the final pond in case of extreme events. This overflow system caused a temporary failure in the system. It is indeed sealed with clay on the edge of the pipe to remain watertight, but a too low water level in the pond at first caused the clay to dry and subsequently retract. When water filled the pond back at a higher level, it started leaking at the joint and the pond drained. More clay was added to solve the problem.

Fig. 4 Hybrid treatment system just after construction. The settling pond is on the lower side of the picture; the constructed treatment wetland is on the upper left side of the picture. A bit further behind, a discharge pond can be seen (*Source* ICube Laboratory)

Table 3 Stormwater treatment system characteristics and hydraulic features (adapted from Schmitt 2014)

		#1	#3
Watershed	Surface (ha)	2.7	1.8
	Active surface (ha)	0.9	0.52
Pond	Size (m × m)	11 × 9	5 × 4.5
	Maximum hydraulic load (m^3/m^2/day)	10	10
	Permanent water volume (m^3)	28	2
	Maximum temporary water volume (m^3)	56	10
CTW	Area (m^2)	90	100
	Surface/watershed active surface (%)	1	2
	Maximum hydraulic load (m^3/m^2/day)	60	30

From the surface to the bottom of the CTWs, the porous medium is distributed as follows: The top layer is made of 20 cm (CTW #1), respectively, 30 cm (CTW #3) of sand (particle size from 0 to 4 mm); for both CTWs, the intermediate layer is made of 25 cm of fine gravel (particle size from 4 to 8 mm), and the drainage layer is made of 25 cm of gravel (particle size from 16 to 22 mm). The choice and the arrangement of these layers are a crucial and delicate point for CTWs, as it controls the infiltration rate and subsequently the hydraulic residence time, determining the absence or prominence of clogging in the system.

Finally, as the system feeding is by essence stochastic, a minimal water level is maintained at the bottom of the CTWs to reduce the water stress that vegetation, especially wetland plants, could endure during long dry periods. The ponds were

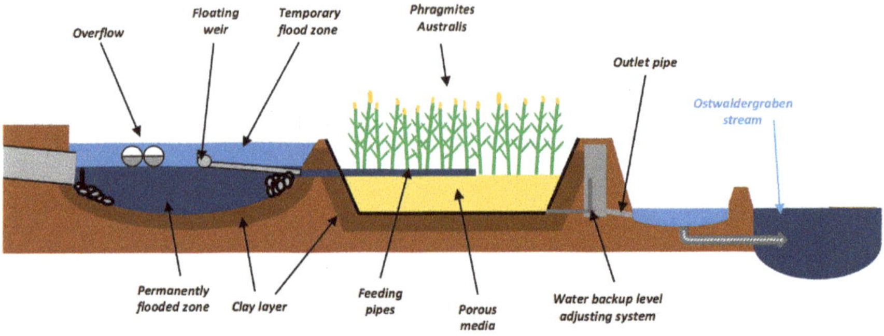

Fig. 5 Side view of the treatment system. From left to right (corresponding also to the water flux), pond, constructed wetland, discharge pond, stream (adapted from Schmitt 2014)

built free of vegetation, while the constructed wetlands were planted with *Phragmites australis* (9 plants/m^2).

4 The Ostwaldergraben's Response: Aftermath of the Restoration Project

4.1 The Creation of a Haven for the Green Toad? A New Face for the Corridor

The operations on the natural environment showed positive changes in a very short time after achievement. For example, the first ponds were created between March and August 2012, and as early as in September immature individuals of green toads were observed in these newly created habitats. In April of the following year (2013), egg clutches of several green toads were observed. The creation of a network of ponds at the upstream site (Bohrie Pond—2008) produced the same trends (Michel and Zrak 2015): a very fast colonization and a rise in the number of use of these environments, at least before the vegetation development (2011). These few results given as examples show the rapidity of spread in new environments by organisms whenever source populations are around.

The notion of a mosaic of habitats has also a central place in this type of project. The amphibian species *Bufotes viridis* acts here as an umbrella species that needs fallow and even cultivated soil to find suitable habitat during its terrestrial phase. This pioneer environmental species needs a constantly rejuvenated environment. Working gravel pits on both sides of the restored site allow the maintenance of pioneer habitats, while the restored site can be considered as a secondary habitat favoring connectivity between subpopulations. To keep this passage interesting and functional for the green toad, vegetation must be managed to avoid too great development and keep this system at a pioneer stage. The maintenance needed to manage the vegetation under and around the power lines pylons also helps rejuvenating the vegetation along the dispersal corridor, in particular on the bare soil surrounding the artificial breeding ponds. The obligatory status of this management compensates the absence of self-sustaining natural processes to maintain each adequate habitat available in the mid- and long-terms.

4.2 *Sustainability of the stormwater treatment: evolution of the hybrid system over 5 years*

The responses of the treatment system were observed from the beginning of operation (back in 2012). The hybrid system was meant to provide regulating services on peak flows and water quality. First, less than 20% of the runoff events discharging into the pond actually discharge from the pond to the CTW. Thus, the sizing of the pond provides a strong buffering effect on runoff water discharge into the stream. Second, the pollution of the stormwater is clearly mitigated for both watersheds: major pollution (COD, TSS, TN, TP) (Schmitt et al. 2015) and micropollution (heavy metals and polycyclic aromatic hydrocarbons—PAHs) drops from the inlet to the outlet of the systems. These performances are sustained in time, as sampling sessions from 2013 to 2017 gave similar results.

The ecosystem services expected for the hybrid systems are well provided. The way the system works is also interesting: For instance, in case of runoff a strong shortcut is created by the floating weir because of its arm's length, which largely reduces the hydraulic residence time in the pond—and thus the efficiency of the settling phase (Laurent et al. 2013). This could be easily enhanced by installing a static weir leaving the whole pond surface available. In the CTW, the flow distribution is heterogeneous unlike what was expected: very little water reaches the end of the longest feeding pipes, which creates a feeding gradient.

The accumulation of sediments progressively fills the pond; the organic layer depositing at the surface of the CTW is around 4 cm after 6 years for the most covered CTW and 0 for the less covered. Ponds are colonized (2 out of 3) by macrophytes or algae and transform slowly into wetlands. In the CTWs, a gradient of organic deposit according to the feeding gradient is observed. In both constructed wetlands, trees and grass are appearing and sometimes taking over on macrophytes.

After 6 years of operation, the pollution mitigation is still working. As the pollution that is treated is mainly due to heavy metals and PAH, that are either non-degradable or highly stable compounds, their removal from stormwater means they migrated from the liquid phase to another phase. These pollutants are then logically found in the solid phase (sediments from the pond and organic matter and sand from the constructed wetland)—in the pond sediments, heavy metals were detected at a few mg/kg_{DM} to more than 2,000 mg/kg_{DM} for zinc, and around 1 mg/kg_{DM} of PAHs—and in the vegetation of the constructed wetland—from 1 to 22 mg/kg_{plant_DM} of heavy metals were measured in the reeds growing in the CTW

—(Schmitt et al. 2015). The contamination level in the sediments will rise steadily with time, which makes it first a sink for pollutants, but could also create a source of pollution under changing physicochemical conditions (Semadeni-Davies 2006).

When we look further ahead, the hydraulic functionality is likely to remain as long as sediments' accumulation is not too important. Yet, as the system keeps on retaining suspended solids from stormwater, accumulation will go on and call for sediment dredging, at least in the pond. With a minimal maintenance, this ecosystem service can be easily sustained over time. It is less crucial in the CTW, as the accumulation of mainly organic matter on its surface is much lower than in the pond, as the latter is the first in the system to retain such pollutants. If we look now from an ecotoxicological perspective on this system, the handling and disposal of these sediments appear to be of utmost importance: The toxicity of stormwater sediments was shown for much lower metal concentrations (Hatch and Allen 1999; Snodgrass et al. 2008). And as the treatment system is bound to stay and continue working, this long-term question is critical for the sustainability of the system. Eventually, the issue of the sediment behavior and fate should be carefully thought after.

4.3 Nature-Based Solutions and Citizen Representations: Sociological Aspects of the Project

To study the way this freshly built socio-ecosystem is perceived by the local residents and surveys were carried out during spring 2017. As a first step, seven people living in close proximity to the site were interviewed about their perception of their neighborhood. These semi-structured interviews were focused on the representation of their living environment and the potential nuisances, the representation of the Ostwaldergraben and its utility and on their own practices in terms of water pollution. This step was meant to define precise questions before individual questionnaires would be created. The individual questionnaires summarized the main issues of the interviews and were addressed to all the inhabitants living in the street right next to the site and to the inhabitants living in the street right behind (147 households). The questionnaire contained 23 questions (including 7 questions on the social characteristics of the respondents). It was self-administrated, and the principle of answering to the questions was built on the Likert scale: Respondents had to choose between five modalities (from 1 = 'not at all' to 5 = 'absolutely'). Of the 147 questionnaires mailed, 66 answers were received (45%—without any reminder). The sample was quite similar to the population living in this area: In our sample, we had 53% men and 47% female, the average age was 52 years ($\sigma = 15.1$) and there was a strong percentage of retired people (30%). Fifty-eight percent of the respondents had a direct look upon the zone. The questionnaire allowed us to test three main hypotheses: (i) The distance to the Ostwaldergraben site influences the residents' representation of the Ostwaldergraben; (ii) the knowledge of the functionality of the site (depollution) modifies the inhabitants' behaviors linked to their own pollution in the rainwater network collection; and (iii) there is a typical profile of inhabitants who show a stronger awareness of the link between pollution in the stormwater network collection and the pollution of the Ostwaldergraben stream.

The results showed that distance to the site influenced the representation of pollution of the Ostwaldergraben: The respondents with a direct look on the site preferentially think that the stream is polluted [F (1.61) = 4.0334, $p = 0.05$]. Likewise, the respondents with direct look on the site wish that the site remains closed to the public [F (1.60) = 7.1361; $p = 0.01$].

To check the link between the respondents' understanding of the site and their own behaviors (hypothesis #2), further questions were asked (Figs. 6 and 7). The answers to the question on the system's functioning (Fig. 6) showed that the reintroduction function is the most understood, followed by the aesthetics enhancement. On the contrary, the regulating service—pollution mitigation—is poorly understood and the gutter products do not seem to be associated with pollution. To precise things a bit more, the question of the origin of the water entering the pond-CTW system was asked (Fig. 7). These answers showed that people think of the system as only designed to treat 'rainfall,' which means actually direct rainfall, as shows the answers to the last question '[…] gutter-discharged products.' People are aware that this system is not meant to treat wastewater as shows the answer to the second question (…domestic wastewater). Eventually, the understanding of the stream restoration conforms to reality but there is a gap between the {pond-CTW} function and its perception from the respondents. Additionally, an analysis of the variance (ANOVA) of the results shows that reading the information panel does not make a significant difference in the understanding of neither the treatment function of the hybrid system nor the origin of the water discharging in the system. We can conclude that the panels failed to explain the function of the treatment system, and that if deeper understanding of the people is wished by the city of Strasbourg, additional measures should be taken.

To understand the weight of the communication and information in the respondents' understanding of the site functionality, we tested the part of the communication implemented by the city of Strasbourg (public meetings, visits of the site once restored and information panels). Ninety-one percent of the respondents did not attend the public meetings, and 82% did not visit the restored site. Apart from these formal information sequences, 55% of the respondents read the information panels that are displayed on the access doors to the site, but the impact of the panels on the level of understanding is clearly questionable. It also seems that the respondents did not feel informed enough on the site and its functionality (52% of the respondents). Only 31% think that they are informed on that point.

Concerning a typical profile of inhabitants that were more aware on the link between the stormwater pollution and the pollution of the Ostwaldergraben site (hypothesis #3), it seems that the representation of a level of pollution was not linked with social characteristics.

We also tested the future of the site, trying to answer the question of the Ostwaldergraben sustainability and a kind of 'appropriation' of the site by the

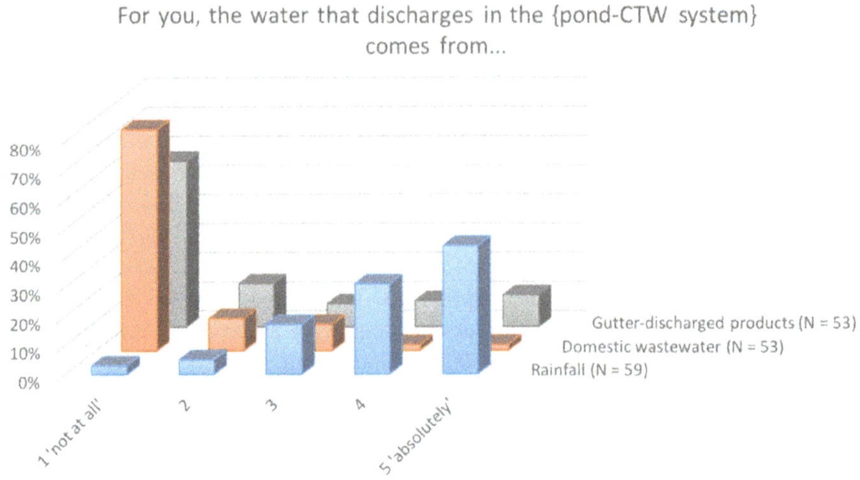

Fig. 6 Results from the questionnaire sent to local communities. The questions were related to the pond and constructed treatment wetland (CTW) in operation phase. Here the origin of the water that flows into the {pond-CTW} system

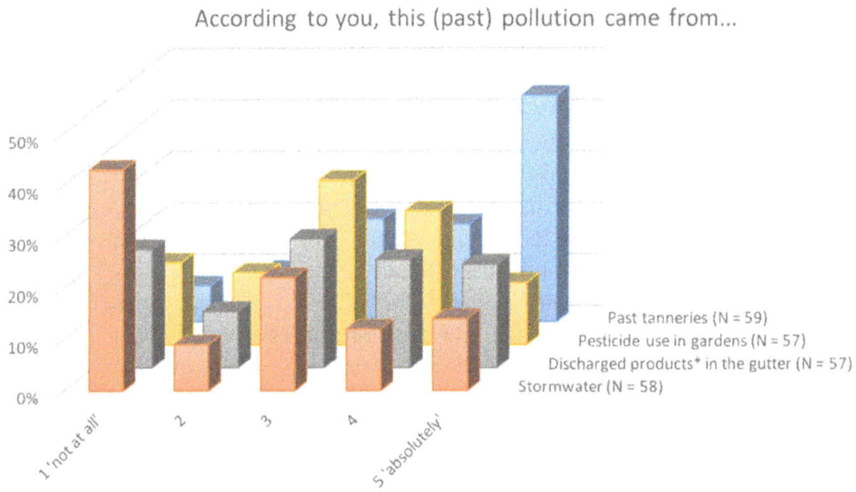

Fig. 7 Results from the questionnaire sent to local communities. The questions were related to the pond and constructed treatment wetland (CTW) in operation phase. Here the origin of the water that flows into the {pond-CTW} system. * paint, oil, water from carwash

inhabitants. A majority of the respondents (56%) wanted it to remain closed, while 30% wanted the opening to be restricted and 14% wanted it to be open. Eventually, we tested the interest of the local community about getting more active in the life of the site and the future vision of the site they had (Table 4). In accordance with the will of having it closed to the public, 46% of the answers about tours open for the

Table 4 Results of the questionnaire sent to residents on the social aspects of the project. The question was: « About the future use of the site, you would like to… » (response expressed in %). 1 = 'not at all,' 5 ='absolutely'

	1	2	3	4	5
Use it for educational purpose (N = 66)	23	3	12	14	48
Get involved in the stewardship of the site (N = 62)	58	10	13	8	11
Have it more used for tours open to the public (N = 65)	46	6	20	8	20
Make it an example for other sites in Strasbourg (N = 65)	20	2	14	11	53

public are 'not at all.' A clear tendency also showed that the inhabitants did not want to get involved in the stewardship of the site (68% of the answers between 1 and 2). They did believe, however, that this system is valuable, in terms of use for educational purpose or as an example to be replicated across Strasbourg.

To summarize, the Ostwaldergraben site seemed mostly 'accepted' and well perceived by the inhabitants. Indeed, the first source of perceived nuisance in the neighborhood was actually youngsters hanging out (20 occurrences at the open question '*In your neighborhood, what kind of nuisances do you perceive (noise pollution, olfactory pollution, visual pollution)*' N = 53). The second nuisance is linked to the amphibians (18 occurrences). Let us remember that the respondents who perceived the most negatively the amphibians are located in one spot, just across the natural pond. Yet, the results showed that the respondents who perceived the amphibians as a nuisance were significantly not favorable to wetland habitats [$F(1.49) = 10.576$; $p = 0.002$].

5 Conclusion: The Trade-Offs of This Project

The target species of this program was *B. viridis*, an endangered amphibian. Many amphibian species have populations structured as patchy networks or metapopulations. Urbanization reduces the ability of these networks of populations to function due to the construction of roads and urban infrastructure that inhibit or discourage amphibian dispersal (Hamer and McDonnell 2008). Stormwater wetlands and their neighboring terrestrial habitats may play an under-appreciated role in the conservation of urban amphibians (Scheffers and Paszkowski 2013). The construction of stormwater ponds is recognized as a useful tool both to mitigate the loss of wetlands, to retain water runoff from impermeable urban surfaces and to treat them, but their ecological value, in particular as breeding habitat for amphibians, remains poorly known (Chester and Robson 2013; Scheffers and Paszkowski 2013). Habitat surrounding stormwater sites also merits attention and preservation considering the importance of small-scale connections between the habitat of immatures and that of adults (Scheffers and Paszkowski 2013).

Controversial views of urban small water bodies are related to their water quality, even if in recent decades the quality of many urban aquatic habitats has been significantly improved. In green toad, heavy metals like copper and lead have been shown to increase the frequency of morphological malformations (Dorchin and Shanas 2010) but *B. viridis* was already found in heavily contaminated water bodies (Adlassnig et al. 2013). Further, analyses are required to investigate the potential cost for the species of a putative lack of water quality.

The great removal efficiency of the treatment system provides benefits, but the cost of the sediments pollution and long-term solution to this is not to be underestimated. Additionally, taking into account local communities advices could help replicating this type of project and improve its acceptability among the population.

Acknowledgements The research on the Ostwaldergraben system is mainly supported by the Rhin-Meuse Water Agency and the French Biodiversity Agency through the LumiEau-Stra project. Additional research support is provided by the ZAEU (French LTSER network). Ideas for this chapter came through collaborations supported by the Urban Sustainability Research Coordination Network, support by the U.S. National Science Foundation (Grant No. 1140070).

References

Adlassnig W, Sassmann S, Grawunder A, Puschenreiter M, Horvath A, Koller-Peroutka M (2013) Amphibians in metal-contaminated habitats. Salamandra 49(3):149–158

Azimi S, Rocher V, Muller M, Moilleron R, Thevenot DR (2005) Sources, distribution and variability of hydrocarbons and metals in atmospheric deposition in an urban area (Paris, France). Sci Total Environ 337:223–239

Barbosa AE, Fernandes JN, David LM (2012) Key issues for sustainable urban stormwater management. Water Res 46:6787–6798

Becker CG, Fonseca CR, Haddad CFB, Batista RF, Prado PI (2007) Habitat split and the global decline of amphibians. Science 318(5857):1775–1777

Chester ET, Robson BJ (2013) Anthropogenic refuges for freshwater biodiversity: their ecological characteristics and management. Biol Cons 166:64–75

Chong MN, Sidhu J, Aryal R, Tang J, Gernjak W, Escher B, Toze S (2013) Urban stormwater harvesting and reuse: a probe into the chemical, toxicology and microbiological contaminants in water quality. Environ Monit Assess 185:6645–6652

Dorchin A, Shanas U (2010) Assessment of pollution in road runoff using a *Bufo viridis* biological assay. Environ Pollut 158(12):3626–3633

Dunning JB, Danielson BJ, Pulliam HR (1992) Ecological processes that affect populations in complex landscapes. Oikos 65(1):169

Erickson AJ, Weiss PT, Gulliver JS (2013) Stormwater treatment processes. Optimizing stormwater treatment practices. Springer, New York, NY, pp 22–34

Fenger J (1999) Urban air quality. Atmos Environ 33:4877–4900

Gosset A, Durrieu C, Orias F, Bayard R, Perrodin Y (2017) Identification and assessment of ecotoxicological hazards attributable to pollutants in 2 urban wet weather discharges. Environ Sci Process Impacts 19:1150–1168

Hamer AJ, McDonnell MJ (2008) Amphibian ecology and conservation in the urbanising world: a review. Biol Cons 141(10):2432–2449

Hatch AC, Allen G Jr (1999) Sediment toxicity and stormwater runoff in a contaminated receiving system: consideration of different bioassays in the laboratory and field. Chemosphere 39 (6):1001–1017

Laurent J, Finaud-Guyot P, Wanko A, Bois P, Mosé R (2013) Hydrodynamic of artificial wetlands at the outlet of urban catchment: complementarity of the systemic approach and computational fluid dynamics tools. Récents Progrès en Génie des Procédés, 104, Ed. SFGP, Paris, France

Michel V, Zrak E (2015) Bilan de dix années de suivi des indicateurs de la biodiversité en Alsace. Les effectifs de crapauds verts, Bufotes viridis. CICONIA, 39 (2–3):144–151

Mitsch WJ (2012) What is ecological engineering? Ecol Eng 45:5–12

Moore TLC, Hunt WF (2012) Ecosystem service provision by stormwater wetlands and ponds: a means for evaluation? Water Res 46:6811–6823

O'Sullivan AD, Wicke D, Hengen TJ, Sieverding HL, Stone JJ (2015) Life cycle assessment modelling of stormwater treatment systems. J Environ Manage 149:236–244

Scheffers BR, Paszkowski CA (2013) Amphibian use of urban stormwater wetlands: the role of natural habitat features. Landsc Urban Plann 113:139–149

Scheyer A, Morville S, Mirabel P, Millet M (2007) Pesticides analyzed in rainwater in Alsace region (Eastern France): comparison between urban and rural sites. Atmos Environ 41:7241–7252

Schmitt N (2014) Characterization of hybrid systems for the treatment of urban stormwater: fate of emerging pollutants. PhD Thesis. University of Strasbourg. HAL-ID: tel-01214516

Schmitt N, Wanko A, Laurent J, Bois P, Molle P, Mosé R (2015) Constructed wetlands treating stormwater from separate sewer networks in a residential Strasbourg urban catchment area: Micropollutant removal and fate. J Environ Chem Eng 3:2816–2824

Semadeni-Davies A (2006) Winter performance of an urban stormwater pond in southern Sweden. Hydrol Process 20:165–182

Snodgrass JW, Casey RE, Joseph D, Simon JA (2008) Microcosm investigations of stormwater pond sediment toxicity to embryonic and larval amphibians: Variation in sensitivity among species. Environ Pollut 154:291–297

The Role of Ecological Wisdom in Brownfields Redevelopment in China

Deepika Kumari, Xin Wang and Jiang Wu

Abstract Brownfields are challenging and common landscape problem especially in industrial and post-industrial cities in many countries. With accelerated population rate and rapid urbanization, many Chinese cities are developing brownfield sites. Such sites inhibit economic growth and use of urban land, and thus, their redevelopment is highly warranted. It has become an important problem that restricts the sustainable development of cities in China. Certainly, brownfield redevelopment is a new term in China and is more challenging compared to other countries worldwide. To address this issue, it is important to coordinate and unify ecological wisdom from other countries where brownfield redevelopment is studied well and applying those technologies to strengthen the current apprehension of brownfield redevelopment. Ecological wisdom aims to increase the awareness of ecological knowledge during brownfield redevelopment to achieve sustainability during this process. This chapter mainly focuses on the concept of brownfield and the causes and formation of brownfield in China, introduces the concept of ecological wisdom to take guidance from brownfield researches in foreign countries, and puts forward for brownfield redevelopment and remediation in China.

Keywords Brownfield · Remediation · Redevelopment · Ecological wisdom Policy

1 Introduction

Brownfields are generally abandoned commercial or industrial resources where redevelopment is usually prevented by feasible environmental pollution and the possible obligation that prevails under Superfund regulations (Kretchik 2002).

D. Kumari (✉) · X. Wang · J. Wu
College of Architecture and Urban Planning, Tongji University,
1239 Siping Road,
Shanghai 200092, China
e-mail: deepikakumari@tongji.edu.cn

V. Achal and A. Mukherjee (eds.), *Ecological Wisdom Inspired Restoration Engineering*, EcoWISE, https://doi.org/10.1007/978-981-13-0149-0_10

The problems associated with brownfields are recently regarded as big as that affects the sustainable development of the society. The brownfield issue is complex and extensive in all countries due to step-by-step, but steady, movement of industries from the city to greenfield areas since the mid-1970s, leaving many big cities with innumerable underutilized or vacant industrial sites (De Sousa 2000). This problem is not only restricted to developed countries. In China, brownfields in the form of idled or abandoned industrial sites and associated land contamination have become an alarming problem in both rural and urban areas. The existence of brownfield brings environmental risks in terms of polluted soil and groundwater, which can affect public health via contaminated food chain and air pollution due to volatilization of pollutants. This chapter aims to provide various definitions of brownfield, with emphasis on brownfield formation in China, associated problems and how ecological wisdom guides us towards brownfield redevelopment.

1.1 Definition of Brownfield

Brownfields are defined by US Environmental Protection Agency (USEPA) as abandoned, idled or underutilized industrial and commercial properties where expansion, redevelopment or reuse may be puzzled by the presence of any type of contaminant (USEPA 1997).

The British National Land Use Database defined brownfields as problematic or contaminated sites in use but with potential for redevelopment (NLUD 2000). Later, the American Planning Association explained brownfields as vacant or underused properties passed over for development because of pollutants or contaminants (APA 2005). One of the most accepted and widely used definitions of brownfield was proposed by Alker et al. (2000) as any site which has previously been used but is not currently fully in use, although it may be partially utilized that may also be derelict, vacant or contaminated. From Chinese urban land use prospect, brownfields are industrial and commercial sites in urban areas which are abandoned or underused due to real or perceived environmental problems and cannot be immediately used without remediation (Cao and Guan 2007).

2 Brownfields Formation in China

Considering environmental concern associated with brownfields, the redevelopment has acquired significant consideration in Western countries since the 1970s (Adams et al. 2010; Bjelland 2004). The US Congress passed the Superfund act in 1980 towards brownfield governance under which the chemical and petroleum industries should provide tax upon releases of hazardous substances that may endanger public health or the environment. In Europe, greening movement plays a big role in the brownfield redevelopment in designing sustainable communities. In Britain itself, more than 19% of brownfield sites were transformed into green spaces (UK DETR 1998). One city that has been particularly proactive in converting brownfields into green spaces over the last decade is Toronto, Canada. Toronto in Canada is known to be very active in getting brownfields converted into green spaces during the last decade and thus described as 'city within a park' (De Sousa 2002, 2003).

However, comparatively redevelopment of brownfield has started attracting consideration of Chinese researchers in recent times (Cao and Guan 2007). The history of Chinese 'brownfield' can be traced back to the Great Leap Forward period 50 years ago or even earlier in the construction period of some high-polluting industries. Due to the lack of awareness of environmental protection in the early industrial planning, lots of factories were built in the centre or surrounding area of the city. The land pollution is getting worse in terms of intensive management and deficient anti-pollution measures. With the industrial restructuring in recent years, large numbers of high-polluting industries are being relocated, which has left numbers of polluted land that cannot be reused, as so-called brownfield. It has grown as a big problem that holds back the urban sustainable development in China.

As per the World Bank Report on May 2005 (Xie and Li 2010), there are minimum of 5000 brownfield sites in China and this number is increasing continuously. Considering serious problem arising from brownfields, the Government of China released the Special Plan for the Revitalisation of Old Industrial-Base Cities (2013–2022) in March 2013 with an aim to redevelop formerly industrial urban areas into sustainable urban land use. A stage model summarizing the process to produce brownfields in China compromised of three steps, namely industrialization, suburbanization and deindustrialization, is presented in Fig. 1 (Liu et al. 2014). The process started during Third Front Construction (1964–1971) of China with formation of industrial-based cities by constituting Industrial Areas, especially in urban fringe areas (Li and Xie 2010). During the late 1970s, China launched a sequence of economic reforms and moved from a planned economy to a market

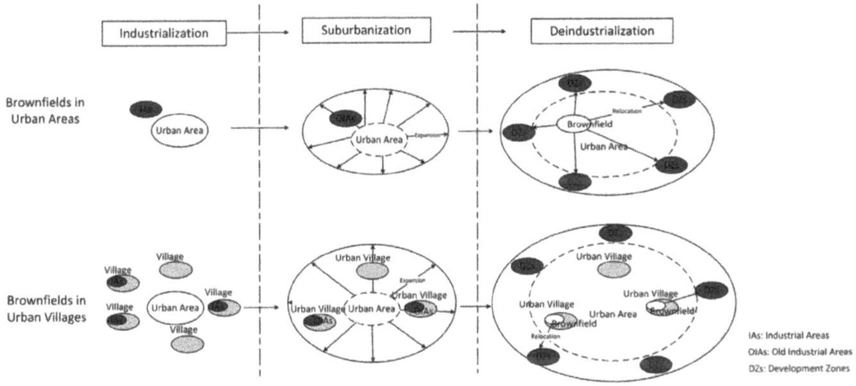

Fig. 1 Stage model of brownfield formation in China (reprinted with permission from Liu et al. 2014)

economy leading to thorough industrialization process (Lin and Ho 2005; Liu et al. 2014). This process was referred to as 'the bottom-up rural industrialization' (Lin 2001, 2007).

Further, a great expansion in urban area was observed in post-reform China where more than 100 Chinese cities were identified as old industrial-base cities according to 'The Special Plan for the Revitalisation of Old Industrial-Base Cities (2013–2022)'. With progressive economic reformation in China, the Old Industrial Areas (OIAs) in the central urban areas are recognized as a big obstacle. Considering environmental issues, these industrial enterprises are required to get relocated. Thus, recently regional deindustrialization was reported in big Chinese cities including Beijing, Shanghai and Tianjin (Wang et al. 2012). In mere five years from 2000 to 2005, more than 140 traditional manufacturing plants were relocated outside the urban area in Beijing, while many such relocations were performed before also (Gao et al. 2010; Li 2011). Such relocation of industries in urban villages leads to big numbers of brownfield sites, which may be polluted, derelict or have been partly redeveloped.

3 Problems with Brownfields

Brownfields in China are mainly contaminated with (1) heavy metals, coming from iron and steel smelting enterprises, mine tailings, e-waste and its recycling process; (2) organic pollutants, such as benzene compounds and halogenated hydrocarbons coming from petroleum, chemical and coking industries; (3) persistent organic pollutants (POPs), coming from history of pesticide use including chlorobenzene, chlordane, DDT and mirex; and (4) compound pollution or landfill containing complex sources of waste, industrial waste, construction waste and other mix contaminants. Brownfields can further be divided into industrial brown, commercial brown, residential brown and public brownfield. Industrial brownfield mainly refers to the brownfield suitable for transformation into industrial land; commercial brownfield mainly refers to the brown land suitable for transformation into commercial places; residential brownfield mainly refers to the brown land suitable for transformation into a residential place; while public brownfield mainly refers to the brown land suitable for transformation into public facilities to facilitate the public daily life.

In 2006, the State Environmental Protection Administration and the Ministry of Land and Resources of China jointly launched the national soil pollution status and prevention measures to comprehensively and systematically grasp the overall situation of the national soil environmental quality, identify the pollutants, extent of soil pollution and put forward the measures of environmental management. The State Environmental Protection Administration and the Ministry of Land and Resources of China issued first national soil pollution investigation bulletin on April 17, 2014 based on a soil survey of about 6.3 million square miles of land in China, covering all the arable land, part of the woodland, grassland, unused land

and construction land, carried out during April 2005 to December 2013 (MEP China 2014). The report said overall 16.1% of China's soil in general is polluted, out of which 1.1% is severely polluted. The main contaminant was identified as inorganic pollutants followed by organic type and then small proportion of compound pollution. This data implicate current situation of brownfield in China. It is suggested that a follow-up survey should be carried out every year, and the updated results can be included in the annual statistical work of the land and redevelopment of remediation process can be adapted based on such data.

In an attempt to harmonize economic and social development with environmental safety, a new environmental protection law (EPL) took effect in China from January 1, 2015 under which harsher penalties are framed for environmental offences such as falsifying environmental data and discharge of contaminants (Zhang and Cao 2015).

4 Ecological Wisdom and Brownfield Redevelopment in China

The brownfield problem exists in all industrialized nation due to continuous shifting of industries from the city to greenfield areas since the mid-1970s, leaving many cities with vacant industrial sites (Gertler et al. 1995). Brownfield redevelopment in China is a big challenge to governments, enterprises and local societies. At the same time, environmental supervision and management of polluted sites has become an important obligation for environmental authorities. Thus, it is necessary to bring the concept of 'ecological wisdom' to brownfields problem in creating criteria under which humans and nature can exist in productive harmony to cover the social, economic and other necessities of present and future generations.

Ecological wisdom is perhaps a very important concept in developing management decisions in the twenty-first century. During brownfield formation, we basically play with nature by making land unfit for any human activity and thus natural processes should be integrated into planning and design of brownfield redevelopment. Ecological factors are superimposed onto the land to determine its capacity to support human activity and its suitability for a particular type of land use (McHarg and Steiner 1998). From Chinese philosophical point of view, ecological wisdom can provide solutions, for example, to brownfields problem by enhancing harmonious coexistence of humans with nature by making adequate decisions and choices in a resilient, systemic manner using available knowledge (Zhang et al. 2016). This could include policy, management, science or engineering. The use of ecological wisdom is undertaken with the long-term goal of developing 'management directives' aimed towards sustainability of ecosystems whether they are human-based such as urban ecosystems, where brownfield redevelopment fits, or nature-based such as natural ecosystems (Patten 2016).

Development of ecological wisdom requires observation over long periods of time, historical knowledge and data collected from research, monitoring and experimental results (Patten 2016). This is exactly the same way brownfields were turned into green spaces in many countries (De Sousa 2003; Burke et al. 2015; Rizzo et al. 2015). In order to treat various contaminants of brownfields, different remediation strategies are generally adapted, as shown in Table 1 during the conversion of brownfields into green space in Toronto (De Sousa 2003). This Toronto case study can provide guidance to brownfield redevelopment in China. After all,

Table 1 Contamination and remediation approaches adapted in different brownfield sites of Toronto, Canada (Reprinted with permission from De Sousa 2003)

Site	Contaminants	Approach
Village of Yorkville Park	Metals, oils, lubricants	Dig-and-dump (removal and disposal of 800 m^3 of contaminated fill at an appropriate facility) 800 m^3 of contaminated fill removed
Parliament Square	Heavy metals, coal tar, polyaromatic hydrocarbons (PAHs)	SSRA Composite geo-membrane/clean soil cap (on top of contaminated soil)
Spadina Gardens/The Music Garden	Heavy metals, PAHs	SSRA Contaminated soil used as roadbed-integrated landscape surfacing/ clean soil cap
Domtar Polyresins	Ethyl benzene, toluene, styrene, PAHs	SSRA Innovative bioremediation approach
Sorauren Park	Heavy metals, hydrocarbons, chlorinated solvents	SSRA Concrete slab preserved to cap contaminants/integrated landscape and clean soil cap
Chester Springs Marsh	Cinder, rubble, heavy metals	SSRA Monitoring of heavy metals is ongoing
Woodbine Park	Heavy metals, hydrocarbons	SSRA Removal of coal and clinker ash Bio-pad treatment of hydrocarbons X-ray fluorescent technology for soil testing and treatment 500,000 m^3 of engineered fill brought to the entire site Demolition of buildings and parking lot
Don Valley Brickworks	Heavy metals, asbestos, PAHs	Site filled with clean soil in anticipation of a residential redevelopment project PAH area has been isolated
Toronto Port Lands	Contaminants and heavy metals	Ongoing
Leslie Street Spit	Heavy metals, PCBs	Clean fill/wetland cap

the beauty of ecological wisdom lies in its endurance over time, efficacy in practice and ability to predict project performance decades, if not centuries, in advance (Martin et al. 2010).

In China, some of the land contamination incidents were published in the media and have captured the attention of the public (Xie and Li 2010). Following this, the State Environmental Protection Administration released the 'Notice on Effective Prevention and Control of Environmental Pollution for Industrial Enterprise Relocations' in 2004. According to it, in the presence of soil pollution, it warranted immediately reporting to the State Environmental Protection Administration, and pollution control countermeasures must be implemented.

Despite the fact that there is no detailed regulation on brownfield redevelopment in China, the Government has actively supported the redevelopment of wasteland to promote densification. Brownfield redevelopment is of great value for local and urban economic development, and living conditions of people as it can produce new employment opportunities, encourage community activity and reduce health risks by creating a harmonious and stable society (Xie and Li 2010). By combining Western conceptions of brownfield redevelopment with Chinese context, ecological wisdom could bring passage to urban sustainability (Xiang 2014). Ecological wisdom views sustainability as a set of practices in brownfield redevelopment that influences both ingenious and interventionist strategies.

Brownfield redevelopment process could be a long step involving many stakeholders for the purposes of planning and preparatory work, stakeholder identification and selection, and feedbacks to stakeholders, among others (Rizzo et al. 2015). And thereafter, it is necessary to implement ecological wisdom for achieving practically feasible brownfield regeneration. Brownfields are contaminated sites and pose carcinogenic and other accumulative health and environmental risks; therefore, remediation of such sites is warranted. Assessment of such sites is required based on identification, environmental risk assessment and remediation based on preparatory investigation, primary investigation, selection of suitable remediation technique and implementation of the most suitable technique.

Due to rapid urbanization, many cities of China require a vast amount of land for residential and commercial purposes; and thus, the redevelopment of brownfields is necessary. Considering the negative impact of brownfield, in Beijing itself, 142 plants were relocated during the period of 2001–2005, making 8.78 million m^2 lands available for reuse (Xie and Li 2010). The poisoning incident of Songjiazhuang Subway Station, Beijing, in 2004 was a turning point regarding the public opinion and environmental management of contaminated sites. Thereafter, to speed up the redevelopment process, *Site Environmental Assessment Guidelines* (2007) developed by Beijing Municipal Environmental Protection Bureau stipulated a stepwise procedure for site environmental assessment that includes contaminant identification, contaminant verification and risk assessment and remediation measures. Directly or indirectly, knowingly or unknowingly, ecological wisdom plays a big role in formulating such guidelines. Brownfield redevelopment should also support sustainability to avoid adverse effects on health and environment. Ecological practice (planning, design, construction and management) has the potential of supporting sustainability (Wang and Xiang 2016).

5 Sustainable Remediation for Brownfield Redevelopment

From the perspective of contamination types, inorganic contaminants get first place followed by organic contaminants and comparatively the proportion of complex pollution is lower in soils of China. The content of cadmium, mercury, arsenic, copper, lead, chromium, zinc and nickel in soil was 7%, 1.6%, 2.7%, 2.1%, 1.5%, 1.1%, 0.9% and 4.8%, respectively, while organic pollutants such as benzene hexachloride, DDT and polycyclic aromatic hydrocarbons were 0.5%, 1.9% and 1.4%, respectively. In the view of the serious soil environmental situation and to strengthen protection of soil environment and pollution control, Ministry of Environmental Protection of People's Republic of China compiled 'Action Plan for Prevention and Control of Soil Pollution' (MEP China 2014).

In order to remediate all such contaminants to redevelop brownfields in a sustainable way, ecological engineering-based solutions to enhance ecosystem resilience on such sites are warranted. Compared with traditional remediation techniques used to remediate or redevelop brownfield sites, biological processes are found to be economic and environmentally friendly. Such processes include phytoremediation, which uses plants to extract, degrade, absorb or immobilize pollutants (Sleegers 2010); and bioremediation including an advance microbially induced calcite precipitation method to immobilize inorganic contaminants (Kumari et al. 2016) for sustainable remediation. The sustainable remediation concept originated around a decade ago (Hou and Al-Tabbaa 2014). In a scheme promoted by Sustainable Remediation Forum in the UK, sustainable remediation is defined as 'remediation that removes or controls uncontrollable risks in a safe and timely manner, and which maximizes the overall environmental, social, and economic benefits of contaminants remediation' (Surf-UK 2010). The remediation work at the 2012 London Olympic Park site has successfully demonstrated sustainable construction practices into brownfield remediation and redevelopment projects (Hellings et al. 2011). The work phases in this site redevelopment involved site investigation, demolition and clearance. Further, contaminated soil was treated by soil washing, complex sorting, ex situ stabilization and bioremediation; thereafter, groundwater was remediated using innovative in situ treatment technologies including in situ bioremediation (Hellings et al. 2011; Hou et al. 2015). Such sustainable remediation practices could be the best option to implement in brownfield sites in China for their redevelopment.

6 Conclusions

In conclusion, given the large number of contaminated sites in China and issues (whether environmental and/or development) associated with it, it is urgently required to create land risk level assessments to clean-up those sites. Also, as brownfield redevelopment is comparatively new in China compared to European countries or USA, the concept of ecological wisdom should be included by taking specific examples of brownfield redevelopment from those countries in the domain of ecological research and planning practices. An appropriate remediation technology depending upon site history, physical factors and type of contaminants is required in order to cope with challenges due to brownfields. China also requires to toughen its legal, organizational and other related frameworks for management of brownfield and moreover to form an implementation policy.

References

Adams D, De Sousa C, Tiesdell S (2010) Brownfield development: a comparison of North American and British approaches. Urban Stud 47:75–104

Alker S, Joy V, Roberts P, Smith N (2000) The definition of brownfield. J Environ Plann Man 43:49–69

APA (American Planning Association) (2005) Legislative priority: brownfields. http://www.planning.org/priorities/brownfields.htm

Bjelland MD (2004) Brownfield sites in Minneapolis-St. Paul: the interwoven geographies of industrial disinvestment and environmental contamination. Urban Geogr 25:631–657

Burke H, Hough E, Morgan DJR, Hughes L, Lawrence DJ (2015) Approaches to inform redevelopment of brownfield sites: an example from the Leeds area of the West Yorkshire coalfield, UK. Land Use Policy 47:321–331

Cao K, Guan H (2007) Brownfield redevelopment toward sustainable urban land use in China. Chinese Geogr Sci 17:127–134

De Sousa C (2000) Brownfield redevelopment versus greenfield development: a private sector perspective on the costs and risks associated with brownfield redevelopment in the Greater Toronto Area. J Environ Planning Manage 43:831–853

De Sousa CA (2002) Brownfield redevelopment in toronto: an examination of past trends and future prospects. Land Use Policy 19:297–309

De Sousa CA (2003) Turning brownfields into green space in the city of toronto. Landsc Urban Plann 62:181–198

Gao BY, Liu WD, Norcliffe G, Du C (2010) Land policy and the geographies of manufacturing industries: a case study of Beijing. Prog Geogr 29:878–886

Gertler M (1995) Berridge Lewinberg Greenberg Dark Gabor Ltd., Municipality of Metropolitan Toronto, 1995. Adapting to New Realities: Industrial Land Outlook for Metropolitan Toronto, Durham, York, Halton, Peel, Hamilton-Wentworth and Waterloo. Municipality of Metropolitan Toronto, Toronto

Hellings J, Lass M, Apted J, Mead I (2011) Delivering London 2012: geotechnical enabling works. In: Proceedings of the institution of civil engineers—civil engineering, 164, pp 5–10

Hou D, Al-Tabbaa A (2014) Sustainability: a new imperative in contaminated land remediation. Environ Sci Policy 39:25–34

Hou D, Al-Tabbaa A, Hellings J (2015) Sustainable site clean-up from megaprojects: lessons from London 2012. Proc Instit Civil Eng-Eng Sustain 168:61–70
Kretchik JT (2002) Brownfields. Chem Health Saf 9:34
Kumari D, Qian X-Y, Pan X, Achal V, Li Q, Gadd GM (2016) Microbially-induced carbonate precipitation for immobilization of toxic metals. Adv Appl Microbiol 94:79–108
Li X (2011) Brownfields in China: how cities recycle industrial land. Massachusetts Institute of Technology
Lin GC (2001) Metropolitan development in a transitional socialist economy: spatial restructuring in the Pearl River Delta, China. Urban Stud 38:383–406
Lin GC (2007) Reproducing spaces of Chinese urbanisation: new city-based and land-centred urban transformation. Urban Stud 44:1827–1855
Lin GC, Ho SP (2005) The state, land system, and land development processes in contemporary china. Ann Assoc Am Geogr 95:411–436
Liu Y, van Oort F, Geertman S, Lin Y (2014) Institutional determinants of brownfield formation in Chinese cities and urban villages. Habitat Int 44:72–78
Martin JF, Roy ED, Diemont SA, Ferguson BG (2010) Traditional ecological knowledge (TEK): ideas, inspiration, and designs for ecological engineering. Ecol Eng 36:839–849
McHarg IL, Steiner FR (1998) To heal the earth: selected writings of Ian L McHarg. Island Press, Washington, DC
MEP China (2014) http://www.mep.gov.cn/gkml/hbb/qt/201404/t20140417_270670.htm
NLUD (National Land Use Database) (2000) Previously developed land: data specification. National Land Use Database, London
Patten DT (2016) The role of ecological wisdom in managing for sustainable interdependent urban and natural ecosystems. Landsc Urban Plan 155:3–10
Rizzo E, Pesce M, Pizzol L, Alexandrescu FM, Giubilato E, Critto A, Marcomini A, Bartke S (2015) Brownfield regeneration in Europe: identifying stakeholder perceptions, concerns, attitudes and information needs. Land Use Policy 48:437–453
Site Environmental Assessment Guidelines (2007) Beijing Municipal Environmental Protection Bureau
Sleegers F (2010) Phytoremediation as green infrastructure and a landscape of experiences. In: Proceedings of the annual international conference on soils, sediments water and energy, 15, pp 132–140
Surf-UK (Sustainable Remediation Forum in the UK) (2010) A framework for assessing the sustainability of soil and groundwater remediation. In: Contaminated land: applications in real environments. Surf-UK, London, UK
Uk DETR (1998) Derelict Land Surveys in 1988 and 1993. Department of Environment, London
US EPA (1997) Brownfields Definition. US EPA Brownfields Homepage, http://www.epa.gov/swerosps/bf/glossary.htm
Wang QS, Wang YX, Du QZ (2012) An analysis of China's current situation of deindustrialization. Contemp Fin Econo 12:5–13
Wang X, Xiang W-N (2016) Ecological wisdom for urban sustainability: Doing real and permanent good in ecological practice. Landsc Urban Plann 155:1–2
Xiang W-N (2014) Doing real and permanent good in landscape and urban planning: ecological wisdom for urban sustainability. Landsc Urban Plan 121:65–69
Xie J, Li F (2010) Overview of the current situation on brownfield remediation and redevelopment in China. The World Bank, Washington DC
Zhang B, Cao C (2015) Policy: four gaps in China's new environmental law. Nature 517:433–434
Zhang L, Yang Z, Voinov A, Gao S (2016) Nature-inspired stormwater management practice: the ecological wisdom underlying the Tuanchen drainage system in Beijing, China and its contemporary relevance. Landsc Urban Plan 155:11–20

Ecological Wisdom-Inspired Remediation Technology for Aquaculture Water Quality Improvement in Ecological Agricultural Park

Zhang Qiu-zhuo, Li Hua, Xu Ya-tong and Tang Jia-wen

Abstract The current ecological situation of aquaculture water quality in Wushe (Songjiang District, Shanghai City) District is not up to the mark: In the present study, characteristics and basic information of ecological agricultural park were first investigated. Ecological remediation integration technology, including aquatic plant purification and aquatic economic crops floating bed, was used to improve the aquaculture water quality in this district. After operating for a period of defined time, effects of the integration technology on quality improvement in the aquaculture water quality at experimental site with removal rates of COD_{Cr}, TN, and TP were 73.5%, 24.9%, and 67.4%, respectively. In addition, the purification effects of ecological remediation integration technology and individual technology were compared. The results indicated that the integration technology was better than individual technology, as TN, TP, nitrite nitrogen, and COD_{Cr} were removed at the rate of 13.7%, 40.1% 78.7%, and 114.4%, respectively, higher than that of individual technology. In addition, water spinach, which was grown on aquatic economic crops floating bed, was harvested at regular time intervals. The total harvest weight of water spinach was 11907 kg, which could achieve good economic benefits. A small parts of the water spinach were transplanted to crab ponds. The transplanted water spinach could not only serve as food source for crab, but also purify the water quality in crab ponds, which could maintain the biological diversity in aquacultural ponds and coincide with production mode of circulation ecological agriculture.

Keywords Aquaculture water · Water quality improvement · Ecological agriculture district · Aquatic economic crops floating bed

Z. Qiu-zhuo (✉) · L. Hua · X. Ya-tong · T. Jia-wen
Shanghai Key Lab for Urban Ecological Processes and Eco-Restoration, School of Ecological and Environmental Sciences, East China Normal University, 200241 Shanghai, China
e-mail: qzhzhang@des.ecnu.edu.cn

V. Achal and A. Mukherjee (eds.), *Ecological Wisdom Inspired Restoration Engineering*, EcoWISE, https://doi.org/10.1007/978-981-13-0149-0_11

1 Introduction

China is a large aquaculture country, with aquaculture production for many years ranked first in the world (Xiang 2001). Shanghai is located in the Yangtze River export, facing the Yellow Sea and the East China Sea, and it is located west of the mainland. So the development of fisheries is also unique. According to the statistics, the total output of Shanghai aquatic products was 338,574 tons, 153,151 tons of fishing products and 200,388 tons of aquatic products in 2009 (Zheng et al. 2011). However, with the rapid development of the intensive aquaculture model, water quality deterioration and wastewater discharge directly restricted the sustainable development of the aquaculture industry. The decomposition of fodder, the discharge of excreta, and the use of chemicals and antibiotics made deeply deterioration of aquaculture water, resulting in ecosystem out-of-balance and water eutrophication (Lai et al. 2009; Soriano et al. 2009). To ensure the development of aquaculture industry while reducing the pollution of aquaculture wastewater on the surrounding environment, aquaculture is to ensure sustainable development of the inevitable requirement.

The fundamental proposition and the highest pursuit of Chinese traditional ecological wisdom is the harmony between human and nature. At present, the world's aquaculture water purification technology and measures are carried out extensive research (Ma et al. 2012; Jiang et al. 2010; Boopathy et al. 2007), including the use of mechanical filtration treatment of physical methods (Palacios and Timmons 2001), the use of ozone purification (Fang 2004) and electrochemical treatment of chemical methods (Mook et al. 2012), the use of biofilm (Boopathy et al. 2007; Gregory et al. 2012; Fontenot et al. 2007), algae (Soriano et al. 2009; Marinho-soriano et al. 2011; Meng 2009), plants (Ma et al. 2011), and other biological methods of treatment. Compared with the traditional physical repair and chemical repair technology, bioremediation technology has been widely recognized because it has many advantages such as low cost, no secondary pollution, no harm to aquaculture function, and no damage to ecological balance. Chemical or physical methods always tend to produce some by-products, the cost is high, and the ecological value is little. Moreover, the plants which were used in biological methods could build green aquaculture system for aquaculture remediation. The wastewater in aquaculture system could provide nutrition for the growth of plants, and the plants could purify wastewater in aquaculture systems at the same time. It was quite accorded with the concept of modern aquaculture nowadays (Yang 2008).

Nowadays, ecological engineering is advancing toward the sustainable development of ecological wisdom. Xiang pointed out the relationship among ecological wisdom, theoretical wisdom, and practical wisdom in his article (Xiang 2016). Practitioners are often aware of the reciprocal relationship between man and nature in ecological practice. Motivated by human beings' enlightened self-interest, ecophronesis is the master skill par excellence of moral improvisation to make, and act well upon, right choices in any given circumstance of ecological practice. Duncan T. Patten demonstrated the interdependence of coupled urban and natural systems, and the potential management role of ecological wisdom in influencing sustainability of holistic human/nature ecosystems, through two urban systems from arid and semiarid western North America, Los Angeles, and Phoenix used as cases (Patten 2016). Since ancient times, ecological wisdom has been fully integrated into the urban construction, especially in the ecological harmony of green ecological projects. Such as McHarg who used a case study on Staten Island in New York City, Li Bing who planned and host the Dujiangyan irrigation system in Sichuan, China, both of their projects achieve a paramount level of "doing real and permanent good in landscape and urban planning" (Xiang 2014). This is the ecological wisdom in sustainable urban development on the one hand reflected. On the other hand, as present construction technologies use too much material and energy and produce huge quantities of greenhouse gases, more and more scientists are inspired by the ecological wisdom embedded in natural habitats such as anthills, coral reefs, silk, and bones; they use biological mineralization technology to prepare biological materials, are fully applied to urban infrastructure, and are conducive to the sustainable development of the city (Achal et al. 2016). Aquaculture industry runs through the process of human development. The use of biological methods to solve the aquaculture pollution problems is in line with the concept of the development of ecological wisdom.

SongjiangWushe Ecological Agriculture Park is located in the southeastern edge of the Yangtze River Delta, east of the Taihu Lake Basin and the upper reaches of the Huangpu River in Shanghai. It is a plain river network tidal area, with a total area of 11.19 km^2. This park was formally established in February 2001 and is a modern agricultural park in Shanghai, the national agricultural standardization production base area. However, the untreated high-load aquaculture drainage in the ecological park constitutes the main source of pollution in the park. Most of the aquaculture drainage flows into the Huangpu River through the outer channel, resulting in water eutrophication and biodiversity loss. How to alleviate the environmental pollution caused by aquaculture and construct the green ecological aquaculture mode according to the local conditions is an urgent problem to be solved in the agricultural ecological park.

2 Materials and Methods

2.1 Location of Pilot-Scale Experiment

Wastewater from aquacultural drainage river, which was located in Songjiang Wushe Ecological Agriculture Park (Shanghai, China), was selected as objective in present study. The length and width of the selected aquacultural drainage river were 600 m and 6.5 m, respectively. The average water depth was 1.2 m, and the highest water depth could reach 1.65 m when breeding peak arrived. The bank of river was totally hard cement, the average length of bank was 4 m, and the slope angle was

Fig. 1 Original appearance of river

45°. There was a water outlet in every aquacultural pond, through which the wastewater in aquacultural ponds could drain away to the selected river. The original appearance of river course and its revetment is shown in Fig. 1.

2.2 *Construction of Aquaculture Water Quality Purification*

Ecological remediation integration technology was selected for pilot-scale experiment to improve aquaculture water quality in ecological agriculture district because of the large drainage flow quantity, high hydraulic loading rate, and characteristics of ecological agricultural park. In the case of high hydraulic loading and short hydraulic retention time, water hyacinth with strong decontaminating ability was used as a pioneer plant to remove part of the pollution, and then the removal rate of pollutants was further improved by floating aquatic economic crops, which is conducive to reduce the breeding pond drainage of external river pollution. In addition, water spinach used on the floating bed could acquire economic benefit especially to local farmers. This cropping pattern was a green and ecological way, which coincides with the theory of ecological agriculture.

River ecological restoration project is divided into two sections: aquatic plant purification and economic crop floating bed area.

2.2.1 Section I—Aquatic Plant Purification

Section I is an aquatic plant purifying zone (Fig. 2), 100 m long, with an area of about 600 m^2, planting decontamination ability of strong aquatic plants—water hyacinth. Water hyacinth plant root is well developed, and it is rapid tiller propagation and extensive management, which is a good water purification plant

Fig. 2 Aquatic plants purification area

(Li 2012; Wang 2012). Due to the strong propagation ability of water hyacinth, netting from the river bottom can prevent the overbreeding of water hyacinth on the growth of aquatic economic crops in Section II.

2.2.2 Section II—Economic Floating Bed Area

Section II is for the economic crops floating bed area, 500 m long, with an area of about 3,000 m^2; the main planting aquatic cash crop is water spinach. The floating bed used was different from other traditional floating beds. The material that was chosen to construct the floating bed was bamboo, which is environmental-friendly. The length and width of bamboo raft were 3–3.5 m and 1–1.2 m, respectively. A small PVC tube (32 mm × 40 mm) was roped every 15–20 cm on bamboo raft, through holes of which the roots of plants could elongate into water. These tubes were fixed by tie-line buckle. Water spinach was put into each PVC tube, and then the settled bamboo raft was kept into selected river, which is shown in Fig. 3. In the late of spinach plant growth, it grew more vigorous, and it can reach 80% coverage of the riparian zone II.

2.3 Layout and Determination of Indicators and Methods of Sampling Points

Demonstration projects and water quality monitoring were divided into two stages: The first stage was from July to August, 600 m of water all the water planted spinach, mainly on economic crops floating bed single technology construction and monitoring. A1, A2, and A3 were chosen as three sampling sites. The second stage was from September to November, 100 m near the water side of the water spinach

Fig. 3 Economic crop on floating bed

replaced by water hyacinth, mainly for the construction of aquatic plant purification area and the purification of aquatic plants, combined purification technology to monitor the effect of water purification. The sampling points in this stage were based on the first three sampling points, and the sampling point a1 was added to the intercept of water hyacinth 100 m to monitor the water purification effect of water hyacinth. The specific arrangement of sampling points is shown in Fig. 4.

One sample per week, the main test indicators are the TN, TP, ammonia nitrogen, nitrate nitrogen, COD_{Cr}, and so on. The determination of water quality indicators were followed "water and wastewater monitoring and analysis methods" (State Environmental Protection Administration editorial committee 2002). Based on the A_1 sampling point for the water inlet quality and the A_3 sampling point for the water outlet quality, the removal of various pollutants was calculated.

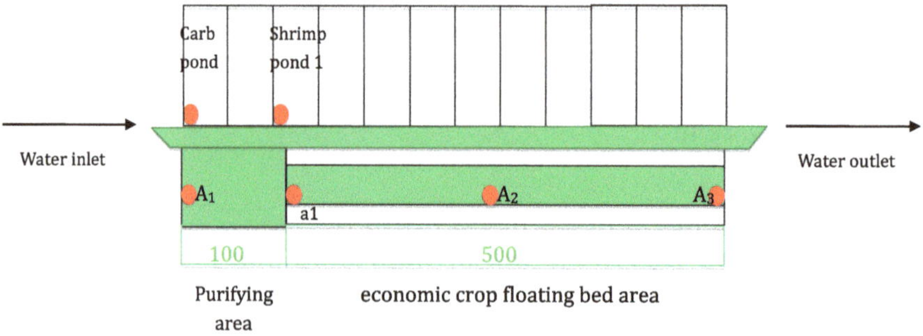

Fig. 4 Location of sampling site

3 Results and Discussion

3.1 *Investigation of Aquaculture in Ecological Agriculture*

3.1.1 Basic Information of Aquacultural Ponds

There were totally 27 different kinds of aquacultural ponds in Songjiang Wushe Ecological Agriculture Park. The length and width of every aquacultural pond were 220 m and 50 m, respectively, and the area is 11,000 m^2. The breeds cultivated were *Penaeus monodon*, *Macro-brachium* species, and the Yangtze River crab; the breeding density for each pond is 1 million larvas. *Penaeus monodon* and *Macro-brachium* species were cultivated from late May to late September, while the Yangtze River crab was cultivated from late March to late November. Three aeration equipments were put in every pond, and they were aerated twice every day when the shrimps were young. The aeration frequency and aeration time were increased as shrimp growing. The aeration frequency in crab pond was less than that in shrimp pond. Basic information of shrimp pool is given in Table 1.

Shrimp pond 1 (*Penaeus monodon* and *Macro-brachium* polyculture shrimp ponds) and shrimp pond 2 (*Penaeus monodon* ponds) were selected as research objects. The changes of the feeding rates of the two shrimps in the whole growth cycle in 2010 are shown in Fig. 5.

It can be seen from Fig. 5 that the quantity of feed addition was increased as shrimp growing. The shrimp grows very fast especially in August and September. With changes in the time during these periods, the quantity of added feed was largest, and pollution intensity reached the maximum value. It suggests that the wastewater in ponds should be drained off and changed by freshwater at that time. In addition, different varieties of shrimp feed requirements are also different. The feed addition to Shrimp pond 1 was generally higher than that of shrimp pond 2.

The main raw materials of the feed added to the shrimp ponds include high-quality fish meal, cuttlefish paste, shrimp powder, beer yeast, phosphate ester, refined fish oil, high gluten, amino acids, minerals, trace elements, stabilizing vitamins, shelling elements, polysaccharides. Moreover, the feed addition mainly consists of crude protein. After adding these feed, they flowed to water body mostly in the form of nitrogen, and only about 1% in the form of phosphorus. Besides, the excrement of shrimp and superfluous of feed could also exert a bad influence on water body in aquacultural ponds. Other inputs included lime, BCDMH and disinfectants. The dosing amount are very small, thus it could not influence the water quality.

Table 1 Basic information of shrimp pool

Breeding density (million/hm^2)	Feeding time (d)	Feeding amount (kg/hm^2)	Lime (kg)	Water quality improver (g)	Harvest (kg/hm^2)
75 ~ 90	113 ~ 150	3300 ~ 7350	7.5 ~ 15	30 ~ 60	3750 ~ 5250

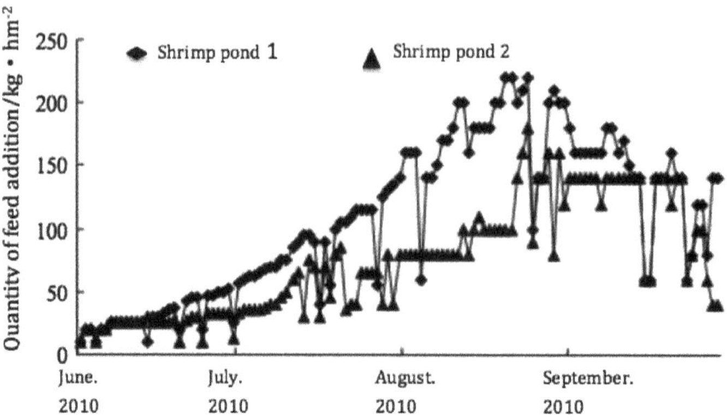

Fig. 5 Dosage of shrimp feed

3.1.2 Water Quality of Aquacultural Ponds

Two different kinds of shrimp ponds and one crab pond were selected as the representatives to investigate water quality of aquacultural ponds in two stages. The first stage was from July to August, and the second stage was from September to November. It was monitored once a week, for a total of 20 times. The average value of water quality is given in Table 2. The results of water quality monitoring in two representative aquaculture ponds showed that there was no large-scale drainage in July, and the change of water quality index was not significant. The peak values of COD_{Cr}, TN, NH_4^+–N, and NO_2^-–N in the period from August to September were increased by 5–30%.

Table 2 Water qualities of aquacultural ponds

Items	Shrimp pond 1		Shrimp pond 2		Carb pond	
	Stage 1	Stage 2	Stage 1	Stage 2	Stage 1	Stage 2
Temperate (°C)	30.3	27.2	30.3	27.3	30.3	27.2
DO (mg L^{-1})	2.95	3.32	3.27	4.58	2.83	3.06
PH	8.0	8.1	8.4	8.7	8.5	8.2
Transparency (cm)	50.7	41.0	30.5	34.0	28.7	34.7
COD_{Cr} (mg L^{-1})	53.17	70.95	52.43	81.60	62.62	57.63
TN (mg L^{-1})	2.37	3.99	2.59	2.92	2.16	2.26
NH_4^+–N (mg L^{-1})	1.01	1.43	0.94	0.89	0.83	0.61
NO_3^-–N (mg L^{-1})	1.17	1.36	1.17	1.20	1.12	1.09
NO_2^-–N (mg L^{-1})	0.05	0.21	0.08	0.22	0.02	0.01
TP (mg L^{-1})	0.21	0.20	0.30	0.27	0.56	0.17

3.1.3 Pollution Intensity Calculation of Aquacultural Ponds

According to the water quality monitoring results, drainage quantity and pattern, pollution intensity of the main water quality index in aquacultural ponds was estimated. And compared with the relevant literature, the results were given in Table 3. The pollution intensity in the present study was also compared with general aquaculture pond in Shanghai. It was found that the intensities of COD_{Cr}, TP, and TN in this research were 3201 kg/ha, 25.7 kg/ha, and 128.6 kg/ha, respectively, all of which were higher than general aquaculture ponds in Shanghai. It may be related to the breeding species and culture density in the region.

3.2 Evaluation of Operation Effect of Demonstration Project

In demonstration project after running a certain stage, we were continuously monitoring its water purification effect. The water quality indicators after the completion of the demonstration project in 2010 were compared with those before the completion of the 2009 demonstration project; we conducted a functional evaluation of the performance of the demonstration project. Moreover, we compared the single crop of floating crops and the combination of aquatic plant purification technology. In addition, we evaluated the economic benefits of this combination of ecological restoration techniques.

3.2.1 Comparison of Annual Water Purification Effect

The water quality of the same aquaculture period (from July to August) in 2009 was compared with that in 2010, and the four indexes of TN, TP, COD_{Cr}, and NH_4^+–N were monitored and analyzed. The results are shown in Figs. 6 and 7.

Table 3 Pollution intensity of aquacultural ponds

Survey region	COD (kg ha^{-1})	TP (kg ha^{-1})	TN (kg ha^{-1})	Remark
Intensive shrimp pond in Songjiang	3201	25.7	128.6	In this research
General aquaculture in Shanghai*	999	4.9	101	Xu (2003)
General aquaculture in Kunshan*	553	2.7	54	Zhang (2002)

**Notes* In pond-intensive, river culture, lake farming collectively referred to as general aquaculture,and one of the most intensive pollution within the pond

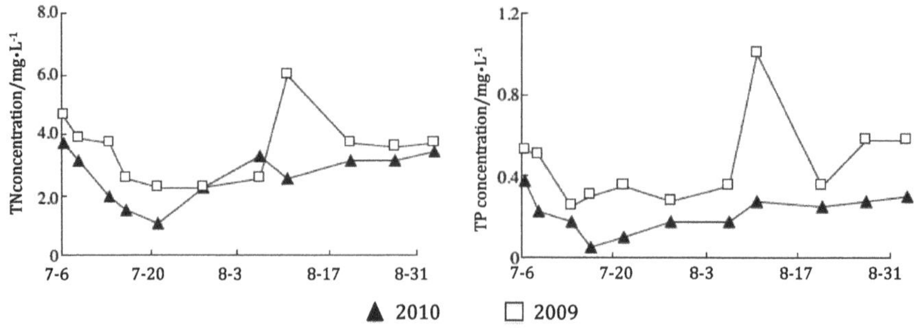

Fig. 6 TN and TP concentrations of river

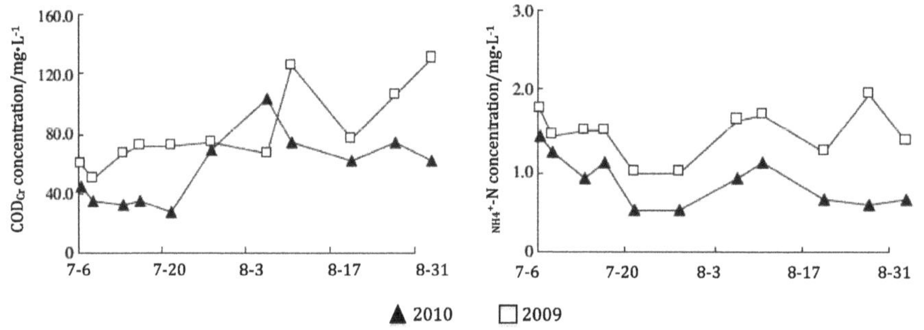

Fig. 7 COD_{Cr} and NH_4^+–N concentration of river

It can be seen from the figure, in July and August, 2009 measured TN, TP, COD_{Cr}, and NH_4^+–N content were higher than in 2010. The average contents of TN, TP, COD_{Cr}, and NH_4^+–N in the period of 2009 were 3.52 mg L^{-1}, 0.46 mg L^{-1}, 81.1 mg L^{-1}, and 1.46 mg L^{-1}, respectively. The average contents of TN, TP, COD_{Cr}, and NH_4^+–N in 2010 were 2.61 mg L^{-1}, 0.21 mgL^{-1}, 55.9 mg L^{-1}, and 0.89 mg L^{-1}, respectively.

According to the preliminary investigation of the basic situation of shrimp farming, conducting a detailed investigation, we can know the scale and variety of farming have not changed in 2009 and 2010, and shrimp pond dosing amount of feed has not changed. And the experimental river channel is a relatively closed shrimp pond drainage channel in the agricultural ecological park,so the surrounding pollution sources and pollution did not change significantly. In addition, through two years of continuous monitoring of river water quality, we found that the inlet water quality is basically the same. It can be seen that the ecological remediation integration technology a good effect on improving the drainage quality of aquaculture in Songjiang Wushe Ecological Agriculture Park, while the pollution source and the amount of shrimp and crab feeding are basically the same circumstances. The combination technique mainly utilizes the stems, leaves, and roots of the planted aquatic plants to absorb, enrich, adsorb, and fix aquaculture water pollutants, and provide habitat for microorganisms to reduce the pollution of aquaculture water bodies.

3.2.2 Comparison of Two Stages of Water Purification Effect

The water purification effect of the two different stages in experimental sites was shown in Figs. 8 and 9. It can be seen from the figure that the removal efficiency of the combined purification technology of aquatic plants (the second stage) was obviously higher than that of the aquatic economic crops floating bed single technology (the first stage). The average removal rates of TN, TP, NO_2^-–N, and COD_{Cr} in the first stage (the tenth week) were 11.2, 27.3, 0.1, −40.9%; the average removal rates of TN, TP, NO_2^-–N, and COD_{Cr} in the second stage (the seventh week) were 40.9%, 67.4%, 78.8%, and 63.5%, respectively.

In the first stage, the single technology of aquatic economic crop floating bed has a certain ability to remove total nitrogen and total phosphorus, which is consistent with the reports of most researchers (Tang et al. 2012; Song et al. 2011). However, the removal of COD_{Cr} was negative in the single-row technique of aquatic cash crops, which may be due to the gradual increase in the amount of feed added to the first stage, and the spinach itself did not have the ability to remove COD_{Cr}..

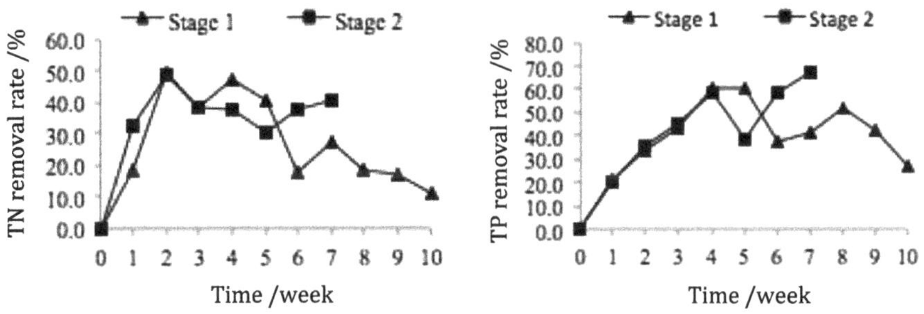

Fig. 8 TP and TN removal rate of two stages

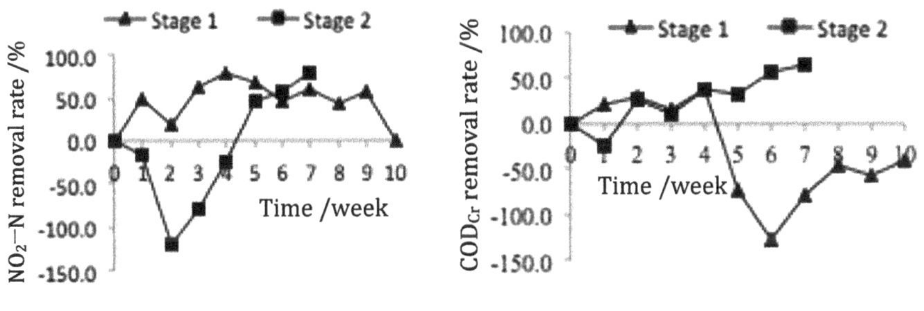

Fig. 9 Nitrite nitrogen and COD_{Cr} removal rate of two stages

The second stage of aquatic plant water hyacinth has the advantages of fast growth rate, large biomass, strong nitrogen, and phosphorus enrichment ability, has better purification ability to polluted water than other aquatic plants, and is widely used in eutrophic lakes and rivers, aquaculture wastewater, and industrial wastewater treatment (Zhang 2011). Water hyacinth mainly accumulates and degrades pollutants in the water through the absorption of stem and leaf. In addition, the root system can also release O_2 to the water body, forming a microbial biofilm on the active surface of the root layer, providing habitat for microorganisms, forming rhizospheric microbial ecosystem, and thus enhancing the purification effect of eutrophic water (Yi et al. 2009). This is the reason why second stage of the pollutant removal rate was higher than the first stage.

3.3 Benefit Analysis of Demonstration Project

Floating bed of aquatic economic crops can also produce some economic benefits in the purification of water. Water spinach, also known as bamboo leafy vegetables, is an annual or perennial herb. Among the many floating plant species, water spinach has the characteristics of strong stain resistance, high survival rate, simple cultivation, and high biomass. It is proved to be an ideal material for floating bed plants. Moreover, it has a good economic value, high nutritional value, and good palatability, and it can be used as a vegetable (Jia et al. 2011). The use of floating beds in aquaculture drainage channels can not only make full use of land, promote product diversity, low investment, and high yield, increase the income of farmers, but also reduce the risk of single cropping patterns.

During the trial period, July and August were the largest seasons for the biomass of the water spinach, at which time it is harvested and harvested every half month. Since November, the water temperature gradually decreased, water spinach began to wilt, and it was the time for its timely harvest and salvage. Aquatic cash crop spinach was harvested in the test period of the demonstration project with weight of about 11,907 kg, an average of 1488 kg per harvest, the specific harvest time and weight as given in Table 4. The economic benefits of water spinach were accounted for, and the results are given in Table 5. From output-to-input ratio, we can find that

Table 4 Harvest of water spinach

Date	Harvest weight (kg)		Total weight (kg)	Dry weight (kg)
	The first time	The second time		
June	2117	1817	3934	248
August	1701	1820	3521	232
September	1150	1239	2389	155
October	1095	968	2063	144
Total	6063	5844	11907	779

Table 5 Economic benefit of water spinach floating bed

Investment (yuan/a)			Crop output (kg/hm^2 a)	Price per unit (yuan/kg)	Income (ten thousand yuan/hm^2 a)	Input–output ratio
Water spinach plant	Frame materials	Expense of labor				
3000	5000	2000	45 000	6	27	1.80

the water spinach floating bed has good promotional value. In addition, we put part of the water spinach transplant in the crab pond in July, not only for crab food, but also for other aquatic animals to provide a certain living environment. This approach is conducive to maintaining the biological diversity of aquaculture systems, improving ecological efficiency, and cleaning the crab pond water quality, which conforms to the cycle of eco-agriculture model.

4 Conclusion

Taking the drainage channel of a shrimp pond in SongjiangWushe Ecological Agriculture Parks, the demonstration site, based on the investigation and research on the characteristics of aquaculture drainage in the park, the eco-floating bed purification zone of economic crops was established according to the local conditions. Combined with the purification of aquatic plants, we proposed an optimized combination technology of aquaculture drainage treatment. This technology can not only alleviate the environmental pollution caused by aquaculture, but also build a green ecological aquaculture model, which can provide reference for aquaculture drainage and treatment in Shanghai and other plain water network areas. The main conclusions are as follows:

(1) The pollution of Songjiang Wushe Ecological Agriculture Park aquaculture drainage was mainly from the intensive culture of shrimp ponds at late cultural phase and the cleaning up stage when the pond was changing water. The average discharge intensities of COD_{Cr}, TN, and TP was 3201 kg ha^{-1}, 128.8 kg ha^{-1}, and 25.7 kg ha^{-1}, respectively, which were higher than these of the average aquaculture in Shanghai suburb.
(2) Comparing the water quality at the same time in 2009 and 2010, the results show that the demonstration project can reduce the pollutants in aquaculture in this area with a large number of drainage. The removal rates of COD_{Cr}, TN, and TP were 73.5%, 24.9%, and 67.4%, respectively, which solved the problem of aquaculture drainage pollution in the park.
(3) By comparing the results of the first and second stages, it was found that the average removal rates of COD_{Cr}, TN, TP, and NO_2^-–N in the second stage of the river culture increased, but the average removal rate was higher in the first phase. The average removal rates of TN, TP, NO_2^-–N, and COD_{Cr} were

increased by 13.7%, 40.1%, 78.7%, and 114.4%, respectively. It shows that the combined effect of aquatic plant purification technology is better than single technology removal effect.

(4) Timely harvesting the water spinach on the aquatic economic crop floating bed, we harvested 11907 kg of water spinach, access to certain economic benefits, and input–output ratio is relatively high. In addition, we put part of the spinach placed in the crab pond, not only to purify crab pond water quality, but also for crab food,which reduces the cost of farming, but also have a certain landscape effects and water purification effect.

References

Achal V, Mukherjee A, Zhang Q (2016) Unearthing ecological wisdom from natural habitats and its ramifications on development of biocement and sustainable cities. Landsc Urban Plann 155:61–68

Boopathy R, Bonvillain C, Fontenot Q et al (2007) Biological treatment of low-salinity shrimp aquaculture waste water using sequencing batch reactor. Int Biodeter Biodegrad 59(1):16–19

Fang SQ, Hu XF, Wu HX (2004) Technology of aquaculture wastewater treatment and application. Chin J Environ Eng 5 (9):51–55

Fontenot Q, Bonvillain C, Kilgen M et al (2007) Effects of temperature, salinity, and carbon: nitrogen ration sequencing batch reactor treating shrimp aquaculture wastewater. Biores Technol 98:1700–1703

Gregory SP, Dyson PJ, Fletcher D et al (2012) Nitrogen removal and changes to microbial communities in model flood/drain and submerged biofilters treating aquaculture wastewater[J]. Aquacult Eng 50:37–45

Jia Y, Li XZ, Tang YY et al (2011) Effects of different cutting regimes on the productivity of Ipomoea aquatic Forsk planted on floating-mats in a eutrophicated river of Shanghai. Chin J Ecol 30 (6):1091–1099

Jiang YF, Liu DH, Sun TX et al (2010) Process characteristics of zeolite media biological aerated filter for treating aquaculture wastewater. Environ Sci 31(3):703–708

Lai HT, Hou JH, Su CI et al (2009) Effects of chloramphenicol, florfenicol, and thiamphenicol on growth of algae *Chlorella pyrenoidosa*, *Isochrysisgalbana*, and *Tetraselmischui*. Ecotoxicol Environ Saf 72(2):329–334

Li CG (2012) A feasibility study on blue algae pollution control by Water Hyacinth in Lake Dianchi. Environ Sci Surv 31(3):64–68

Ma M, Zhu CX, Liang H L et al (2011) Repairing effect of several plants on the aquaculture wastewater of fisheries. Environ Sci Technol 34 (6):18–22

Ma HF, Li X, Hu HY et al (2012) Growth removal of nitrogen and phosphorus, and lipid accumulation property of *Scenedesmus sp.*LX1 in aquaculture wastewater. Environ Sci 33 (6):1891–1896

Marinho-soriano E, Azevedo C AA, Trigueiro TG et al (2011) Bioremediation of aquaculture wastewater using macroalgae and Artemia. Int Biodeterior Biodegrad 65:253–257

Meng FP, Gong YY, Ma DD (2009) Recent progress in treatment of aquaculture wastewater based on microalgae: a review. *Acta Microbiologica Sinica* 49(6):691–696

Mook WT, Chakrabarti MH, Aroua MK et al (2012) Removal of total ammonia nitrogen (TAN), nitrate and total organic carbon (TOC) from aquaculture wastewater using electrochemical technology: A review. Desalination 285:1–13

Palacios GL, Timmons MB (2001) Determining design parameters for recovery of aquaculture wastewater using sand beds. Aquacult Eng 24:289–299

Patten DT (2016) The role of ecological wisdom in managing for sustainable interdependent urban and natural ecosystems. Landsc Urban Plann 155:3–10

Song C, Chen JZ, Ge XP et al (2011) The control of nitrogen and phosphorus to tilapia fishpond by floating-bed-grown water spinach (Ipomoea aquatic). Chin Agric Sci Bull 27(23):70–75

Soriano EM, Nunes SO, Carneiro MA et al (2009) Nutrients' removal from aquaculture wastewater using the *macroalgae Gracilariabirdiae*. Biomass Bioenerg 22(3):327–331

State Environmental Protection Administration editorial committee (2002) Water and wastewater monitoring method, 4th edn. China Environmental Science Press, Beijing, pp 223–281

Tang YY, Li XZ, Zhou YQ et al (2012) Effect of Ipomoea aquatic Floating-bed on the quantity and distribution of nitrogen cycling bacteria and nitrogen removal. Acta Ecologica Sinica 32 (9):2837–2846

Wang ZX, Calderon MM (2012) Environmental and economic analysis of application of water hyacinth foreutrophic water treatment coupled with biogas production. J Environ Manage 110:246–253

Xiang JH (2001) Disease occurrence and control strategies of mari culture organisms. Maritime Press, Beijing

Xiang WN (2014) Doing real and permanent good in landscape and urban planning: Ecological wisdom for urban sustainability. Landsc Urban Plann 121(2014):65–69

Xiang WN (2016) Ecophronesis: the ecological practical wisdom for and from ecological practice. Landsc Urban Plann 155:53–60

Xu ZX (2003) Planning theory and practice of river pollution control. China Environmental Science Press, Beijing

Yang DJ, Jing YX (2008) Advances in plants in aquaculture wastewater treatment. Ecol Sci 27 (6):522–526

Yi Q, Kim Y, Tateda M (2009) Evaluation of nitrogen reduction in water hyacinth ponds integrated with waste stabilization ponds. Desalination 249(2):528–534

Zhang W (2002) Conditional of pollution and countermeasures to breeding of Kunshan City. Pollut Control Technol 15(2):60–62

Zhang YY, Zhang ZY, Wang YL et al (2011) Research on the growth characteristics and accumulation ability to N and P of *Eichhorniacrassipes* in different water areas of Dianchi Lake. J Ecol Rural Environ 27(6):73–77

Zheng JM, Zhang XG, Huang T (2011) The influence of aquaculture products government regulation on aquaculture farmers economic benefits: based on shanghai case. Shanghai J Econ 3:92–99

Wetlands Restoration Engineering in the Metropolitan Area

Limin Ma, Rohit Kumar Jha, Prantor Kumar Mondal and Weiwei Li

Abstract Wetlands are one of the most endangered environments with global change on earth. Wetlands play an important to the ecosystem because they act as buffers between surface runoff and rivers, and decrease flooding peaks by as much as 60%. Wetlands covered by an acre can store more than one million gallons of floodwater, so they can help save lives and property. Moreover, they also fight against pollution by supplying clean water to our lakes, streams, and rivers. Whenever rain and melt waters flow through farms and cities, they can pick up pollution. If those waters flow through wetlands, the wetland naturally filters out most of the pollutions. By the time, it enters lake or river, it is clean. Thus, wetlands act like a natural waste treatment plant. Wetlands also provide a habitat for huge number of plant, animal, and endangered species. Wetlands have a lot of diversity through their various nature, which attracts a different field of scientists and provide an occupation. Wetlands provide various kinds of study and research area for wetland ecologists, aquatic biologist, botanists, ecologists, and hydrologists. Wetlands include opportunity for resource managers, water quality manager, and wetland program coordinator.

Keywords Wetlands · Eco-restoration · Ecological wisdom · Metropolitan Vegetation

1 Introduction

The wetland ecotone between water and land is a variable ecosystem, an important area for natural protection, and research on global change. The wetland is an ecosystem which has various functions and values (Guntenspergen and Stearns 1985; Hruby 1999). Wetlands have a special structure and a function of purifying

L. Ma (✉) · R. K. Jha · P. K. Mondal · W. Li
College of Environmental Science and Engineering, Tongji University, 1239 Siping Road Shanghai, Shanghai 200092, China
e-mail: lmma@tongji.edu.cn

V. Achal and A. Mukherjee (eds.), *Ecological Wisdom Inspired Restoration Engineering*, EcoWISE, https://doi.org/10.1007/978-981-13-0149-0_12

wetlands that make it very special, and a group of people and scientists called it "the kidney of landscape". The wetland plays a very vital role to maintain the environmental regularity with various aspects such as regulating weather, controlling erosion of the soil, promoting accretion and making land, and decomposing pollutants. The huge attention has been paid to wetland studies in recent years, by investigating different fields. Many of the researchers have covered wetland identification, classification, modeling, assessment, and the relationship to the ambient environment.

The Ramsar Convention is a 1971 international agreement which was signed in Ramsar, Iran. It lays out an outline for various national actions and international cooperation for the conservation and wise use of wetlands and their various resources (*Fact Sheet on Wetland Values and Functions: Flood Control*, https://www.ramsar.org/sites/default/files/documents/tmp/pdf/info/values_floodcontrol_e.pdf). The definition of wetlands under this agreement is quite broad which includes both natural and human-made wetlands and ranging to 6 m below low tide along ocean shorelines. Approximately, 124 million ha (hectares) of wetlands in 1421 different sites around the world have been selected as Wetlands of International Importance (http://www.ramsar.org), whereas only 19 sites are in the USA which covers 1,192,730 ha and less than 1% of the total US land. The U.S. Fish and Wildlife Service (FWS) definition (Fish et al. 1979) is much narrower, but still includes shallow aquatic systems: "Wetlands are lands transitional between terrestrial and aquatic systems where the water table is usually at or near the surface or the land is covered by surface water. For purposes of this classification wetlands must have one or more of the following three attributes: (1) at least periodically, the land supports predominantly hydrophytes (plants that grow in water); (2) the substrate is predominantly undrained hydric soil (wet and periodically anaerobic); and (3) the substrate is non-soil and is saturated with water or covered by shallow water at some time during the growing season of the year."

As in the world's major river delta (Yangtze Delta), the natural wetlands in Shanghai are important for they have abundant biological resources and various ecological functions. Its great diversity and presence of key habitats make a region of important ecological and socioeconomic value in China. Wetlands are becoming more significant for the development of resources and the protection of ecological environment in the city of Shanghai.

2 Ecosystem of Wetlands

Wetlands are categorized by the occurrence of macrophytes which include representatives of various flowering plants, pteridophytes, and bryophytes (mosses) and a few large algae (e.g., Chara, Nitella). All of these plants float freely on the water surface or remain suspended in the water column, whereas mainstream plants are rooted in or attached to the substratum with their shoots being either wholly emergent or submerged above the water surface. But in the deep open water systems such as lakes, reservoirs, and rivers, domination of different algae such as phytoplankton and some filamentous algae can be seen. These algae also occur commonly within the macrophyte-dominated wetlands. Demonstration of how wetlands works is shown in Fig. 1.

The diversity and plenty of macrophytes in a wetland is closely related with its water regime. The depth, duration, frequency, and amplitude—these four components of the water control various life processes throughout the lifecycle of various macrophytes (as also the fauna) in the wetland ecosystems. The water regimes are also additionally influenced by the flow velocity and wave action in large water bodies. Macrophytes provide substrate for a large diversity of periphyton which create the food cycle of several grazing fish and invertebrates. Some macrophytes are operated by the ducks for feeding, nesting material, and roosting at different levels of the plant canopy. And submerged macrophytes provide food and refugia for many faunas. Most of the macrophyte production turns into detritus where microbes start complex food webs, enhance secondary production, and enrich the biodiversity in the wetland ecosystems. In lakes, reservoirs, and rivers, the food chains are sustained by allochthonous inputs. And wetland fauna also requires different water regimes and food sources at different times of their life cycle. Numerous fish, amphibia, reptiles, and invertebrates complete breeding, nourishing, and other activities by seasonally migrating between macrophyte-dominated wetlands and open water systems. Some terrestrial insects also depend upon shallow water habitats for their breeding and larval stages.

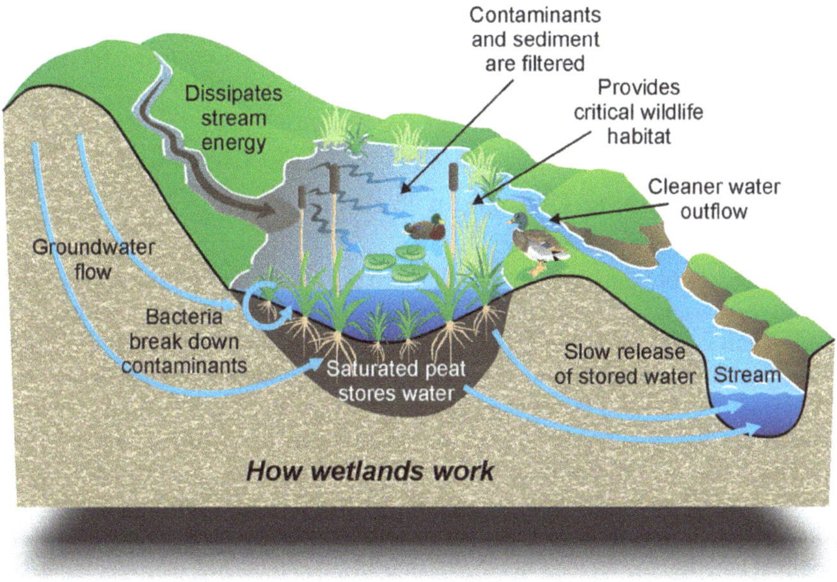

Fig. 1 Illustrate that how wetlands work

The faunal diversity of wetlands can be divided into five major categories such as: (1) residents in the wetlands; (2) regular migrants from terrestrial uplands (particularly large herbivores); (3) regular migrants from deepwater habitats; (4) regular migrants from other wetlands (e.g., waterfowl) and occasional visitors; and (5) those indirectly dependent on wetland biota. Thus, wetlands are "hotspots" of biodiversity and often support thousands of species.

Wetland ecosystems are considered as natural wealth. Worldwide, they provide us with numerous goods and services which worth trillions of US dollars every year and mostly they are entirely free which makes a vital contribution to human health and well-being. As the global population estimated to increase to nine billion by 2050, then there will be accumulative pressure on aquatic resources and the threats by climate change; thus, the need to maximize these benefits has never been greater or more urgent.

The profiles the "ecosystem services"—the benefits people obtain from ecosystems provided by wetlands that cover:

- Biological habitats
- Reservoirs of biodiversity
- Wetland products
- Flood control
- Groundwater replenishment
- Shoreline stabilization and storm protection
- Sediment and nutrient retention and export
- The source and sink of carbon
- Water purification
- Cultural values
- Recreation and tourism
- Climate change mitigation and adaptation.

3 Effect of Climate Change on Wetland

The occurrences of climate change will disturb the hydrology of various individual wetland ecosystems mostly through changes in precipitation and temperature regimes with great global variability. Climate change is known as a major threat to the survival of species and integrity of ecosystems worldwide (Hulme 2005).

Wetlands can be used by regular changes in hydrology; it has direct and indirect effects of changes in temperatures, as well as land use change (Ferrati et al. 2005). In addition, impacts transforming on projected changes in extreme climate events (STRP 2002) include: alteration in base flows; transformed hydrology (depth and hydroperiod); in wildlife heat stress has improved; prolonged range and activity of some pest and disease vectors; flooding, landslide, and mudslide damage; increased soil erosion and increased flood.

Somehow, the approach of assessment of climate change and the tremendous impact on wetlands, these ecosystems require monitoring broad range of their spatial location in a watershed within a specific region. Because human populations are increasing (filling wetlands on the upland side) and sea levels are rising (eliminating wetlands along the ocean edge), results to global warming have specific concern in regard to coastal wetlands. However, coastal populations are increasing at two times the average rate and at worlds scale 21% of human population exists within 30 km of the coast (Nicholls et al. 1999).

By rising sea level, coastal wetlands can be overawed with poor sedimentation. Another important concern is that canals have been dredged and spoils have been piled alongside, the problems of levees. From their sediment-rich water source and negatively affected march plant growth, the spoil banks have isolated the wetlands. The loss of vegetation further harms the capacity of coastal wetlands to combat rising sea level (Turner 1997, 2001; Turner and Boye 1997).

More delicately, by way of the coastline subsides, saline water creeps inland, emphasizing freshwater wetlands and shifting. Shifts in the relative area of tidal water and marsh vegetation can change the amount of marsh-edge habitat, which is available for shellfish and finfish (Delaney et al. 2000).

Every year, the area of coastal wetlands is eliminating due to the development at a rate of 1%. There is one kind of prediction that global sea level rise of 20 cm by the 2018s would result in considerable damage, despite the fact is 1-m rise would eliminate 46% of the world's coastal wetlands (Nicholls et al. 1999). In addition, coastal wetlands would experience increased flooding. Their classical model shows geographically different impacts, with wetland loss most extensive along the Mediterranean, Baltic, and Atlantic coasts, plus the Caribbean islands and coastal flooding greatest for wetlands in the southern Mediterranean, Africa, and South and Southeast Asia. Through flooding small islands of the Caribbean, Indian Ocean and the Pacific Ocean would receive the largest impacts of flooding. In 2004, tsunami devastated small islands and coastal areas in Indonesia and Sri Lanka.

4 Hydrological Change of Wetlands

Wetlands are identified through the nature of soil, vegetation, and hydrology of certain area or field. Out of these, hydrology is considered as a primary factor in the development and plays a vital role for long-term survival of all wetlands as well.

In addition, hydrology includes both the hydrologic cycle (HC) and the hydrologic budget (HB). The HC is known as the movement of water between the atmosphere and earth. Generally, HC is made up of precipitation, evaporation, runoff, and infiltration of water hooked on the ground. The HB applied to many water wetlands made up of six parameters is arranged mathematically as:

$$\text{Outflow} = \text{Precipitation} + \text{Groundwater} + \text{Surface water} - \text{Evaporation} - \text{Transpiration}$$

Basically, the outflow dependent upon the precipitation of the watershed of the wetland, surface run-off and flooding, and groundwater, which is either discharged or percolates downward by the soil profile. Groundwater flows by the internal hydrogeologic framework of the wetland and may act as a source or sink for surface water. In some wetland systems, the HB can be manipulated.

Wetlands are called to perform "hydrological functions," "to act as a sponge," soaking up water during wet periods and releasing it during dry periods. Wetlands are the significant part for complete HC—they soak up runoff and filter and release floodwaters slightly. This maintains flows in dry by alleviating the flow of flood.

Wetland hydrology determines:

- The amount of nutrients entering and leaving a wetland
- The chemistry of soil in a wetland
- The chemistry of water in a wetland
- The plants and animals in a wetland
- The productivity of a wetland.

Inside the wetlands, living creatures such as plants, animal, microbes depend on water and its different characteristics like its volume, depth, permanence, temperature, and chemistry. In addition, ecosystem of wetlands is directly related to long-term fluctuations in water level over months and years. There are several factors of wetland; drainage is one of the primary factors of its drying up. While wetlands become too dry, the number of plants reduces, then the wetland plants are taken over by dry land plants, particularly weeds. Peat soils also break down when they dry up and are exposed to the air. There is another factor like flooding which can change the hydrology of wetland which is artificially flooded and may not give plants potential seasonal variations that are evidences for growth and flowering. However, permanent flooding will reduce the health and the diversity of wetland.

4.1 Human Factors Behind the Wetland Degradation

In the last few years, rapid wetlands degradation has been noticed all around the world due to human activities. In desire of fast economic growth, humans are unable to show ecofriendly and sustainable behavior toward wetlands/environment. There are uncountable activities, which have a negative impact on wetlands in direct or indirect ways; somehow, most of the activities range from agriculture, construction, water diversion, and so on. From the estimated value, it is stated that about 5% of agricultural land worldwide (264 million ha) is irrigated, through South Asia (35%), Southeast Asia (15%), and East Asia (7%) showing a high dependence on irrigation.

Distribution of irrigated area in world is not the same; China and India both have 39% of the global irrigated area, and Western Europe and USA have 13%, while sub-Saharan Africa and Oceania have less than one percent of their agricultural land irrigated (Wood et al. 2000).

Evaluations similarly indicate in order to the portion of cropland that is irrigated has increased by nearly 72% from 1996. On the one hand, for human use around 70% of water has been reserved from the freshwater system; on the other hand, only 30–60% is subsequently used downstream, making irrigation the largest net user of freshwater.

Millennium Ecosystem Assessment (MEA Board 2005) quantified that the degradation and loss of inland wetlands and species have been determined by several kinds of causes such as infrastructure expansion (dams, dikes, and levees), pollution from different factors, overharvesting, land conversion, water withdrawals, and the introduction of invasive alien species while several studies made on wetlands loss in place in China significant ecosystem service which reduce the risks of living and working in coastal landscapes. However, filling and draining have a negative impact on wetland vegetation and maintain hydric soils; it put effect on waster regime as well.

For example, if a wetland is lost, most if not all of its wetland functions are also lost. Human activities proposed to reduce damage to life and properties from extreme climate situation have inadvertently increased the susceptibility of coastal areas to climate change by changing the natural hydrologic functions of wetlands. Regular disturbances straightly can change the specific structure and behavior of wetlands.

4.2 Invasive Species

The most wetlands in China are under great threats from reclamation, water diversion, dam construction, pollution resource overuse, biological invasion, and desertification and climate change. It is specified that 127 alien invasive species have been spotted in China, together with 11 plants and 10 animals that are found in wetlands (Li and Xie 2002). The Environmental Protection Administration of China stated 16 notorious invasive species, with 5 wetlands species. There will be a huge direct economic loss annually 7.0 thousand million dollars (USD), for invasive of 16 species which is accredited to the 5 species that invaded wetlands (SEPA China 2005). For example, alligator weed (*Alternanthera philoxeroides*) was first chronicled in the suburbs of Shanghai in 1892, and water hyacinth (*Eichhornia crassipes*) was reported as a garden flower into China in 1901 (Li and Xie 2002). Both of these species inherent to South America were used as feed from 1950 to 1980 and presently occur in most lakes and rivers in eastern China (Fig. 2).

Smooth cordgrass (*Spartina alterniflora*) which was born in Atlantic Coast of North America was introduced in the year 1979 to reduce coastal erosions from tides, to protect coastal dikes, and it is presently spread over 1120 km^2 of coastal

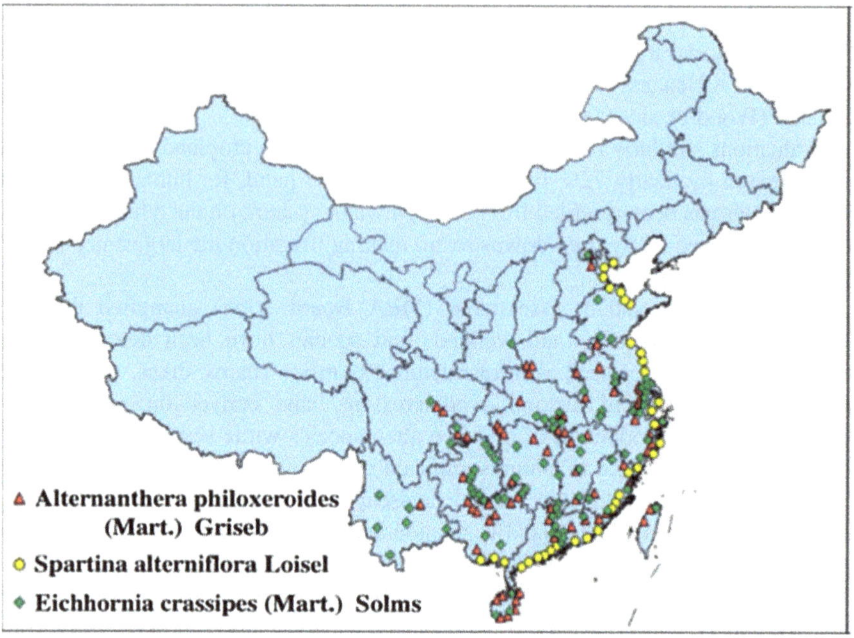

Fig. 2 Distribution of three important invasive plant species in China's natural wetlands. The locations for the smooth cordgrass are based on Ref. 5, whereas the locations of both alligator weed and water hyacinth came from (Li and Xie 2002), with minor corrections based on SEPA China 2005 and SFA China 2005

areas. It was projected that this three invasive species may have caused a total annual economic loss of 2.0 thousand million dollars (USD) (Li and Xie 2002; SFA China 2005; SEPA China 1998) in terms of congested coastal waterways and economic species loss from habitat conversion and ecosystem collapses (Li and Xie 2002; An 2003).

4.3 *Invasive Species Potentially Associated with Wetlands in the Project Area*

Common Name ***Latin Name***

Autumn olive *Elaeagnus umbellata*
Purple loosestrife *Lythrum salicaria L.*
Common reed *Phragmites australis*
Buckthorn *Rhamnus spp.*
Multiflora rose *Rosa multiflora*
Asiatic bittersweet *Celastrus orbiculatus*

Japanese knotweed *Polygonum cuspidatum*
Japanese barberry *Berberis thunbergii*
Tatarian honeysuckle *Lonicera tatarica*
Reed canary grass *Phalaris arundinacea L.*
Privet *Ligustrum vulgare*
Spurge (leafy) *Euphorbia esula L.*

4.4 Wetlands at Risk in Metropolitan Area

Urbanization is one of the major causes of degradation of coastal wetlands. Urbanization has great impact on wetlands in several direct or indirect ways. Urbanization in the metropolitan area results comes from the countryside and has growth in fertility rate, and these had given rise to provide accommodation/employment for the teaming population. For example, degradation of wetlands can be caused directly by habitat loss, suspended solids additions, hydrologic changes, and changed water quality due to constructions. In addition, indirect impacts include transformation in hydrology and sedimentations which noticeably vary wetlands quality. It has noteworthy impacts on the structure and function of coastal wetlands as well, mostly via modifying the hydrological and sedimentation systems, and the dynamics of nutrients and chemical pollutants. In recent years, urban system became a renewed, focused study area, as urbanization remains a challenge for development and putting awesome pressure on social, economic, and environmental sustainability (Pickett et al. 2001). In developing countries, urbanization has associated of some points such as national population growth, rural–urban migration, convergence in rural and urban lifestyle, and the political and economic process with globalization (Cohen 2004). However, in the current scenario urban areas have an account for 3% of total earth's surface, and the association of ecological footprint with urban enlargement has vital environmental consequences especially on wetland ecosystem. Generally, wetlands are seen in human-dominated landscapes like agricultural and urban regions. Several studies show that due to human activities negative effect on wetlands species and ecosystem functioning can be expected in such areas (Ehrenfeld and Schneider 1991).

5 Wetlands in Shanghai

The Yangtze River is the longest river in Asia and fourth largest river in the world in terms of water discharge and sediment load (Yang et al. 2007). Shanghai is located in the Yangtze River estuary in eastern China with 31°41′ north latitude and 121°29′ east longitude. Shanghai has been considered as one of the most populated cities (more than 24 million in 2017) in the world. Shanghai is bordering on Jiangsu and Zhejiang province on the west and the city faces East China Sea on the east and Hangzhou Bay on the south. As a whole, the municipality consists of a peninsula situated between the Yangtze River and Hangzhou Bay, Mainland China's third largest island (Chongming).

The vast majority of Shanghai's 6797 km^2 land area is flat alluvial plains, with a few hills to the southwest (average elevation, 4 m). Shanghai has humid subtropical climate, and it experiences four district seasons with temperature averages between 4.2 °C in January, 27.9 °C in July, and 16.1 °C annually. Widely regarded as the hub of China's modern economy, the city also serves as one of the most important cultural, commercial, financial, industrial, and communication centers in China.

Shanghai is well known for its rich aquatic resources and various wetlands types associated with estuaries, rivers, canals, streams, and lakes. The area comprises two internationally important wetland sites, two nationally important wetland sites, four national and provincial nature reserves, two national wetland parks, and four major drinking water reservoirs.

Shanghai contained total of 376970.6 ha of wetlands, with 99.4% of accuracy in 2013. Many pertinent characteristics and trends emerged in the distribution of wetland (Table 1) and wetland areas between 2003 and 2013 (Table 2).

In Shanghai, wetlands were found in every physiographic region. Longitudinal associations with land use types, there are the various wetland categories and type in the city (Fig. 3a). Figure 3b illustrates the marine wetland, and estuarine wetlands are the most common wetlands category and type.

Table 1 Area and proportion of wetlands in Shanghai, 2013

Wetlands		Area (ha)	Proportion (%)
Marine/coastal wetlands	Permanent shallow marine waters	3250.5	0.86
	Rocky marine shores	39.4	0.01
	Intertidal mud flats	43611.0	11.57
	Intertidal marshes	17794.5	4.72
	Intertidal forested wetlands	237.7	0.06
	Estuarine waters	218565.4	57.98
	Sand shores	13474.7	3.58
Riverine wetlands	Permanent rivers	7241.5	1.92
Lacustrine wetlands	Permanent freshwater lakes	5795.2	1.54
Palustrine wetlands	Permanent freshwater marshes	9051.5	2.40
Constructed wetlands	Water storage areas and ponds	7821.0	2.07
	Canals and drainage channels, ditches	28525.1	7.57
	Aquaculture ponds	21563.1	5.72
All wetlands		376970.6	100

Table 2 Changes in wetlands area for selected wetlands categories in Shanghai, 2003–2013

Wetland category	Area (ha)			
	2003	2013	Change, 2003–13	Change (%)
Marine/coastal	305421.4	250902.1	−54519.3	−17.85
Riverine	7190.7	7241.5	50.8	0.71
Lacustrine	6803.1	5123.6	−1679.5	−24.69
Constructed wetlands (water storage areas and ponds)	299.0	5927.9	5628.9	1882.58
All wetlands for selected category	319714.2	269195.1	−50519.1	−1340

5.1 *Method*

During the study and monitoring about wetlands, the study method has a very magnificent role. It is divided into the following steps.

- Wetlands scope and classification
- Data collection and processing
- Wetlands interpretation, identification, and delineation
- Data verification and accuracy assessment.

5.1.1 Water in Wetland

Since 1979, China become world's one of the fastest growing economy. But hasty growth of China's development has marked water resources. The Huai, Hai, and Huang basins of Yellow River supply just under half of the population, so far the area accounts for less than 8% of national water resources, and out of these, more than 70% water are polluted (Graymore et al. 2001).

In China, due to the highest level of urbanization Shanghai became a leading city. At the end of 2005, the non-agricultural population to the total population as the evaluate standard for the level or urbanization was 84.4% and the national level was only 42.9%. The YRD (Fig. 4) is not only China's longest river delta; it also considered as world's one of the largest deltas. It comprises the Shanghai metropolitan area, southern Jiangsu province, and northern Zhejiang province, which covers 1% of total area of China with 99,600 km^2 and 6% of the national population around 74 million. It is considered as one of most developing areas of China. National Bureau of Statics of China till 2007, the GDP of this area has been reached 4667 billion RMB, which accounts for 19% of the national total and increases at a rate of 15% per annum.

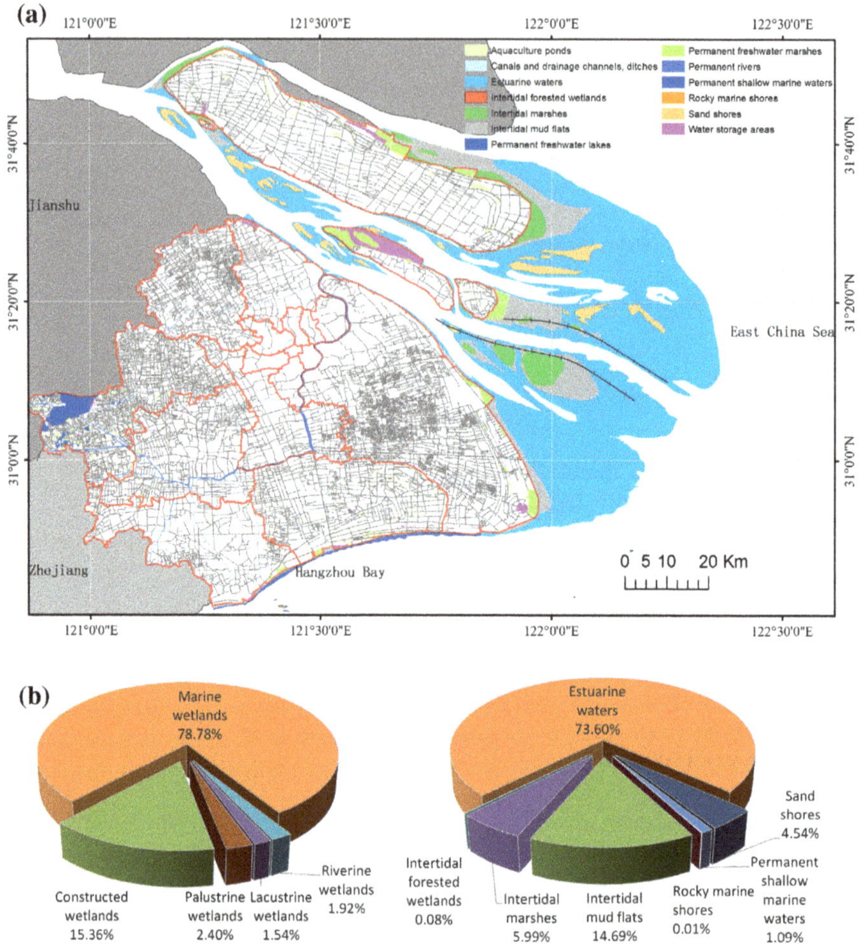

Fig. 3 **a** Type of special distribution of wetlands habitats in Shanghai in 2013. There were 5 categories and 13 types of wetlands (excluding paddy fields) in Shanghai. Estuarine and coastal wetlands were the major category, including forested wetlands, intertidal marshes, intertidal mud flat, rocky marine shores, sand shores and permanent shallow marine waters. **b** Marine wetland and estuarine waters are the most common wetlands category and type (proportions of several wetland categories and types in Shanghai)

5.1.2 Source of Water Pollution

There are many major and minor source of water pollution. Out of these, there are some main source of water pollution of wetlands near metropolitan cities such as industrial waste and municipal waste (Chen et al. 2002; Davidson and White 2005), more fertilizer and pesticides application (Wang et al. 2004), and the byproduct of animal husbandry (Gu et al. 2008). The relative contributions of the various sources are known.

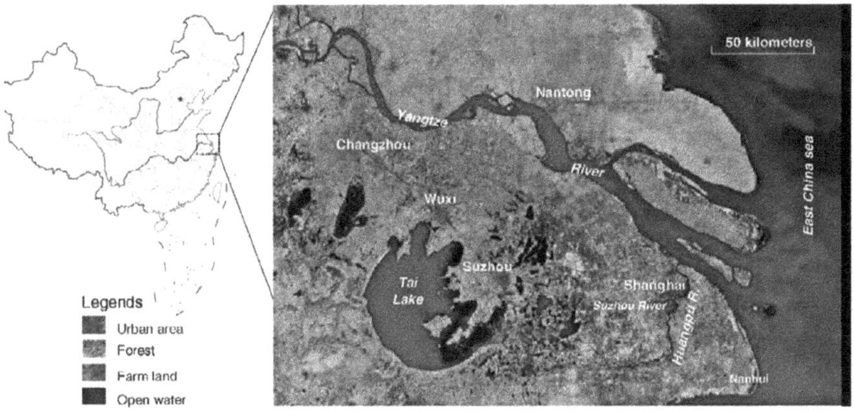

Fig. 4 Location of Yangtze River Delta

5.1.3 Wetlands in the Treatment System

Constructed wetlands has shown its effective potential as a kind of treatment system, and it has become a permissive alternative for the sake of clean water in Yangtze River delta. In the year of 2007, the purification efficiency was measured by Ruan et al. (2007) with subsurface flow and integrated vertical flow wetlands on the muddy flat of a heavily polluted river. The influent concentrations of COD_{Mn} (analyzed with $KMnO_4$ method) and NH_4^+–N of 3.11–117.28 mg/L and 1.30–24.71 mg/L, respectively, both systems showed a >60% reduction in COD_{Mn} and >85% in NH_4^+–N, with better effluent water quality (<14.1 for COD mg/L and <1.0 mg/L for NH_4^+–N With using *Ipomoea aquatica*-dominated treatment system >93% of NH_4^+–N and TP could be removed from heavily polluted stream water (Cao et al. 2006), for the sake of N and P removal in different wetland plants; the performance of *Canna indica* was best (Li et al. 2009). The performance of different wetland treatment systems in the Yangtze Delta is illustrated in Table 3, and in total, these small-scale studies come out with some positives, but application at larger scales is needed. In the region lands are intensively used, the cultivated lands are more than 33%, and due to this, the land is copped twice or more times in a year, so the multi-crop index is >200%; because of the above causes, establishment of new wetlands is difficult especially with the purpose of water purification (Li et al. 2009),

Areas for placement of constructed wetlands including roadsides, river/canal sides, and green space around buildings are important. Such kinds of place are very limited and insufficient to treat the huge amounts of polluted water. Therefore, establishing small constructed wetlands near the outlets of small enterprises or residential complexes could be become an alternative. These techniques are based on the principles of ecological engineering; in some parts, Chinese centuries-old knowledge has been applied (Mitsch and Jørgensen 2003). There are some

Table 3 Efficiency of water quality improvement experiments in Yangtze Delta (Li et al. 2009)

Site			Input (mg/L)	Output (mg/L)	Percent removal (%)
Tai Lake (multi-measure)		TN	4.28	2.14	50
		NH_4^+–N	1.52	0.42	72
		TP	0.161	0.125	22
		Chl-a	0.036	0.036	–
Mesocosm experiments in Hangzhou (*Azolla filiculoides*)		NH_4^+–N	1.058	0.05	95
		NO_3^-–N	3.523	0.16	95
		PO_4^{3-}–P	1.6	0.016	99
Tridal wetland treatment system in Shanghai (multi-species)		TN	60.2 ± 2.0	3.6 + 0.2	94
		NH_4^+–N	9.5 ± 0.6	0.149 ± 0.015	98
		NO_3^-–N	0.088 ± 0.006	0.077 ± 0.017	13
		TP	1.153 ± 0.121	0.267 ± 0.015	77
		DIP[a]	0.166 ± 0.016	0.002 ± 0.000	99
		COD	1180.1 ± 59.6	88.9 ± 5.2	92
Mesocosm experiments in Shanghai	*Ipomea aquatica*	TN	11.24	2.1	81
		NO_3^-–N	6.41	0.65	90
		NH_4^+–N	3.26	0.39	88
		TP	1.55	0.44	95
		DP[b]	0.66	0.12	82
Oenanthe javanica		TN	18.01	3.10	83
		NO_3^-–N	9.41	1.26	87
		NH_4^+–N	8.04	1.12	86
		TP	1.76	0.09	95
		DP[b]	0.89	0.09	90

[a]DIP: diluted inorganic phosphorus;
[b]DP: diluted phosphorus

alternatives with huge prospective such as grow floating mats of vegetation in rivers/canals and lakes/ponds (Hubbard et al. 2004; Headley et al. 2008) or growing plants along the banks, attached with very well-managed biomass collection (Mander et al. 2005; Verhoeven et al. 2006; Sollie et al. 2008; Maddison et al. 2009)

Although wetlands in Yangtze River Delta have been used for thousands of years, and about 80% of area occupied, unfortunately wetlands in Yangtze River Delta rarely have been considered as a treatment system for polluted water. However, convincing the local societies to use their paddy fields and lakes/canals as treatment system for water purification remains a challenge.

5.2 Vegetation Community Wetland

According the change of surface reflection, the wetlands vegetation development and community succession can be investigated by remote sensing technology. The scientists have studied and established the various approximate relationships between vegetation coverage and spectral response, since man-made earth resources observation satellites has been launched (Bannari et al. 1995). While using different bad combinations, vegetation indices become simply effective and experimental measurements on earth surface for vegetation features. The response from vegetation to the electromagnetic spectrum is influenced by a few responsible factors such as the differences in chlorophyll content, nutrient levels, water content, and underlying soil characteristic (Sivanpillai and Latchininsky 2007). The southern part of Yangtze River Delta also called as "Shuixiang" in Chinese which means "land of water," is famous with its low land and abundant water. In this region, there are huge number of lakes and ponds with crisscrossing and canals connecting.

In internal areas of the delta, countless water plants, such as water bamboo (*Zizania latifolia* (Griseb Stapf), lotus (*Nelumbo nucifera* Gaertn. cv. "Fenginye"), water shield (*Brasenia schreberi* J. F. Gmel.), water caltrop (*Trapa japonica* Flerow), and chufa (*Eleocharis tuberosa* Roem.), are harvested as vegetables for the local people. Water hyacinth (*Eichhornia crassipes* (Mart.) Solms), *Alternanthera philoxeroides* (Mart.) Griseb., and water ferns (*Azolla imbricata* (Roxb.) Nakai and *Azolla filiculoides* Lam.) are plants introduced to control pollution, but they turned out to be a annoyance because of superabundant biomass (Zhong et al. 2011). There is another type of artificial wetlands such as rice paddies, which has existence here from centuries. Along the coast, *Phragmites australis* (Cav.) Trin ex steudel, *Scirpus mariqueter* Tang et Wang, and *Spartina alterniflora* Loisel are the dominant plant species that stabilize the shoreline and provide habitat for wild animal (Sun et al. 2002; Xingzhong et al. 2005). These ordinary groups, along with the sediments on tidal flats, contribute to the sequestration of nutrients from rivers to shallow sea water.

5.3 Wetland Birds and Their Habitat

A lot of birds live in the wetlands of the Shanghai, at least for parts of each year, and some of these are rare. There are 150 species of ,birds and these comprise 60% of the number of wetland bird species in China. The birds can be classified into four groups: summer migrants, winter migrants, migrants, residents. The birds are principally snipes (56 species), ducks (33 species), and storks (27 species).

In all kind of wetlands, coastal wetlands are important for waterbirds. It has potential in breeding, stopover, and wintering stopover and wintering habitats for waterbirds. World's half population is living in coastal regions, and it has a great pressure from anthropogenic impacts (Adam 2002).

Because of loss and degradation, in the last few decades globally coastal wetlands become phenomenal, and determinate effects on waterbirds have been observed throughout each and every stage of their life cycle (Schekkerman et al. 1994; Weber et al. 1999; Erwin et al. 2003). In China, coastal wetlands are able to afford habitats for more than 200 waterbirds species (including millions of migratory shorebirds along the East Asian–Australian Flyway and hundreds of thousands of wintering waterfowl) (Barter 2002). The anthropogenic activities have an adverse impact on waterbirds and their habitats in wetlands, and also their surrounding areas (Erwin et al. 2003). However, the effects of these habitats change on the waterbirds community in China are largely unknown.

In the year of 2009, Zhijun Ma has reported a survey in regard to waterbird population changes in wetlands at Chongming Dongtan in the Yangtze River Estuary, China, time period from 1980 to 2000. His study describes that between 1988 and 2000 total of 124 species were recorded (Table 4).

5.3.1 Wetland Habitat Change and Waterbird Population Trends

There are a number of factors which has contributed to the population trends of waterbirds. Dramatic change in area and type of wetlands is one of the most responsible factors, although reduction and change of habitats might result in the variation in waterbird communities as recommended by another study of avian population trends.

Various studies have revealed that artificial wetlands can provide wintering, stopover, and even breeding habitats for waterbirds (Sánchez-Guzmán et al. 2006; Yasué et al. 2007; Rendón et al. 2008). In previous study, it has been found that the artificial aquaculture pounds at Dongtan in 2000s provide habitat for approximately 60% waterbird and become one of the most important habitats.

Table 4 Area (thousands of hectares) of wetlands habitats for waterbirds at Chongming Dongtan 1988 and 2002

Wetland category	1988	2002	Change in area	%Change
Industrial zones	22.20	11.86	−10.34	−46.6
Sea bulrush	4.67	1.50	−3.17	−67.8
Common reed habitat	4.87	1.14	−3.73	−76.5
Smooth cordgrass habitat	0	1.14	+1.14	–
Bare intertidal zone	12.66	8.08	−4.58	−36.2
Aquaculture ponds	1.14	3.29	+2.15	±188.86
Farmland (paddy fields)	3.47	7.01	+3.54	+102.0
Coastal shallow water zone	14.50	14.50	0	0
Total	41.31	36.66	−4.65	−11.3

Recent study has illustrated that aquacultural ponds only roosting habitats for most waterfowl, the major waterbird group there. The majority of waterbird still depends on bare intertidal and sea bulrush habitats for good food (Ma et al. 2004).

5.4 Basic Principles of Ecological Restoration

Ecological restoration is the practice of renewing and restoring of an ecosystem that has been degraded, damaged, and destroyed by active human intervention and action. There are some keywords those might be able to define all basic ecological concepts for restoration such as: disturbance, succession, and fragmentation.

Disturbance—It is an ordinary and even essential system in many ecosystems; the progressively austere disturbance of ecosystem by humans has degraded, damaged, and even destroyed ecosystems. It is a major change of environmental system.

Succession—It is the method of change in an ecological community over time. After a disturbance, an ecological community generally changes from a simple level of organization to a more multifarious level. Many ecological communities recover from slight disturbances on their own, but with severe or constant disturbance, restoration may be needed to assist with restoring ecological successional processes.

Fragmentation—It comes when ecosystems are divided into small, isolated fragments and can only support small populations. These small populations are much more vulnerable to extinction. Sometimes, restoration projects can overcome this by simply adding area or developing habitat corridors that link isolated fragments. Increasing habitat connectivity is an important goal in ecological restoration.

Ecological restoration is determined to the recovery of an ecosystems, although it concerns at first health (functional processes such as water filtration, sequestration of carbon dioxide), integrity (species composition and community structure), and sustainability (resistance and resilience to disturbance).

Ecological restoration becomes a global issue from the years and also for upcoming years, but the first and important concern here is that HOW?

There are some following points and guidelines that might be helpful for ecological restoration.

- Assess the site
  - What are the present conditions of site?
  - Identify the responsible factors for disturbance?
  - Identify the methods for stopping or reversing the disturbance?
- Develop projects goal
  - Use reference sites (nearby sites in natural condition)
  - Consult historical sources that describe the pre-disturbance community

- Remove source of disturbance
  - Remove toxic material
  - Remove causes of erosions
  - Reduce overpopulation of species
  - Eradicate invasive species
- Restore processes/natural disturbance cycle such as flood and fire cycles
  - Restore tidal flow
  - Rehabilitate substrates
  - Restore soil texture or chemistry, water regimes or water quality
- Restore vegetation
  - Direct re-vegetation or seeding of a site using native species suited to local environmental conditions. Where possible a variety of sources within the local region should be used to ensure genetic diversity.
- Monitor and Maintain
  - Monitoring restoration sites is critical to determine whether goals are being met. Ideally, restoration projects will eventually achieve a self-sustaining ecosystem without further need for intervention.

5.5 Case Study

A study on the impacts of the Yangtze River South Water Conservancy on wetland along the riverbank provides us a full understanding of the advantages and disadvantages of water conservancy facilities, and a new idea for the construction of levees in the future, with more comprehensive understanding on the development and protection of wetlands. It further provides data and theory support for wildlife protection and management.

River embankment, along the rivers, channels, lakes, coastal, or flood diversion area, the edge of the reclamation area construction, is retaining structures. The embankment can resist the flooding, block the tide and wave, can protect the residents and industrial and agricultural production safety. It is the world's first widely adopted flood control measure. Embankment can protect the lives of residents, while the construction of the levee to a certain extent changed the original ecological environment. For example, the construction of the levee reduce the water seepage performance of riverside environment, this will lead a reduction of underground water level and a rise of the water level outside the embankment. The change of hydrological conditions make a series of changes in wetland ecosystem near the embankment.

In certain extent, the construction of the levee has caused great damage to the river wetlands. With the construction of the levee, the riverside wetlands have been destroyed, and the area has shrunk dramatically. In recent years, put forward to build ecological river construction ideas. The construction of ecological rivers, the increase of riverside wetlands and floodplains, the restoration of riparian waterside plant communities, the maintenance of biological diversity have become the international trend of river construction and development. It also shows that the international water conservancy project with ecological construction is in the ascendant, becoming the trend. Therefore, the study on the influence of water conservancy facilities in the Yangtze River to the riverbank beaches will provide valuable experience and theoretical support for ecological river construction in China. In addition, through the investigation of the wild animal and plant resources along the Yangtze River in Shanghai and its evolution with the embankment construction, it will provide firsthand data and the most advanced theoretical support for wildlife conservation and management, enrich the basic data of wildlife protection and management, and increase the management experience of wildlife protection.

5.5.1 Brief Introduction of Wetland Ecological Restoration Project of Shanghai Baoshan of Yangtze River

Shanghai Baoshan Yangtze River coastal wetland ecological restoration project, located in Baoshan District, Shanghai Luojing Town (E 121° 22′16.37″, N 31° 29′ 35.14″), before the repair its three sides are surrounded by the Yangtze River dam, one side by the river. A barren Yangtze River riverside beach (Fig. 5) covers an area of about 10 ha. This beach is located near the mouth of the Yangtze River. Due to the erosion of the Yangtze River and the tide, the area of soil erosion is serious, and a lot of garbage from the river rushed to the shore. There is only a small amount of reed and scallions scattered in the area, and the wetland ecology is seriously degraded. After the field survey it was found that the region is an important bird habitat of Shanghai green wings duck (Latin name) and crested (Latin name) habitat. As the site being located in the metropolitan area of Shanghai, the strong human activities, industrial production and port operations made the birds difficult to find a suitable habitat in the region. In order to protect birds, the wetland ecological restoration of the area decided to provide a good habitat for birds. The restoration plan is to divide the ecological restoration area from the Yangtze River by building a permeable dam (dam length of 156 m, the height of the dam body is 4.05 m, and the head of the dam is 3.4 m). Through the height of the dam, control the water level of the repair area, forming a closed connection with the river water (high-level connectivity). According to the protection of birds living habits, build a two ha (depth of 1.5–2 m) of the closed surface and build several bird islands in the repair. At the same time, some fish and crabs were delivered in the water, to attract birds for the purpose of providing a safe habitat and foraging environment for waterfowl. To establish the wetland ecosystem quickly, wetland vegetation area is optimized and the common wetland vegetation is planted. At the same time, it will establish isolation systems around the repair area to avoid interference with human activities. After one year of engineering restoration, the ecological and ecological environment of the wetland in the region has improved completely. The water quality is gradually restored, the wetland vegetation system has been well developed, the bird activity in the region is gradually restored, the landscape effect of the beach has also been greatly improved, and the soil quality is gradually restored (Fig. 5). The wetland restoration project has achieved the desired effect.

Through the summary of the successful experience of this project, it is found that ecological restoration in a metropolitan area with strong population and strong human activities should follow the following principles:

(1) To comply with the basic principles of environmental ecology
(2) To repair the target clear, to create a habitat of protected species
(3) Make full use of human activities in the area of human activities
(4) The repair area is not necessarily great, but do well, and the effective isolation of human activities can also achieve good results.

In the year of 2013, our research team first time visited this wetland and collected the sample from five locations, and in the year of 2016, our research again visited this wetland. Findings from 2013 and 2016 are mentioned (Table 5).

Fig. 5 Baoshan Wetland Research Area

Table 5 Showing the data of Baoshan Wetland in the year of 2013 and 2016

Sampling site	2013			2016		
	TN (mg/L)	TP (mg/L)	COD (mg/L)	TN (mg/L	TP (mg/L)	COD (mg/L)
1	2.42	0.067	5.47	2.5	0.329	6.42
2	2.2	0.127	18.57	1.45	0.104	20.16
3	0.9	0.026	0.47	0.52	0.067	8.36
4	2.2	0.142	104.82	2.26	0.178	50.32
5	0.5	0.023	2.69	0.64	0.059	66.37
6	0.54	0.057	52.77	2.43	0.269	30.44
7	0.26	0.019	44.7	0.47	0.049	42.35

6 Development and Conversion Continue to Pose Major Threats to Wetlands, Despite Value and Importance

Local planning authorities have stated that they are planning a huge suburban wetland project to improve the quality of Shanghai's tap water. The district's planning administration bureau stated the Qingpu District is planning to develop a 15-sq-km wetland around Dianshan Lake which will be source of about 30% of Shanghai's drinking water and can act as a filter for the water flowing into the city. According to an official surnamed Zhang, to build a harmonious ecological system the district government will also encourage farmers living around the lake to farm "properly". Furthermore, to improve the ecological situation a 5.5-hectare forest of pond cypress which is a water-preserving plant will be grown along the wetland, as its details are yet to be finalized, and he did not give the schedule for the project. Last month, local authorities found out the blue-green algae boom in Tai Lake in neighboring Jiangsu Province and so they have been keeping a close eye on the water quality in Dianshan Lake. A thorough inspection has been done by the district's environment monitoring station, and they affirmed that no algae were found in Dianshan Lake.

Cui Liping, Vice Director of the Shanghai Landscaping Administration Bureau (SLAB), stated that completion of the wetland project will improve the water quality of Dianshan Lake significantly. "Shanghai's water source is being threatened, and we have to move quickly," Cui said. "It's crucial to protect the wetlands, which are an integral part of our water resources." Building nature reserves could increase the natural wetlands reserves from 25 to 30%, Cui said.

This wetland project is a part of that plan. "In Shanghai approximately, 30% of the soil lands are wetlands," the SLAB said. While the city's forest coverage was 11.63% in 2006, its green space coverage was 37.3%. Despite the government's efforts to reverse the effects of industrialization, some people stay cautious about the city's water quality. Visitors in Shanghai have complained about the tap water due to its strong smell of bleach. A local primary school teacher named Mao Kejia, said her family always drink purified water and never drink tap water. "It seems to me the water quality is getting worse and worse," she said.

Fig. 6 A bird's eye view of the future suburban wetland project in Shanghai's Dianshan Lake

"The problems are serious and it takes time before those pollution-tackling measures work" (Fig. 6).

7 Protection of Wetlands and Conclusions

Addressing a recent forum on the Yangtze River held in Changsha, the capital of central China's Hunan Province, Zhu Lieke, Deputy Head of the State Forestry Administration, said China has made an inventory of 173 wetlands, most of which are in northeast China and the Yangtze River Valley (YRV). Thirty of the country's wetlands are listed in the international wetland catalogue, and one-third of them are situated along the Yangtze.

Phenomena such as the rapid drop in the number of lakes and fast shrinkage in lake area got worse as China's economy tears through resources, said Zhu, who warned that wetlands in the YRV face unprecedented ecological threats. "The problems that plague wetlands in the YRV include pollution, ecological degradation, and dwindling water resources," said Zhu. "The protection of our wetlands is urgent." The 6,300-km-long Yangtze, the country's longest, originates in the Tanggula Range on the Qinghai-Tibet Plateau and passes through Qinghai, Tibet, Sichuan, Yunnan, Chongqing, Hubei, Hunan, Jiangxi, Anhui, Jiangsu, and Shanghai before emptying into the East China Sea. Wetlands in the YRV include salty plateau lakes and plateau marshlands, the galaxy of lakes on the middle reaches of the Yangtze, and the coastal wetland near Chongming Island at the estuary of the river.

Dongting Lake, which exists as an important wetland and also flows into Yangtze River, for instance, is shockingly polluted. Marine life has already divested, and people are badly affected by schistosomiasis disease. The water area of Dongting Lake has shrunk from 4,350 sq km in 1949 to present 2,625 sq km as a result of silting and land reclamation for farming. According to Zhu, People's Republic of China has already launched back to back three programs to protect the wetlands in the YRV, including the national program for conservation of wildlife, plants, and nature reserves, and the program to protect the Sanjiangyuan wetland in Qinghai Province. But much remains to be done. More efforts to protect wetlands, there are some guidelines and restoration principles mentioned below:

- Preserve and protect aquatic resources
- Restore ecological integrity
- Restore natural structure and function
- Work within the watershed/landscape context
- Understand the potential of watershed
- Address ongoing causes of degradation
- Develop clear achievable and measurable goals
- Focus on feasibility
- Use references site
- Anticipate future changes
- Involve a multidisciplinary team
- Design for self-sustainability
- Use passive restoration when appropriate
- Restore native species and avoid non-native species
- Use natural fixes and bioengineering
- Monitor and adopt where changes are necessary.

References

Adam P (2002) Saltmarshes in a time of change. Environ Conserv 29(01):39–61

An S (2003) Ecological engineering of wetlands. Chemical Industry Press, Beijing, pp 22–35

Bannari A, Morin D, Bonn F, Huete A (1995) A review of vegetation indices. Remote Sens Rev 13(1–2):95–120

Barter MA (2002) Shorebirds of the yellow sea: importance, threats and conservation status. Wetlands International Global Series 9, International Wader Studies 12, Wetlands International of Oceania, Canberra

Board Millenium Ecosystem Assessment (2005) Millenium ecosystem assessment—ecosystem and human well-being: synthesis. Phys Teach 34(9):534

Chen Z, Xu S, Xu Q, Hu X, Yu L, Chen Z, Xu S, Xu Q, Hu X, Yu X (2002) Surface water pollution in the Yangtze River Delta: patterns and countermeasures. Pedosphere 12(2):111–120

Cohen B (2004) Urban growth in developing countries: a review of current trends and a caution regarding existing forecasts. World Dev 32(1):23–51

Davidson B, White R (2005) Bio-economic strategy to combat non-point pollution in China. Pedosphere 15(2):156–163

Delaney T, Webb JW, Minello TJ (2000) Comparison of physical characteristics between created and natural estuarine marshes in Galveston Bay, Texas. Wetlands Ecol Manag 8:343–352
Ehrenfeld JG, Schneider JP (1991) Chamaecyparis thyoides wetlands and sub-urbanization: effects on hydrology, water quality and plant community composition. J Appl Ecol 28:467–490
Erwin RM, Allen DH, Jenkins D (2003) Created versus natural coastal islands: atlantic waterbird populations, habitat choices, and management implications. Estuaries Coasts 26(4):949–955
Ferrati R, Canziani GA, Moreno DR (2005) Esteros del Ibera: hydrometeorological and hydrological characterization. Ecol Model 186(1):3–15
Fish U, Service W, Cowardin LM (1979) Classification of wetlands and deepwater habitats of the United States, fish and wildlife service. US Department of the Interior, Washington, DC
Graymore M, Stagnitti F, Allinson G (2001) Impacts of atrazine in aquatic ecosystems. Environ Int 26(7):483–495
Gu P, Shen RF, Chen YD (2008) Diffusion pollution from livestock and poultry rearing in the Yangtze Delta, China. Environ Sci Pollut Res 15(3):273
Guntenspergen G, Stearns F (1985) Ecological perspectives on wetland systems. In: Ecological considerations in wetlands treatment of municipal wastewaters. Van Nostrand Reinhold Company, New York, pp 69–97, 177, Ref. NSF Grant DEB 791 2516
Headley T, Tanner CC, Council AR (2008) Application of floating wetlands for enhanced for stormwater treatment: a review, Auckland Regional Council
Hruby T (1999) Assessments of wetland functions: what they are and what they are not. Environ Manage 23(1):75–85
Hubbard R, Gascho G, Newton G (2004) Use of floating vegetation to remove nutrients from swine lagoon wastewater. Trans ASAE 47(6):1963–1972
Hulme PE (2005) Adapting to climate change: is there scope for ecological management in the face of a global threat? J Appl Ecol 42(5):784–794
Li X, Mander Ü, Ma Z, Jia Y (2009) Water quality problems and potential for wetlands as treatment systems in the Yangtze River Delta, China. Wetlands 29(4):1125–1132
Li Z, Xie Y (2002) Invasive alien species in China. China Forestry Publishing House, Beijing, pp 24–34
Ma Z, Li B, Zhao B, Jing K, Tang S, Chen J (2004) Are artificial wetlands good alternatives to natural wetlands for waterbirds?–A case study on Chongming Island, China. Biodivers Conserv 13(2):333–350
Maddison M, Mauring T, Remm K, Lesta M, Mander Ü (2009) Dynamics of Typha latifolia L. populations in treatment wetlands in Estonia. Ecol Eng 35(2):258–264
Mander Ü, Hayakawa Y, Kuusemets V (2005) Purification processes, ecological functions, planning and design of riparian buffer zones in agricultural watersheds. Ecol Eng 24(5):421–432
Mitsch, WJ, Jørgensen SE (2003). Ecological engineering: a field whose time has come. Ecol Eng 20:363–77
Nicholls RJ, Hoozemans FMJ, Marchand M (1999) Increasing flood risk and wetland losses due to global sea-level rise: regional and global analyses. Glob Environ Change 9:S69–S87
Pickett STA, Cadenasso ML, Grove JM, Nilon CH, Pouyat RV, Zipperer WC, Constanza R (2001) Urban ecological systems: linking terrestrial ecological, physical, and socioeconomic components of metropolitan areas. Annu Rev Ecol Syst 32:127–157
Rendón MA, Green AJ, Aguilera E, Almaraz P (2008) Status, distribution and long-term changes in the waterbird community wintering in Doñana, south–west Spain. Biol Cons 141(5):1371–1388
Ruan Xh, Wu Jq, Sun M (2007) Analysis on efficiency of polluted river water purification using SFCW and IVFCW. China Water Wastewater 23(11):17
Sánchez-Guzmán J, Morán R, Masero J, Corbacho C, Costillo E, Villegas A, Santiago-Quesada F (2006) Identifying new buffer areas for conserving waterbirds in the Mediterranean basin: the importance of the rice fields in Extremadura, Spain. Biodiversity and Conservation in Europe, Springer, pp 91–102

Schekkerman H, Meininger PL, Meire PM (1994) Changes in the waterbird populations of the Oosterschelde (SW Netherlands) as a result of large-scale coastal engineering works. Hydrobiologia 282(1):509–524
Sivanpillai R, Latchininsky AV (2007) Mapping locust habitats in the Amudarya river delta, Uzbekistan with multi-temporal MODIS imagery. Environ Manage 39(6):876–886
Sollie S, Coops H, Verhoeven JT (2008) Natural and constructed littoral zones as nutrient traps in eutrophicated shallow lakes. Hydrobiologia 605(1):219–233
State Environmental Protection Administration of China (1998) National Report of Biodiversity of China. China Environmental Science Press, Beijing, pp 430. (In Chinese)
State Environmental Protection Administration of China (2005) Environmental Report. (http://www.zhb.gov.cn/)
State Forestry Administration of China (2005) Wildlife and Wetland Reserves Information. (http://www.cnwm.org.cn/wildlife/index.asp)
STRP (Scientific and Technical Review Panel of the Ramsar Convention on Wetlands) (2002) New guidelines for management planning for Ramsar sites and other wetlands. In: "Wetlands: water Life, and culture" 8th meeting of the conference of the contracting parties to the convention on wetlands (Ramsar, Iran, 1971). Valencia, Spain, 18–26 Nov 2002
Sun S, Cai Y, An S (2002) Differences in morphology and biomass allocation of Scirpus mariqueter between creekside and inland communities in the Changjiang estuary, China. Wetlands 22(4):786–793
Turner RE, Boyer ME (1997) Mississippi River diversions, coastal wetland restora-tion/creation and an economy of scale. Ecol Eng 8:117–128
Turner RE (1997) Wetland loss in the northern Gulf of Mexico: multipleworking hy-potheses. Estuaries 20:1–13
Turner RE (2001) Estimating the indirect effects of hydrologic change on wetland loss: If the earth is curved, then how would we know it? Estuaries 24:639–646
Verhoeven JT, Arheimer B, Yin C, Hefting MM (2006) Regional and global concerns over wetlands and water quality. Trends Ecol Evol 21(2):96–103
Wang D, Liu Q, Lin J, Sun R (2004) Optimum nitrogen use and reduced nitrogen loss for production of rice and wheat in the Yangtse Delta region. Environ Geochem Health 26(2): 221–227
Weber TP, Houston AI, Ens BJ (1999) Consequences of habitat loss at migratory stopover sites: a theoretical investigation. J Avian Biol:416–426
Wood S, Sebastian K, Scherr S (2000) Pilot analysis of global ecosystems: agroecosystems, a joint study by International Food Policy Research Institute and World Resources Institute. International Food Policy Research Institute and World Resources Institute, Washington DC
Xingzhong Y, Hong L, Jianjian L (2005) Effect of Scirpus mariqueter vegetation on salt marsh benthic macrofaunal community of the Changjiang estuary. J Coast Res: 73–78
Yang S, Zhang J, Xu X (2007) Influence of the three gorges dam on downstream delivery of sediment and its environmental implications, Yangtze River. Geophys Res Lett 34(10)
Yasué M, Patterson A, Dearden P (2007) Are saltflats suitable supplementary nesting habitats for Malaysian Plovers Charadrius peronii threatened by beach habitat loss in Thailand? Bird Conserv Int 17(03):211–223
Zhong L, Deng J, Song Z, Ding P (2011) Research on environmental impacts of tourism in China: progress and prospect. J Environ Manage 92(11):2972–2983

Modern and Emerging Methods of Wastewater Treatment

Neetu Sharma, Abhinashi Singh and Navneet Batra

Abstract The world is progressing at a rapid pace, and this progression is exerting a lot of pressure on natural resources. The resources are limited, but demand has been increased to all time higher. The nations are facing many issues including climate change, extinction of natural flora and fauna. Another big issue which is common to all the nations is management of wastewater being generated from domestic, industries, and agricultural sources. Inflow of wastewater into rivers, seas, and oceans harms the aquatic environment in many ways such as acidification, nutrient enrichment, high level of growth, and oxygen depletion. This leads to the extinction of rich flora and fauna of aquatic systems and also has indirect impact on human health by means of biomagnifications. Over the years, different natural methods are being used to treat wastewater such as wetlands and ponds with specific aquatic plants. But due to certain limitations, new and advanced methods have emerged which are much more efficient in treating wastewater with different types of contaminants. Methods like activated sludge, oxidation ditches, UASB, activated carbon, nanoparticles, carbon nanotubes, and microbial fuel cells are currently being employed at various levels to treat wastewater. These advanced methods are having more efficiency but on the same hand have less sustainability. The other major drawback associated with these systems is that their establishment leads to forfeiture of natural resources which are valuable to us. The treatment and disposal of wastewater are a major challenge for protection of public health and environmental resources. Thus, principle of ecological wisdom should be employed to design new treatment strategies and also to strengthen the available natural treatment methods in order to achieve the goal of sustainable development.

Keywords Wastewater · Remediation · Ecological wisdom · Wetland
Microbial fuel cell

N. Sharma (✉) · A. Singh · N. Batra
Department of Biotechnology, GGDSD College, Sector-32-C, Chandigarh 160030, India
e-mail: neetusharma4@gmail.com

V. Achal and A. Mukherjee (eds.), *Ecological Wisdom Inspired Restoration Engineering*, EcoWISE, https://doi.org/10.1007/978-981-13-0149-0_13

1 Introduction

Water is very essential for supporting life on earth. Although seventy percent of earth is covered with water, less than one percent is actually available. Rest of the water is either frozen or available in the form of salt water. Clean and potable water is the basic requirement of every human being, but at the same time access to this is the major concern and challenge of the modern world. Some regions are having rich source of water bodies and receive regular rainfalls, but on the other hand some regions are lacking in this. This poses a major problem in managing water distribution across the different regions of the world. The rapid growth of human civilization across the planet in the form of population increase, urbanization, and industrialization resulted in rise of the demand for clean and potable drinking water on the one hand and the generation of wastewater on the other hand. Both in the industrialized world and developing world, human activities have resulted in the contamination of water resources through release of hazardous chemicals from industries (Table 1), municipal waste, agricultural runoff, etc., putting an extra burden on available water resources and generating large amount of wastewater (Qu et al. 2013). This wastewater coming from different anthropogenic sources is discharged in the same form into different water bodies such as rivers, estuaries, seas, and oceans where they contaminate them and deteriorate the aquatic flora and fauna to a large extent. Over the years, various policies, standards, and methods have been framed and implemented all over the world related to the treatment and release of wastewater into water bodies and its recycling in order to meet the current

Table 1 Common contaminants present in industrial effluents

Sr. No.	Industry	Type of waste generated
1.	Construction	Ignitable/toxic debris, lead pipes, petroleum distillates, and asphalt wastes
2.	Laboratories	Unused reagents, reaction products, experimental samples, and used solvents
3.	Leather	Chromium, acid and alkaline wastewater, ammonium sulfate, calcium hydroxide, hydrogen sulfide, alcohols (methanol, ethanol, propanol, etc.)
4.	Pesticide	Used/unused pesticides, solvent wastes, ignitable wastes, contaminated soil (from spills), contaminated rinse water, empty containers
5.	Textile	Hydrogen peroxide, sodium silicate, and organic stabilizer Alkali and sodium hydroxide, tetrachloroethylene, trichloroethylene, chlorobenzene, toluene
6.	Printing	Waste ink with chromium, barium, and lead content; waste ink contaminated with cleaning solvents, acid plate etching chemicals for metallic lithographic plates Waste photochemical solutions from fixer and rinse water and from alkaline or acid process baths

demand of potable water. But on the other side beside all the rules and regulations, there is another harsh reality that a large amount of wastewater is either left untreated or treated using inefficient methods and released into the environment which ultimately leads to environmental degradation. Wastewater treatment methods already in use have proven to be beneficial in the past, but with increasing population and human activities there is an urgent need to reconsider and modify the present policies and methods and devise new technologies to treat wastewater in order to minimize the risk toward the environment. Another important benefit from the treatment of wastewater is that it can be used for drinking purposes which will substantially result in reducing the stress of growing water need on natural freshwater sources. According to World Health Organization 2015, it has been estimated that approximately 1.1 billion people lack supply of potable drinking water so treating wastewater and making it fit for drinking will help to reduce the demand of potable drinking water.

1.1 General Composition of an Untreated Wastewater and Its Impact on Water Bodies

It is very difficult to define the wastewater. Considering the water in its natural state, i.e., water from streams, lakes, and rivers, they also contain different concentrations of minerals, elements, suspended and dissolved matter. The water being used for one purpose may not be fit for other purpose. So in order to define the purity of water, there are list of standards set for different categories of water by different countries. For example, there are norms set for water used for public supply, irrigation water, and effluents to be discharged by the industries. In order to ensure supply of clean water, amendments are regularly made on the basis of databases available from several water and health regulatory authorities across the globe. Any deterioration in the quality of water due to interference of human activities is water pollution. Anthropogenic and natural waste both contribute toward shift in natural composition of water. Wastes released into water bodies of one country do not remain limited to certain boundaries, and this leads to development of dispute among neighboring nations. Thus, it is a global problem and hence needs international cooperation. The increasing demand for sustainable use of resources has been felt by every segment of the society.

Only 20% of the total wastewater generated undergoes adequate treatment before being released into the environment (UNESCO 2012).Wastewater generally comes from different anthropogenic sources which contribute to different types of contaminants and pollutants present in it including organic, inorganic, and biological origin (Das et al. 2014).

Organic contaminants and pollutants such as detergents, dyes, hydrocarbons, biodegradable compounds are present in water coming from sewage systems. Inorganic contaminants and pollutants include heavy metals, viz., cadmium, copper, nickel, zinc. Biological contaminants are also present in wastewater such as microorganisms (pathogenic and nonpathogenic) bacteria, viruses, fungi (UN Water 2015).

All of the three broad categories of contaminants and pollutants mentioned above if left untreated or inadequately treated directly affect and impact the environment in harmful manner leading to environmental deterioration and related problems. The degree of impact depends on the type, form, and concentration of the contaminants and pollutants.

As mentioned above, wastewater is a complex mixture of various contaminants and pollutants being released from industries, domestic, and agricultural sources and these contaminants and pollutants if left unprocessed will contaminate the water bodies and pose a serious threat to the environment of the water bodies receiving this wastewater which will further effect the state of available drinking water. Biological contaminants whether pathogenic or nonpathogenic present in wastewater coming from different sources can directly or indirectly contaminate groundwater and may lead to infectious outbreaks. Table 2 illustrates a list of microorganisms commonly present in water and the diseases caused by them.

Water bodies receiving improperly treated wastewater from different sources are a big challenge because of the long-term impact it could have on the aquatic ecosystem.

1. **Impact on river water**: All around the globe, the major source of freshwater for humans is river water, and any kind of deterioration will directly affect the human population. As river water is a major source of freshwater, but it is also acting as the major disposal site untreated wastewater generating from various sources. All these are directly affecting the ecosystem of rivers which in turn affects the environmental cycles. The wastewater alters the ecosystem of rivers. High amount of contaminants and pollutants leads to nutrient enrichment in rivers which leads to eutrophication (Kanu and Achi 2011). High level of growth in river water results in oxygen depletion which hampers the growth of natural aquatic organisms

Table 2 Microorganisms present in water and diseases caused by them (UN Water 2015)

Microorganism	Disease
Campylobacter spp.	Gastroenteritis
Escherichia coli	Gastroenteritis
Enteroviruses	Gastroenteritis
Salmonella typhi	Typhoid
Shigella spp.	Dysentery
Vibrio cholera	Cholera
Norovirus	Gastroenteritis
Parvovirus	Gastroenteritis
Rotavirus	Gastroenteritis
Entamoeba histolytica	Amoebiasis
Fasciola	Fascioliasis
Schistosoma	Schistosomiasis

and may result in the depletion of aquatic life. Disposal of industrial and mining wastewater containing large concentration of heavy metals like arsenic, cadmium, lead in rivers can lead to accumulation of these metals which can further result in biomagnification of these metals in organisms. Industrial wastewater also contains acid waste which acidifies the river water and changes its pH and temperature which deteriorates the sensitive aquatic ecosystem (Akpor and Muchie 2011; UN Water 2015). Long exposure of rivers to untreated wastewater can permanently change their structure and properties which affects the normal growth and survival of aquatic life. The presence of large number of suspended solids in rivers reduces the penetration of sunlight and hence reduces photosynthesis which again hampers the growth of aquatic plants and microbes that maintain and stabilize the rivers (Akpor and Muchie 2011).

2. **Impact on marine environment**: Release of untreated wastewater into the marine ecosystem such as seas and oceans has an adverse effect on the growth and survival rate of marine life. The extent of deterioration depends upon the concentration, solubility, and persistence of contaminants and pollutants present in wastewater such as heavy metals, pharmaceuticals, insecticides, pesticides, toxic chemicals, microorganisms, fossil fuels, and their residues. All of them can have alarming impact on the marine life. These can accumulate or bioconcentrate in the organisms. Wastewater being discharged into marine waters impairs the aquatic life due to increased toxicity. This increased toxicity can cause severe and permanent damage to marine life and ultimately disturb the ecological cycle (Ren et al. 2017).

To study the serious impact of these contaminants and pollutants on the aquatic life, a number of studies have been conducted on one of the most diverse as well as the most sensitive parts of marine environment, the coral reefs. The results of various studies have shown that the link between wastewater and marine ecosystem degradation is very serious and is posing a severe risk to the aquatic life. Sewage wastewater, industrial wastewater, and agricultural runoff are affecting the coastal areas and coral reefs to very large extent and are also altering the overall water quality of seas and oceans which in turn is disturbing the growth and survival rate of aquatic life. Another important point observed was that polluted sites spread their pollution to unpolluted sites through water currents in seas and oceans, thus affecting distant areas. Another problem which is having a big impact is the decalcification of calcium carbonate structures present in aquatic system. In marine ecosystem, biocalcification is an important process in which a variety of calcium carbonate structures are formed which helps many marine organisms to grow. Calcium carbonate structures deposition around other organisms provides a range of benefits like protection from predators, unique habitat, and mechanical support (Andersson and Gledhill 2013). Coral reef is one such ecosystem that grows in shallow ocean waters with the help of calcium carbonate structures around them which further helps in providing habitat to other organisms. Acidification of oceans is causing erosion and decalcification of these structures resulting in the weakening of coral structures. Naturally, there is a balance between calcification and decalcification process which maintains the coral reef ecosystem but due to the acidification of oceanic waters. Ocean acidification means the pH of marine water is shifting toward low alkaline condition which is posing a great threat to aquatic life (Abbasi and Abbasi 2011). One possible reason for acidification could be the discharge of industrial wastewater containing hazardous heavy metals and other chemicals. These acids can contaminate water and persist there for a long time period, altering the average pH of the oceanic waters. To analyze the influence of wastewater discharge on coral reefs, study was conducted around the coastal areas of Thailand by Reopanichkul et al. (2010). Samples collected from different coastal areas receiving untreated sewage in Phuket, Thailand, were analyzed. The study concluded that the coastal waters receiving untreated wastewater have high nutrient content which leads to algal bloom and depletes the oxygen level in water resulting in an unbalanced growth of aquatic organisms and also altering the percentage of coral reef communities. With increasing pollution, the percentage of coral cover and fish population decreases. It was observed that coastal waters with poor quality and high nutrient content have higher amount of dead coral and at the same time percentage of algae increases.

Another important aspect which is being considered for deterioration of aquatic life is nutrient enrichment of marine water. Agricultural runoffs and untreated sewage are contributing to the nutrient enrichment of marine water in an immense way and affecting the aquatic life. Nutrient enrichment causes a steep rise in the concentration of inorganic compounds like nitrate, ammonium, and phosphate. Long-term persistence causes a chemical imbalance of marine waters and increased algal growth, reducing oxygen levels and photosynthetic activity of aquatic plants, thus affecting their growth. Nutrient enrichment also results in bleaching of coral reefs resulting in the loss of coral covers globally (Wiedenmann et al. 2013).

2 Wastewater Treatment Strategies

With the rapid urbanization, there is also an increase in the exploitation of natural resources. The people from different parts of world showed concern in this area and organized various meetings and frame agendas for protection of environment. Different scholars showed deep concerns for the continuous depletion of natural resources and offered solutions by framing different principles. One such important concept is ecological wisdom which aims at exploring the available ecological knowledge and applying those principles for designing new projects in order to achieve a goal of sustainable development without doing any harm to the environment and also compromising the needs of coming generations. One such term ecosophy having a similar meaning to ecological wisdom was first used by Norwegian philosopher Arne Naess in 1973 which means the human responses toward Mother Nature (Wang et al. 2016). The principle of ecological wisdom can also be applied in designing wastewater treatments strategies using soil, plants, and microorganisms. Many such methods are currently being used such as wetlands, aquatic ponds, oxidation ponds, trickling filters, activated sludge method. Some of these methods used natural system, while others are used in constructed systems in a modified and controlled form. Treatment refers to as reducing the number of contaminants or pollutants in the wastewater before it can be released into the environment with least adverse impacts. Treatment process varies greatly and depends on factors like complexity of technology, cost, affordability, efficiency, and type of wastewater involved such as agricultural, industrial, or urban. There are various standards and guidelines framed by respective countries to maintain quality of water and for industries regarding permissible levels of toxic components in their effluents. The drinking water standards are mainly categorized into two: primary and secondary. Primary standards are enforceable by law and are meant to regulate potential harmful substances like organic, inorganic chemicals, microbes in order to ensure safe water supply to the public. Secondary standards on the other hand are not enforceable by law, mainly include color, odor, and hence are not a matter of major concern. The technology is changing with a rapid pace and is indirectly hampering the natural resources including water. The by-products and waste generated by these industries are adding xenobiotic components which are difficult to get degraded by natural means. Hence, there is a need to get advanced treatment techniques to tackle this continuously changing waste matter.

The type of wastewater treatment strategies depends upon the type of environment in which the wastewater would be released later on and what type of impact it could have on that particular environment (Laugesen et al. 2010). The wastewater treatment schemes are classified into different categories depending on the process involved and approach used to meet the required standards. The simplest type of classification is based on natural and advanced methods. But the worldwide accepted and most studied approaches categorized the treatment scheme into three main classes: primary, secondary, and tertiary schemes (Fig. 1). These schemes further consist of different types of physical, chemical, and biological treatment methods. With the increasing interest and research in this area, the components of above schemes are expanding at a faster pace and more novel options are available nowadays. The primary schemes emphasize physicochemical and mechanical approaches and do not have much impact on biological oxygen demand (BOD) and chemical oxygen demand (COD) of effluent. It is mainly aimed to reduce the bulkiness of the effluent by removing the contaminants which may hamper the further operation facility of wastewater treatment plant. The primary process mainly comprises of screening the effluent using bar screens, comminutors to chop solids present in effluent, grit chambers to remove gritty materials like eggshells, sands followed by primary clarifiers where effluent is given sufficient detention time to carry out gravity-aided separation. Chemicals may be added to clarifiers to increase the rate of settling to increase the competence of the process. Primary treatment is meant to remove only those pollutants which have the tendency to float or settle under the influence of gravity. Hence, there is a need to remove those pollutants which are in suspended form and are mainly comprised of oxygen-demanding substances. The process is followed by secondary treatment which is popularly known as biological treatment. The most commonly used biological treatment processes are trickling filters which is a fixed growth system and activated sludge system which is suspended growth system. Both the systems are used successfully

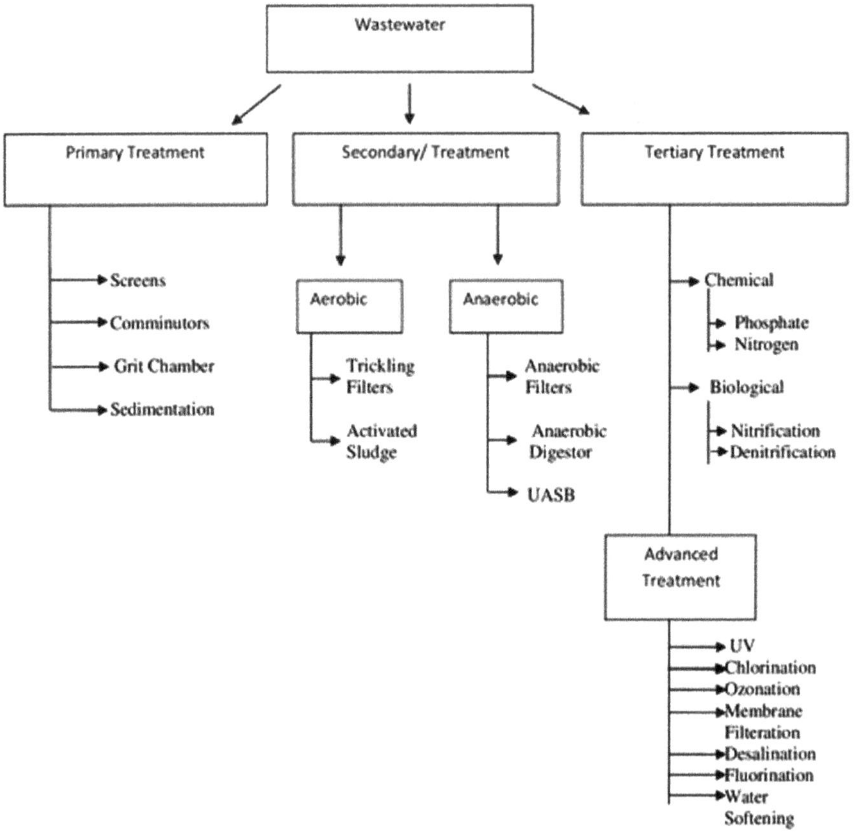

Fig. 1 Wastewater treatment scheme

for the effective removal of total suspended solids and dissolved BOD. The biological treatment depends on the operation of microbial community. Hence, desired results can be achieved by providing optimum growth conditions required for the present community. Several parameters are taken care to monitor the effect of the process like hydraulic retention time, residence time, food-to-microbe ratio, rate of organic load, recirculation ratio. The activated sludge process is the most researched area and is one of the oldest and main methods being used by the wastewater treatment industry.

The various modifications have been made in activated sludge method to have better outcomes. The efficiency of activated sludge system can be increased by several modifications in conventional type of facility like step aeration, extended aeration, mechanical aeration, and contact stabilization. With the advancement in technology, it is possible to modify the process by judiciously augmenting the sludge with acclimatized microbes and optimization of other parameters related to the process. The use of Anammox technology holds a great promise with better removal efficiency and increases energy production which can be utilized to recover the amount spent on running the facility (Kartal et al. 2010). The developed countries have better treatment options with centralized conventional activated sludge treatment (CAS) facilities. These systems have edge over conventional plants in terms of handling high organic load, treatment facilities and enhanced removal of total suspended solids (TSS), biological oxygen demand (BOD), and chemical oxygen demand (COD) substances (van Loosdrecht and Brdjanovic 2014). Apart from producing energy, these systems have the facilities to recover valuable resources from the wastewater. In the last few years, the research has been done in this area and thus it is possible to recover many valuable products, viz., cellulose fibers, bioplastics, biopolymers (Guest et al. 2009; Lin et al. 2010; Ruiken et al. 2013). The secondary effluent thus obtained can be further treated by advanced techniques like tertiary treatment followed by membrane filtration and disinfection to recover good-quality potable water. The secondary treatments were also categorized into aerobic and anaerobic processes. The most popularly used aerobic processes are already discussed above like trickling filters, activated sludge, rotating biological contactors, oxidation ditches. Among anaerobic methods are anaerobic filters, anaerobic digestors, upflow anaerobic sludge blanket reactor. The anaerobic methods have certain advantages over aerobic methods in terms of less biomass production and more energy production in the form of methane gas having high calorific value (Metcalf and Eddy 2003). The tertiary treatment scheme popularly known as effluent polishing is employed to remove plant nutrients like phosphorous and nitrogen which are responsible for eutrophication of water bodies. The tertiary treatment schemes employ physicochemical and biological process. The removal of nitrogen can be carried out by biological process which employs nitrification and denitrification with use of additional aeration and nitrifying and denitrifying bacteria. Another method involves ammonia stripping where pH of water is first raised to alkaline range leading to formation of ammonium ions followed by passing water at high speed through long towers which results in removal of nitrogen in the form of ammonia gas. The phosphorus removal is based on chemical treatment aided by addition of alkaline chemical, alum which results in precipitation (Nathanson 2000).

2.1 Natural Processes for Treatment of Wastewater

Natural treatment strategies involve the appropriate use of natural systems such as soil, plants, microorganisms where contaminants and pollutants are removed without the aid of any kind of chemicals. Natural treatment makes use of the following strategies to counter wastewater problem.

2.1.1 Wetlands

Wetlands comprise marshy areas comprising of swampy forest areas, salt marshes which can be natural, artificial, temporary, or permanent. The naturally occurring wetlands are rich in microflora and can be used for treatment of wastewaters. The constructed wetlands are engineered wetlands and are seen as future of wastewater treatment industry. Constructed wetlands are the areas with a certain depth having water in them and supporting the growth of various life forms including microorganisms and various plant forms like reeds, sedges. They have the edge over currently available options as they are cost-effective and easy to operate and maintain. The approach is aimed at utilizing the natural potential of microbes and plants to remove pollutants. The plant–microbe interactions played a major role in remediation of pollutants. The roots of plants provide surface for growth of bacteria, supply them sufficient nutrients for growth, and also enrich water bodies with oxygen fixed via photosynthesis. The bacteria residing in the roots carry out the removal or degradation of contaminants present in wastewater. These systems resemble oxidation ponds to some extent in their functioning. The most widely used wetland systems are: surface water flow system and subsurface water flow systems. Surface systems have depth of approx two feet having impervious layer of soil at bottom (Sanks 1976). Plants are rooted in layer of either sand or other materials at the base of the ground or remain in the floating stage. A range of vegetation can be used for surface type like emergent, submerged, or floating plants (Tilley et al. 2002). Pretreated water is employed as a source of effluent at surface wetland systems. In contrast to subsurface system, the land requirement is more in this case. This type of system is more suitable for removal of suspended solids. On the other hand, subsurface water flow systems have a base made up of impermeable material like sand and gravel. The water to be treated flows through the roots and stems of the plants growing in surface systems while in case of subsurface systems, wastewater flows through the plant material and is mainly aimed for removal of dissolved contaminants from wastewater. The subsurface type is further divided into two types: horizontal and vertical flow types. The different species of macrophytes belonging to emergent submerged and free-floating type are used in constructed wetlands. Among them, emergent plants are used widely in surface and subsurface flow wetlands. The most popular type of species used is *Phragmites australis, Cyperus papyrus, Typha latifolia, Scirpus validus* (Vyzmal 2013). In tropical and subtropical countries, use of ornamental species, e.g., *Iris pseudacorus,* has also been reported (Yan and Xu 2014). The example of type of plants and microbial species is given in Table 3. The wetlands can serve as future of wastewater industry as they are cost-effective, eco-friendly and have more potential to degrade/remove toxins which remain untreated in conventional wastewater facilities. One such contaminant is the pharmaceutical waste generated from domestic sources. The reason for this is the lack of any guidelines and separate collection system for expired drugs. The major attention was paid to the degradation of xenobiotic compounds and effluents generated from various industries. This part of the waste remains unnoticed as they are discharged along with regular domestic

Table 3 Plant and microbial community of wetlands (Zhang et al. 2016; Lamers et al. 2012; Pester et al. 2012; Stahl et al. 2007; Kolb and Horn 2012; Vymazal 2011; Fester 2013)

Type	Example
Species	
Emergent	*Phragmites* spp., *Typha* spp., *Scirpus* spp. *Iris* spp., *Juncus* spp., *Eleocharis* spp.
Submerged	*Hydrilla verticillata, Ceratophyllum demersum, Vallisneria natans*
Floating-leaved plants	*Nymphae tetragona, Nymphoides peltata, Marsilea quadrifolia*
Free-floating plants	*Salvinia natans, Lemna minor, Hydrocharis dubia*
Methanotrophs	*Methylococcus* spp., *Methylocystis* spp.
Denitrifiers	*Bradyrhizobium japonicum, Achromobacter xylosoxidans, Pseudomonas, Azospirillum lipoferum*
Nitrifying bacteria	*Nitrosomonas* spp., *Nitrosospira* spp., *Nitrobacter* spp., *Nitrospira* spp.
Sulfur-oxidizing bacteria	*Beggiatoa* spp., *Thiobacillus* spp., *Desulfovibrio* spp., *Desulfotomaculum* spp.
Iron-reducing bacteria	*Gallionella ferruginea*
Fungi	*Funneliformis mosseae, Rhizophagus irregularis, Chaetomium* spp., *Aspergillus* spp., *Trichoderma* spp., *Cladosporium* spp., *Emericallopsis* spp.

waste. The main problem associated with them is that they vary a lot in terms of their chemical nature, i.e., lipophilic, hydrophilic, hydrophobic, oxidizable, and reducible. They are considered as pseudo-persistent pollutants as they are being added to the environment at a continuous rate and thus pose an immediate threat to aquatic species and thus are difficult to remove by any single approach (Fent et al. 2006). They are more toxic as they remain there in effluent due to lack of any treatment scheme meant for their removal. Hence, wetlands can be used to remove these contaminants by the use of microbial and phytoremediation potential of microbes and plants. The main problem associated with use of constructed wetlands is the requirement of large area of land which is more as compared to conventional treatment methods. This is an efficient method for wastewater treatment in rural or remote areas having a small population (Wu et al. 2015).

2.1.2 Aquatic Plant Ponds

This kind of treatment makes use of aquatic plants of floating type including water hyacinth, duckweeds (Starkl et al. 2013), water clover, and Elodea. The depth of water in the pond is 1.6 to 6.0 ft. In general, removal of contaminants and pollutants from environment using plants is called phytoremediation. Specifically, floating aquatic plants grow rapidly in water bodies, help in the removal of metals, nitrogen, total suspended solids due to their high uptake and accumulation ability, and also reduce biological oxygen demand (BOD) significantly (UN Economic and Social Commission for Western Asia 2003). Furthermore, the plants provide shading to water surface which reduces the growth of algae. Using aquatic plants such as water hyacinth is a cost-effective and at the same time an eco-friendly method as compared to other technologies which involve use of chemicals (Rezania et al. 2015). In a study conducted by Rezania et al. (2015) the potential of water hyacinth was explored to treat wastewater coming from domestic sources. Experimental setup was established with all the parameters under analysis which were monitored for 3 weeks. Ten different parameters were analyzed including dissolved oxygen (DO), biological oxygen demand (BOD), chemical oxygen demand (COD), and total suspended solids (TSS). The results showed significant decrease of 38 to 96% in BOD, COD, and TSS when compared to different water quality standards such as Interim National Water Quality Standards (INWQS) and Water Quality Index (WQI) resulting in an improvement in the water quality. The above study suggested that the use of water hyacinth is a better method due to its biosorption and phytoremediation ability and also being cost-effective. Different aquatic plants have been tested for the removal of heavy metals present in wastewater like copper, cadmium, and selenium. In addition to this aquatic plant, duckweed is also a popular choice for treatment of wastewater due to its phytoremediation potential and its ability to convert nutrients present in wastewater to useful compounds. Zhao et al. (2014) studied this aspect of duckweed plant in comparison with water hyacinth in two pilot-scale plants for wastewater treatment in China. Both the plants were filled with duckweed and water hyacinth, and continuous monitoring of wastewater samples was done. In the study lasted for more than a year, it was found that the amount of total nitrogen and total phosphorus recovered was comparable to that of water hyacinth and wastewater nutrients were effectively converted to useful compounds.

2.2 *Emerging Methods*

Traditional methods such as chlorination and filtration can remove physical particles and other contaminants but generate a toxic sludge leading to its difficult disposal (Ferroudj et al. 2013; Anjum et al. 2016).

2.2.1 Activated Carbon

The activated carbon (AC) is the most widely used material in various fields due to its excellent adsorbent properties. These properties are contributed by finely developed internal pore structure, high surface area, different functional groups, viz., carbonyl, phenols, carboxyl (Bhatnagar et al. 2013). The functional groups present on the surface of activated carbon can be modified as per the requirement by various surface treatments like acid (Shen et al. 2008a, b) and base treatment

(Kierzek et al. 2006), impregnation (Bhatnagar et al. 2013), microwave (Ania et al. 2005), ozone (Zaror 1997), chemical treatment using cationic and anionic surfactants (Gao et al. 2009). The use of activated carbon for the removal of pollutants has been reported by Yang (2003). The type of modifications used for improving the efficiency of activated carbon along with their mechanism (Bhatnagar et al. 2013) is explained in Table 4. The potential of activated carbon and various modified methods have been reported widely for the removal of pollutants from the wastewater. Further studies are required to target specific type of pollutant by using selective modified activated carbon in wastewater by selecting specific industrial effluent. The main constraint in using activated carbon is its high cost, but more research in this field and using economically feasible method can make it a suitable treatment alternative for wastewater treatment.

Table 4 Types of modifications in activated carbon and their mechanism to treat wastewater

Treatment	Mechanism	References
Acid-modified AC	Oxidation of surface with nitric acid, sulfuric acid, and other oxidizing agents to increase the removal of cationic metals and dyes	Wang and Zhu (2007)
Base-modified AC	Treatment in hydrogen, ammonia, or inert atmosphere at elevated temperature which results in production of positive charge on the surface of activated carbon which helps in removal of negatively charged contaminants	Menéndez et al. (1996), Jurewicz et al. (2003)
Impregnation	Treatment with metals like iron, copper, silver to increase its adsorption capacity	Miyanga et al. (2002), Yeddou et al. (2011), Vaughan and Reed (2005)
Ozone treatment	Treatment with ozone and high temperature increase the strength of carbon and also modified its acid-base character	Bhatnagar et al. (2013)
Microwave treatment	Treatment induces enlargement of micropores and results in increased adsorption property	Liu et al. (2010)
Plasma treatment	Addition of oxygen free radicals to the surface of activated carbon on exposure to plasma under vacuum or in controlled environment containing air results in increase of adsorption	Bhatnagar et al. (2013)
Biological treatment	Trapping of bacteria in the activated carbon matrix to produce biological activated carbon (BAC)	Nathanson (2000)

2.2.2 Wastewater Treatment Using Nanoparticles

Nanoparticles are the materials having unique properties and a size of the nanoscale range of 1–100 nm which provides an advantage of large surface area (Zhang et al. 2008). Environmentalists are seeing a huge number of possibilities in nanoparticles for treating environmental pollution. One such area is wastewater treatment using metals-based nanoparticles. A large number of metals are being used to synthesize nanoparticles, e.g., TiO_2, iron oxide, gold and silver nanoparticles. These nanoparticles are proved to be very much effective in removing heavy metal contamination and to some extent radioactive compounds present in wastewater (Qu et al. 2013; Gehrke et al. 2015). These nanoparticles are advantageous as they have little adverse impact on the environment, low solubility and have a high surface area for adsorption, making it a very promising, efficient, and safe technology for wastewater treatment (Gupta et al. 2015; Anjum et al. 2016). Adsorption and removal of contaminants using nanoparticles are completed by adsorption on the outer surface and then its intra-particle diffusion along the walls (Qu et al. 2013; Trivedi and Axe 2000). Moreover, these nanoparticles can be compressed without undergoing any significant change in surface area for adsorption which provides an extra-added advantage for the industries which can produce them in the form of fine powder and pellets too (Qu et al. 2013; Gehrke et al. 2015). Also as the size of particles decreases, their adsorption capacity increases to great extent. It has been observed that a particle of 300 and 20 nm adsorbs the same amount of material but when the particles are compressed below 20 nm their adsorption capacity increases several times resulting in higher amount of contaminant removal from the wastewater. For example in case of wastewater having contamination of As^{3+}, nanoparticles of size below 20 nm adsorb three times more arsenic than 300 and 20 nm (Auffan et al. 2008; Kunduru et al. 2017). Huge amount of time and resources whether it is expertise or financial is being invested over the years to analyze and understand the effect of silver nanoparticles on the wastewater. Silver nanoparticles have various applications ranging from antimicrobial activity, therapeutics, drug delivery systems to treat wastewater treatment (Ahamed et al. 2010). Silver nanoparticles due to antimicrobial activity have been used to remove microorganisms present in wastewater. They are very stable and can be used at a wide range of pH (Kunduru et al. 2017).

Zinc oxide nanoparticles having a porous structure have been used to remove different types of heavy metal contamination like cadmium, mercury, nickel, lead, arsenic (Singh et al. 2013). All these heavy metals are adsorbed on the surface of nanoparticles by developing different interactions and are removed from wastewater. Manganese oxide nanoparticles similar to other nanoparticles offer a large surface area for adsorption of metal contaminants in wastewater. Manganese oxide-based nanoparticles are very much effective in the adsorption and removal of arsenic from wastewater (Wang et al. 2011; Anjum et al. 2016). Our planet has various forms of iron oxides present in its crust such as magnetite, hematite, maghemite. Iron oxides are excellent materials to be used for synthesis of nanoparticles and prove to be a novel and innovative technology for the treatment of wastewater. These particles possess unique properties such as size of the nanoscale range, high surface area for adsorption of material, surface modification, environment-friendly, and magnetism (Xu et al. 2012; Anjum et al. 2016). Iron oxides have great affinity for toxic metal contaminants like Cr^{3+}, Co^{2+}, Cd^{2+}, As^{3+}, Pb^{2+} (Gupta et al. 2015). As iron oxide-based nanoparticles possess magnetic properties, so these can be easily separated and recovered from the mixture under the influence of magnetic field (Kunduru et al. 2017). Due to their magnetism potential, they are also known as magnetic nanoparticles. Magnetic nanoparticles have proven to be a promising in situ technology for water treatment as these can be directly applied on the contaminated site and contaminants can be removed easily without any kind of harm to the environment (Jung et al. 2004). Gehrke et al. (2015) have reviewed that extensive studies related to environmental impact and toxicity of nanoparticles have already been carried out and those results can be employed for wastewater treatment. Showing the effectiveness and efficiency of magnetic nanoparticles in a study conducted by Hu et al. (2005) demonstrated that magnetic nanoparticles and their property to adsorb heavy metals can be explored for the removal of Cr^{6+} from wastewater where Cr^{6+} was easily removed and recovered by using magnetic field. However, simply adsorbing the contaminant onto the nanoparticles does not always result in an efficient removal. Various other factors like pH, temperature, concentration also affect the efficiency of nanoparticles. Metal-based nanoparticles can be easily regenerated by simply changing the pH of the solution without affecting their ability to adsorb pollutants on their surface (Hu et al. 2006; Sharma et al. 2009; Qu et al. 2013).

2.2.3 Carbon Nanotubes and Wastewater Cleansing

Nanotechnology has diverse applicability in multidisciplinary fields including medical science, drug delivery systems, food industry, electronics, and environment. Many nanostructures have been created using nanotechnology such as nanotubes, nanoadsorbents, nanoparticles, zeolites which have remarkable properties and are being used efficiently in various domains (Das et al. 2014). Carbon nanotube is one such nanostructure which is capable of treating and purifying wastewater to a great extent. Carbon nanotubes work on the surface phenomenon of adsorption in which molecules or compounds form a layer around the adsorbing material by forming different interactions. Carbon nanotubes are nanoscale structures having large surface area and possess adsorption sites for the adsorption of contaminants present in wastewater. These are cylindrical structures composed of graphite which is an allotropic form of carbon. It can be of two types, single-walled carbon nanotubes and multi-walled carbon nanotubes. Single-walled carbon nanotubes are composed of single layer of graphite sheet, whereas multi-walled nanotubes have multilayered graphite sheets (Gehrke et al. 2015; Das et al. 2014). These nanotubes are efficient in adsorbing metal ions including toxic heavy metals and organic contaminants even pharmaceutics, insecticides, dyes, etc., inorganic and biological contaminants. These contaminants adsorb onto the carbon nanotubes by developing different types of interactions such as hydrogen bonding, electrostatic interactions, van der Waals interactions with the variety of functional groups present on the adsorbing matrix such as –OH, COOH. Carbon nanotubes have been widely used for adsorbing heavy metals like Cu^{2+}, Cd^{2+}, Pb^{2+} (Qu et al. 2013). Major advantages that carbon nanotubes hold on as compared to other adsorbing technologies are high surface area, reusability, and multiple interactions making functional groups which helps to remove a variety of contaminants making it a diverse material. Carbon nanotubes are very efficient in removing heavy metal contamination from wastewater. Studies in the past have shown that carbon nanotubes with their surface being oxidized with various functional groups are very much efficient in adsorbing and removing heavy metals like cadmium, zinc, copper, lead (Lu et al. 2006; Qu et al. 2013). The hydrophobic nature of carbon nanotubes due to graphene makes them easy to adsorb and remove aromatic organic compounds (Das et al. 2014). Another potential application of carbon nanotubes has been explored of their use as a photocatalyst along with other nanomaterials to make a hybrid to remove contaminants. Photocatalysts use light energy and then covert it into chemical energy which further starts a series of chemical reactions to degrade pollutants (Lee and Park 2013). These can be used in combination with titanium oxide nanomaterials, zinc oxide nanomaterials, etc. This combination of carbon nanotubes and nanomaterials enhances the properties of carbon nanotubes manyfolds. It has increased surface area for adsorption, electrical conductivity, photocatalysis, and overall stability making them an excellent material for water purification. Carbon nanotube and nanomaterial hybrid remove contaminants and pollutants by making interactions with aromatic rings of grapheme in carbon nanotubes and contaminants. Carbon nanotubes are most commonly combined with titanium oxide due to its physical and chemical stability, inexpensiveness, and easy availability. These hybrids are synthesized by doping the metal nanomaterial over the carbon nanotube surface through different interactions including noncovalent and covalent interactions making a uniform surface. Photocatalysis is the phenomenon in which due to energy gap electrons travel from one energy level to other. This occurs when photons from light source hit carbon nanotube and metal hybrid and the electrons getting excited shifts from higher energy state to the lower energy state which is electron deficient creating a hole in higher energy level. These holes start a cascade of reaction in which they react with water and synthesizing hydroxyl radicals that degrades the pollutant adsorbed on carbon nanotube and metal hybrid (Das et al. 2014; Upadhyay et al. 2014). It can be used even in low light proving to be a great advantage, but studies are needed for establishing the photocatalysis technology on a large scale for wastewater treatment and its nonfunctionality in visible light (Lee and Park 2013). Nevertheless, carbon nanotubes have a great applicability in wastewater treatment, but disadvantage of its high production cost does not allow its use in industries on a large scale (De Volder et al. 2013; Kunduru et al. 2017). Carbon nanotubes can be regenerated by changing the pH to acidic range without losing their adsorbing capacity (Kunduru et al. 2017).

2.2.4 Microbial Fuel Cells

Wastewater which is considered as useless can actually be a very valuable resource if handled and treated suitably. To explore this unexplored potential of wastewater, environmentalist and scientists are trying to focus on microbial fuel cell technology. Microbial fuel cell is a relatively new technology whose functionality includes a combination of both biological and electrochemical processes (Ahn and Logan 2010).

Microbial fuel cell technology explores the ability of microorganisms which can colonize around the electrode either anode or cathode and transfer electrons from one electrode to other while treating the contaminants present in wastewater. Microbial fuel cell technology due to the following potential advantages (Fig. 2) can prove to be a promising technology in the field of wastewater treatment including clean generation of energy in the form of electricity and biomass. Furthermore, the technology produces less environmental impact and proves to be a sustainable method for treatment of wastewater. In addition to this, the technology is highly sensitive in generating and collecting real-time data (Li et al. 2014).

Microbial fuel cell technology is being tested for energy production from the organic and inorganic wastes present in wastewater by converting them into electricity.

The major shortcoming of the microbial fuel cell technology is its operating cost and efficiency in energy production. An extensive research input is still needed so that the technology can be applied on a large scale with minimum energy consumption and maximum energy production so that a sustainable wastewater

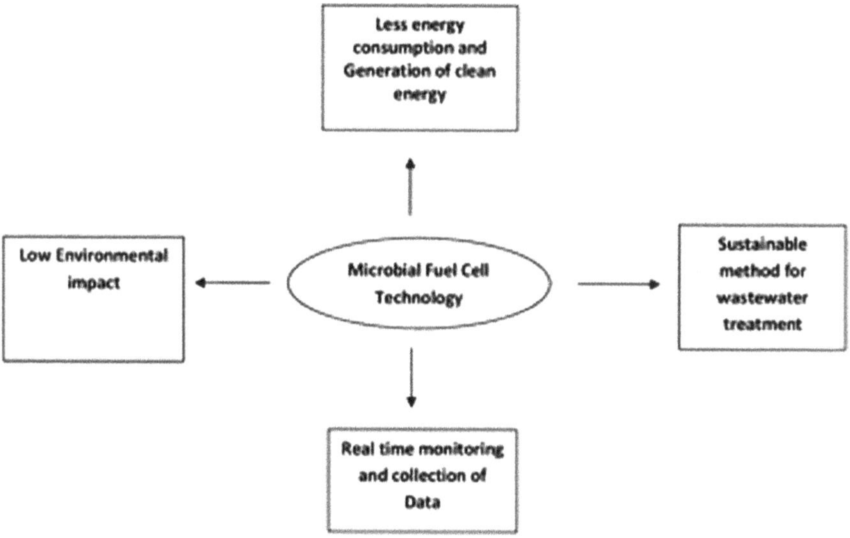

Fig. 2 Advantages of microbial fuel cell technology

treatment can be achieved. Li et al. (2014) reviewed that microbial fuel cell technology works better on the part that it consumes less energy as compared to other treatment processes as it can work well at ambient temperatures and hence less energy requirement will be there for temperature maintenance. Moreover, some studies have been conducted and results obtained were really positive in showing that the technology can be used to produce energy from wastewater. For example, energy of 0.026 kW h m^{-3} has been recovered from wastewater (Zhang et al. 2013a; Ge et al. 2013; Li et al. 2014).

This technology is being tested continuously with different types of contaminants. One such study conducted by Adelaja et al. (2017) targets the efficiency of microbial fuel cell technology against hydrocarbons like phenanthrene and benzene. Hydrocarbon degradation process was tested for 155 days with microbial fuels cells designed according to ex situ and in situ conditions. In situ study showed removal of phenanthrene and benzene and electricity generation up to 6.75 mWm^{-2}. In case of ex situ method, efficient removal of hydrocarbons occurred suggesting that microbial fuel cell technology can prove to be promising technology for cleanup of effluents coming from refinery or wastewater contaminated with hydrocarbons. Removal of inorganic compounds like nitrogen can also be done using this technology.

Microbial fuel cell technology is applied to treat and clean environment's essential part of water. The technology is very much able to efficiently remove various contaminants such as organic, sulfur-based contaminants, and various metals while maintaining the quality standards of water so that it can be reused for different anthropogenic purposes. In addition to efficiency, microbial fuel cell technology releases less CO_2, thus contributes lesser greenhouse gases, and also produces less sludge (Li et al. 2014). Microbial fuel cells produce electricity directly from organic material in which the microbial biofilm oxidizes organic waste and transfers electrons to the electrode. Different studies have been conducted in the recent times to check the applicability of the technology even with some modifications or in combination with other techniques making a hybrid all over the world. For example, in a study conducted by Sevda et al. (2013) microbial fuel cells combined with air cathode have been used against molasses mixed sewage water collected from a treatment plant in Belgium for electricity generation. Original wastewater and half diluted and centrifuged wastewater sample were used, and system was run for 160 days. Maximum electricity generation of about 762 mV and a significant COD cleanup of approximately 60% were recorded with original wastewater followed by half diluted and centrifuged wastewater samples. Results of the study showed that air cathode microbial fuel cell technology is a promising technique for both electricity generation and wastewater treatment. A review of Pandey et al. (2016) focuses on the ability of some microorganisms to clean wastewater and also at the same time produces electricity using different substrates adding to the diverse treatment potential of microbial cell technology as different types of contaminants are each day added to the water through anthropogenic activities. However, this technology is very much at its primary stage so its efficiency and environmental impact are yet to be established and a scale-up of this technique is needed. One major drawback of this technology is the operational cost involved.

2.2.5 Molecularly Imprinted Polymers (MIPs)

The available treatment methods are unable to remove certain type of pollutants which are present in very low concentration, thus escape the treatment process, and began to accumulate in the environment. As discussed in wetland treatment, certain pharmaceutical components belong to this category. In order to degrade such components, specific approach is necessary. One such selective emerging technique is use of MIPs. The template is linked to the monomers carrying specific functional groups to form a complex which will be polymerized along with a selective cross-linker to form molecularly imprinted polymers. The removal of template results in production of cavities having complimentary shape and functional groups (Shen et al. 2012; Advincula 2011; Huang et al. 2015). They have the edge over other techniques in terms of specificity, cost-effectiveness, stability at high temperature and pressure (Bui and Haupt 2010). MIPs are reported to be utilized for various quantification, separation, and catalysis studies (Byun et al. 2010; Sueyoshi et al. 2010; Strikovsky et al. 2003). The adsorption potential of MIPs was reported for the removal of different pollutants present in trace amounts like Diclofenac (Dai et al. 2011), polyaromatic hydrocarbons (Krupadam et al. 2010), and estrogen compounds (Zhongbo and Hu 2008). Another application of MIPs is their use as catalysts. Shen et al. (2008b) reported the use of MIP-coated TiO_2 as photocatalyst for successful removal of 2-nitrophenol and 4-nitrophenol which was more selective and efficient than conventional TiO_2. The currently available MIPs still suffered from many drawbacks like binding affinity, template removal (Huang et al. 2015) which can be overcome by using MIP-based composites using quantum dots, nanomaterials, and magnetic particles (Chen et al. 2014; Mehdinia et al. 2013). These composites possess the property of both MIP and the other associated matrix. Moreover, the adsorption sites are present at the surface of polymers which gives it better affinity and selectivity toward the corresponding analyte. Composite of Fe_3O_4 and MIPs was successfully used for the determination of estrogens and BPA in water sample by Guo et al. (2011) and Zhang et al. (2013b) who studied the removal of estrogens using MIP/TiO_2 nanotubes. More such studies are required to exploit the potential of MIPs in treatment of wastewater to target specific contaminants which cannot be removed by other available approaches.

2.2.6 Fenton's Process in Combination with Cavitation

Different oxidation processes have been used to remove different types of pollutants present in wastewater. It can degrade or convert pollutants with long-chain chemistry into short-chain compounds which are easy to handle conventional methods. (Gogate and Pandit 2004a, b; Bagal and Gogate 2014). Fenton's process is one such oxidation method which is being used to treat wastewater with toxic organic and inorganic contaminants. It produces hydroxyl radicals from hydrogen peroxide. Ferrous ion acts as a catalyst in the process. This process is advantageous in having a range of pollutants that can be removed like pesticides, insecticides, pollutants from refinery, rubber, and plastic industries (Pera-Titus et al. 2004). However, for an efficient process to be carried out different parameters have to be strictly monitored like concentration of hydrogen peroxide, catalyst amount, pollutant concentration, temperature, and pH. pH must be maintained within the range of 2–4 to gain maximum efficiency of the process. Maintenance of all these parameters and high cost of hydrogen peroxide sometimes prove to be disadvantageous, and also its applicability to most of the compounds like carbon tetrachloride and acetic acid is still not well established. Taking the above disadvantages into consideration, some modifications in Fenton's process are being studied to improve the efficiency of removal of contaminants. One such modification is to use Fenton's process in combination with cavitation process. In this combination, Fenton's process is used with ultrasound waves to generate a large number of hydroxyl radicals which will improve the degradation and conversion rate of organic compounds present in wastewater (Bagal and Gogate 2014). Cavitation in addition to ultrasound waves' effect also removes the contaminants with the help of pyrolysis as some organic compounds are not affected by hydroxyl radicals.

3 Conclusion

Currently, there is a wide choice of methods available for the removal of contaminants from the wastewater ranging from principles based on either natural or anthropogenic and combination of these two methods. The promising development has been made in this area, and more research is being focused on this area with increasing concerns raised by different sectors of the world toward protection of the environment. The current study aimed to review the modern and traditional methods available till date for the treatment of wastewaters. The main problem associated with most of the contaminants is their accumulation in environment. Such contaminants remain there in the treated effluent due to lack of necessary treatment technique in conventional methods. One of the reasons for the accumulation is that sometimes they are modified to some secondary form and hence difficult to evaluate their presence and treatment. The basic understanding of the nature of contaminants present in various effluents will further help in increasing the efficiency of the available techniques. The efficacy of treatment techniques can be increased by looking for the lacunae present in the available methods, and required amendments should be made, also by adopting hybrid methods to have better results. The other important area of research is related to movement of contaminants, possible transformations, accumulation in the different environmental conditions. The developments made in the wastewater studies and technologies worldwide should be shared on global platform in order to develop an informative database which will help in boosting the research and development of eco-friendly and innovative methods with minimal drawbacks. The regulatory agencies should be strengthened, and strict regulation should be made for the defaulters. The water management is the most critical issue and hence can be solved by collective efforts of all the sectors, i.e., public, industrial, research, governmental and non-governmental organizations. There is a need to regularly assess the efficacy of available methods and also to add innovative treatment schemes on the same time to have desired results in this area.

References

Abbasi T, Abbasi SA (2011) Ocean acidification: the newest threat to the global environment. Crit Rev Environ Sci Technol 41(18):1601–1663

Adelaja O, Keshavarz T, Kyazze G (2017) Treatment of phenanthrene and benzene using microbial fuel cells operated continuously for possible in situ and ex situ applications. Int Biodeterior Biodegradation 116:91–103

Advincula RC (2011) Engineering molecularly imprinted polymer (MIP) materials: developments and challenges for sensing and separation technologies. Korean J Chem Eng 28(6):1313–1321

Ahamed M, AlSalhi MS, Siddiqui MKJ (2010) Silver nanoparticle applications and human health. Clinica chimica acta 411(23):1841–1848

Ahn Y, Logan BE (2010) Effectiveness of domestic wastewater treatment using microbial fuel cells at ambient and mesophilic temperatures. Biores Technol 101(2):469–475

Akpor OB, Muchie B (2011) Environmental and public health implications of wastewater quality. Afr J Biotech 10(13):2379–2387

Andersson AJ, Gledhill D (2013) Ocean acidification and coral reefs: effects on breakdown, dissolution, and net ecosystem calcification. Annu Rev Mar Sci 5:321–348

Ania CO, Parra JB, Menendez JA, Pis JJ (2005) Effect of microwave and conventional regeneration on the microporous and mesoporous network and on the adsorptive capacity of activated carbons. Microporous Mesoporous Mater 85(1):7–15

Anjum M, Miandad R, Waqas M, Gehany F, Barakat MA (2016) Remediation of wastewater using various nano-materials. Arab J Chem

Auffan M, Rose J, Proux O, Borschneck D, Masion A, Chaurand P, Hazemann JL, Chaneac C, Jolivet JP, Wiesner MR, Van Geen A (2008) Enhanced adsorption of arsenic onto maghemites nanoparticles: as (III) as a probe of the surface structure and heterogeneity. Langmuir 24(7): 3215–3222

Bagal MV, Gogate PR (2014) Wastewater treatment using hybrid treatment schemes based on cavitation and Fenton chemistry: a review. Ultrason Sonochem 21(1):1–14

Bhatnagar A, Hogland W, Marques M, Sillanpää M (2013) An overview of the modification methods of activated carbon for its water treatment applications. Chem Eng J 219:499–511

Bui BTS, Haupt K (2010) Molecularly imprinted polymers: synthetic receptors in bioanalysis. Anal Bioanal Chem 398(6):2481–2492

Byun HS, Youn YN, Yun YH, Yoon SD (2010) Selective separation of aspirin using molecularly imprinted polymers. Sep Purif Technol 74(1):144–153

Chen H, Kong J, Yuan D, Fu G (2014) Synthesis of surface molecularly imprinted nanoparticles for recognition of lysozyme using a metal coordination monomer. Biosens Bioelectron 53:5–11

Dai CM, Geissen SU, Zhang YL, Zhang YJ, Zhou XF (2011) Selective removal of diclofenac from contaminated water using molecularly imprinted polymer microspheres. Environ Pollut 159(6): 1660–1666

Das R, Hamid SBA, Ali ME, Ismail AF, Annuar MSM, Ramakrishna S (2014) Multifunctional carbon nanotubes in water treatment: the present, past and future. Desalination 354:160–179

De Volder MF, Tawfick SH, Baughman RH, Hart AJ (2013) Carbon nanotubes: present and future commercial applications. Science 339(6119):535–539

Eddy MI (2003) Wastewater engineering: treatment and reuse, 4th edn. McGraw Hill Higher Education

Fent K, Weston AA, Caminada D (2006) Ecotoxicology of human pharmaceuticals. Aquat Toxicol 76(2):122–159

Ferroudj N, Nzimoto J, Davidson A, Talbot D, Briot E, Dupuis V, Bée A, Medjram MS, Abramson S (2013) Maghemite nanoparticles and maghemite/silica nanocomposite microspheres as magnetic Fenton catalysts for the removal of water pollutants. Appl Catal B 136: 9–18

Fester T (2013) Arbuscularmycorrhizal fungi in a wetland constructed for benzene-, methyl tert-butyl ether- and ammonia-contaminated groundwater bioremediation. Microb Biotechnol 6(1):80–84

Gao R, Hu Z, Chang X, He Q, Zhang L, Tu Z, Shi J (2009) Chemically modified activated carbon with 1-acylthiosemicarbazide for selective solid-phase extraction and preconcentration of trace Cu (II), Hg (II) and Pb (II) from water samples. J Hazard Mater 172(1):324–329

Ge Z, Li J, Xiao L, Tong Y, He Z (2013) Recovery of electrical energy in microbial fuel cells: brief review. Environ Sci Technol Lett 1(2):137–141

Gehrke I, Geiser A, Somborn-Schulz A (2015) Innovations in nanotechnology for water treatment. Nanotechnol Sci Appl 8:1

Gogate PR, Pandit AB (2004a) A review of imperative technologies for wastewater treatment I: oxidation technologies at ambient conditions. Adv Environ Res 8(3):501–551

Gogate PR, Pandit AB (2004b) A review of imperative technologies for wastewater treatment II: hybrid methods. Adv Environ Res 8(3):553–597

Guest JS, Skerlos SJ, Barnard JL, Beck MB, Daigger GT, Hilger H, Jackson SJ, Karvazy K, Kelly L, Macpherson L, Mihelcic JR (2009) A new planning and design paradigm to achieve sustainable resource recovery from wastewater

Guo W, Hu W, Pan J, Zhou H, Guan W, Wang X, Dai J, Xu L (2011) Selective adsorption and separation of BPA from aqueous solution using novel molecularly imprinted polymers based on kaolinite/Fe_3O_4 composites. Chem Eng J 171(2):603–611

Gupta VK, Tyagi I, Sadegh H, Shahryari-Ghoshekandi R, Makhlouf ASH, Maazinejad B (2015) Nanoparticles as adsorbent; a positive approach for removal of noxious metal ions: a review. Sci Technol Dev 34:195–214

Hu J, Chen G, Lo IM (2005) Removal and recovery of Cr (VI) from wastewater by maghemite nanoparticles. Water Res 39(18):4528–4536

Hu J, Chen G, Lo IM (2006) Selective removal of heavy metals from industrial wastewater using maghemite nanoparticle: performance and mechanisms. J Environ Eng 132(7):709–715

Huang DL, Wang RZ, Liu YG, Zeng GM, Lai C, Xu P, Lu BA, Xu JJ, Wang C, Huang C (2015) Application of molecularly imprinted polymers in wastewater treatment: a review. Environ Sci Pollut Res 22(2):963–977

Nathanson JA (2000) Basic environmental technology: water supply, waste management, and pollution control. Prentice-Hall

Jung JY, Chung YC, Shin HS, Son DH (2004) Enhanced ammonia nitrogen removal using consistent biological regeneration and ammonium exchange of zeolite in modified SBR process. Water Res 38(2):347–354

Jurewicz K, Babeł K, Źiółkowski A, Wachowska H (2003) Ammoxidation of active carbons for improvement of supercapacitor characteristics. Electrochimica Acta 48(11):1491–1498

Kanu I, Achi OK (2011) Industrial effluents and their impact on water quality of receiving rivers in Nigeria. J Appl Technol Environ Sanit 1(1):75–86

Kartal B, Kuenen JG, Van Loosdrecht MCM (2010) Sewage treatment with anammox. Science 328(5979):702–703

Kierzek K, Kowalczuk A, Machnikowska H, Machnikowski J (2006) In: Carbon 2006. Aberdeen

Kolb S, Horn MA (2012) Microbial CH_4 and N_2O consumption in acidic wetlands. Front Microbiol 3(78):10–3389

Krupadam RJ, Khan MS, Wate SR (2010) Removal of probable human carcinogenic polycyclic aromatic hydrocarbons from contaminated water using molecularly imprinted polymer. Water Res 44(3):681–688

Kunduru KR, Nazarkovsky M, Farah S, Pawar RP, Basu A, Domb AJ (2017) Nanotechnology for water purification: applications of nanotechnology methods in wastewater treatment. In Water Purification, pp 33–74

Lamers LP, Van Diggelen JM, Op Den Camp HJ, Visser EJ, Lucassen EC, Vile MA, Jetten MS, Smolders AJ, Roelofs JG (2012) Microbial transformations of nitrogen, sulfur, and iron dictate vegetation composition in wetlands: a review. Front Microbiol 3:156

Laugesen CH, Fryd O, Koottatep T, Brix H (2010) Sustainable wastewater management in developing countries: new paradigms and case studies from the field. ASCE Press, Virginia, USA

Lee SY, Park SJ (2013) TiO_2 photocatalyst for water treatment applications. J Ind Eng Chem 19(6):1761–1769

Li WW, Yu HQ, He Z (2014) Towards sustainable wastewater treatment by using microbial fuel cells-centered technologies. Energy Environ Sci 7(3):911–924

Lin Y, de Kreuk M, Van Loosdrecht MCM, Adin A (2010) Characterization of alginate-like exopolysaccharides isolated from aerobic granular sludge in pilot-plant. Water Res 44(11): 3355–3364

Liu QS, Zheng T, Li N, Wang P, Abulikemu G (2010) Modification of bamboo-based activated carbon using microwave radiation and its effects on the adsorption of methylene blue. Appl Surf Sci 256(10):3309–3315

Lu C, Chiu H, Liu C (2006) Removal of zinc (II) from aqueous solution by purified carbon nanotubes: kinetics and equilibrium studies. Ind Eng Chem Res 45(8):2850–2855

Mehdinia A, Kayyal TB, Jabbari A, Aziz-Zanjani MO, Ziaei E (2013) Magnetic molecularly imprinted nanoparticles based on grafting polymerization for selective detection of 4-nitrophenol in aqueous samples. J Chromatogr A 1283:82–88

Menéndez JA, Phillips J, Xia B, Radovic LR (1996) On the modification and characterization of chemical surface properties of activated carbon: in the search of carbons with stable basic properties. Langmuir 12(18):4404–4410

Miyanaga S, Hiwara A, Yasuda H (2002) Preparation and high bacteriostatic action of the activated carbons possessing ultrafine silver particles. Sci Technol Adv Mater 3(2):103–109

Pandey P, Shinde VN, Deopurkar RL, Kale SP, Patil SA, Pant D (2016) Recent advances in the use of different substrates in microbial fuel cells toward wastewater treatment and simultaneous energy recovery. Appl Energy 168:706–723

Pera-Titus M, García-Molina V, Baños MA, Giménez J, Esplugas S (2004) Degradation of chlorophenols by means of advanced oxidation processes: a general review. Appl Catal B Environ 47(4):219–256

Pester M, Knorr KH, Friedrich MW, Wagner M, Loy A (2012) Sulfate-reducing microorganisms in wetlands–fameless actors in carbon cycling and climate change. Microbiol Wetlands:45

Qu X, Alvarez PJ, Li Q (2013) Applications of nanotechnology in water and wastewater treatment. Water Res 47(12):3931–3946

Ren Z, Chon TS, Xia C, Li F (2017) The monitoring and assessment of aquatic toxicology. Biomed Res Int 2017:9179728

Reopanichkul P, Carter RW, Worachananant S, Crossland CJ (2010) Wastewater discharge degrades coastal waters and reef communities in southern Thailand. Mar Environ Res 69(5): 287–296

Rezania S, Ponraj M, Talaiekhozani A, Mohamad SE, Din MFM, Taib SM, Sabbagh F, Sairan FM (2015) Perspectives of phytoremediation using water hyacinth for removal of heavy metals, organic and inorganic pollutants in wastewater. J Environ Manage 163:125–133

Ruiken CJ, Breuer G, Klaversma E, Santiago T, Van Loosdrecht MCM (2013) Sieving wastewater–cellulose recovery, economic and energy evaluation. Water Res 47(1):43–48

Sanks RL (1976) Land treatment and disposal of municipal and industrial wastewater. Ann Arbor Sci

Sevda S, Dominguez-Benetton X, Vanbroekhoven K, De Wever H, Sreekrishnan TR, Pant D (2013) High strength wastewater treatment accompanied by power generation using air cathode microbial fuel cell. Appl Energy 105:194–206

Sharma YC, Srivastava V, Singh VK, Kaul SN, Weng CH (2009) Nano-adsorbents for the removal of metallic pollutants from water and wastewater. Environ Technol 30(6):583–609

Shen W, Li Z, Liu Y (2008a). Surface chemical functional groups modification of porous carbon. Recent Pat Chem Eng 1(1):27–40

Shen X, Zhu L, Liu G, Yu H, Tang H (2008b) Enhanced photocatalytic degradation and selective removal of nitrophenols by using surface molecular imprinted titania. Environ Sci Technol 42 (5):1687–1692

Shen X, Zhu L, Wang N, Ye L, Tang H (2012) Molecular imprinting for removing highly toxic organic pollutants. Chem Commun 48(6):788–798

Singh S, Barick KC, Bahadur D (2013) Fe_3O_4 embedded ZnO nanocomposites for the removal of toxic metal ions, organic dyes and bacterial pathogens. J Mater Chem A 1(10):3325–3333

Stahl DA, Lpy A, Wagner M (2007) Molecular strategies for studies of natural population of sulphate reducing microorganisms. In: Barton LL, Hamilton WA (eds) I Sulphate-Reducing Bacteria. Cambridge University Press, Cambridge, pp 39–115

Starkl M, Amerasinghe P, Essl L, Jampani M, Kumar D, Asolekar SR (2013) Potential of natural treatment technologies for wastewater management in India. J Water Sanit Hygiene Dev 3(4): 500–511

Strikovsky A, Hradil J, Wulff G (2003) Catalytically active, molecularly imprinted polymers in bead form. React Funct Polym 54(1):49–61

Sueyoshi Y, Fukushima C, Yoshikawa M (2010) Molecularly imprinted nanofiber membranes from cellulose acetate aimed for chiral separation. J Membr Sci 357(1):90–97

Tilley DR, Badrinarayanan H, Rosati R, Son J (2002) Constructed wetlands as recirculation filters in large-scale shrimp aquaculture. Aquacult Eng 26(2):81–109

Trivedi P, Axe L (2000) Modeling Cd and Zn sorption to hydrous metal oxides. Environ Sci Technol 34(11):2215–2223

UNESCO (2012) Managing water under uncertainty and risk. The United Nations World Water Development Report 4. United Nations Educational, Scientific and Cultural Organization, Paris

UN Economic Social Commission for Western Asia (2003) Governance for sustainable development in the Arab region: institutions and instruments for moving beyond and environmental management culture. United Nations, New York

UN Water (2015) Wastewater management-a un-water analytical brief. New York

Upadhyay RK, Soin N, Roy SS (2014) Role of graphene/metal oxide composites as photocatalysts, adsorbents and disinfectants in water treatment: a review. Rsc Adv 4(8): 3823–3851

van Loosdrecht MC, Brdjanovic D (2014) Anticipating the next century of wastewater treatment. Science 344(6191):1452–1453

Vaughan RL, Reed BE (2005) ModelingAs (V) removal by a iron oxide impregnated activated carbon using the surface complexation approach. Water Res 39(6):1005–1014

Vymazal J (2011) Plants used in constructed wetlands with horizontal subsurface flow: a review. Hydrobiologia 674(1):133–156

Vymazal J (2013) Emergent plants used in free water surface constructed wetlands: a review. Ecol Eng 61:582–592

Wang HQ, Yang GF, Li QY, Zhong XX, Wang FP, Li ZS, Li YH (2011) Porous nano-MnO_2: large scale synthesis via a facile quick-redox procedure and application in a supercapacitor. New J Chem 35(2):469–475

Wang S, Zhu ZH (2007) Effects of acidic treatment of activated carbons on dye adsorption. Dyes Pigm 75(2):306–314

Wang X, Palazzo D, Carper M (2016) Ecological wisdom as an emerging field of scholarly inquiry in urban planning and design. Landscape Urban Plan 155:100–107

Wiedenmann J, D'Angelo C, Smith EG, Hunt AN, Legiret FE, Postle AD, Achterberg EP (2013) Nutrient enrichment can increase the susceptibility of reef corals to bleaching. Nature Clim Change 3(2):160–164

World Health Organization (WHO) (2015) Drinking water. Fact sheet no. 391

Wu H, Zhang J, Ngo HH, Guo W, Hu Z, Liang S, Fan J, Liu H (2015) A review on the sustainability of constructed wetlands for wastewater treatment: design and operation. Biores Technol 175:594–601

Xu P, Zeng GM, Huang DL, Feng CL, Hu S, Zhao MH, Lai C, Wei Z, Huang C, Xie GX, Liu ZF (2012) Use of iron oxide nanomaterials in wastewater treatment: a review. Sci Total Environ 424:1–10

Yan Y, Xu J (2014) Improving winter performance of constructed wetlands for wastewater treatment in northern China: a review. Wetlands 34(2):243–253

Yang RT (2003) Adsorbents: fundamentals and applications. John Wiley & Sons

Yeddou AR, Chergui S, Chergui A, Halet F, Hamza A, Nadjemi B, Ould-Dris A, Belkouch J (2011) Removal of cyanide in aqueous solution by oxidation with hydrogen peroxide in presence of copper-impregnated activated carbon. Miner Eng 24(8):788–793

Zaror CA (1997) Enhanced oxidation of toxic effluents using simultaneous ozonation and activated carbon treatment. J Chem Technol Biotechnol 70(1):21–28

Zhang D, Luo J, Lee ZMP, Gersberg RM, Liu Y, Tan SK, Ng WJ (2016) Ibuprofen removal in horizontal subsurface flow constructed wetlands: treatment performance and fungal community dynamics. Environ Technol 37(12):1467–1479

Zhang F, Ge Z, Grimaud J, Hurst J, He Z (2013a) Long-term performance of liter-scale microbial fuel cells treating primary effluent installed in a municipal wastewater treatment facility. Environ Sci Technol 47(9):4941–4948

Zhang W, Li Y, Wang Q, Wang C, Wang P, Mao K (2013b) Performance evaluation and application of surface-molecular-imprinted polymer-modified TiO_2 nanotubes for the removal of estrogenic chemicals from secondary effluents. Environ Sci Pollut Res 20(3):1431–1440

Zhang L, Gu FX, Chan JM, Wang AZ, Langer RS, Farokhzad OC (2008) Nanoparticles in medicine: therapeutic applications and developments. Clin Pharmacol Ther 83(5):761–769

Zhao Y, Fang Y, Jin Y, Huang J, Bao S, Fu T, He Z, Wang F, Zhao H (2014) Potential of duckweed in the conversion of wastewater nutrients to valuable biomass: a pilot-scale comparison with water hyacinth. Biores Technol 163:82–91

Zhongbo Z, Hu J (2008) Selective removal of estrogenic compounds by molecular imprinted polymer (MIP). Water Res 42(15):4101–4108

The Role of Microbes to Improve Crop Productivity and Soil Health

Akhilesh Kumar and Jay Prakash Verma

Abstract Soil microbes are the most important candidature for enhancing soil fertility and health. The plant growth promoting microbes and arbuscular mycorrhizae (AM) are used for enhancing plant growth and yields of agricultural crops under normal and stress conditions. It improves plant growth on various physiological parameters of plant in response to external stimuli by a number of different mechanisms. The mechanisms involved in growth promotion include plant growth regulators, production of different metabolites, and conversion of atmospheric nitrogen into ammonia, etc., by direct and indirect ways. In addition, it also provides resistance against biotic components (pathogens) through induced systemic resistance (ISR) and systemic acquired resistance (SAR). Plant microbe's interaction contributing in plant growth promotion and disease control under changing environment and enabling more sustainable agriculture without compromising ecosystem functioning. Plant growth regulators maintain beneficial plant–microbe interactions such as interaction between plant growth promoting rhizobacteria (PGPR) and arbuscular fungi. The microbial diversity in rhizospheric soil maintains soil health and productivity. Thus, the inclusive use of plant growth promoting rhizobacteria helps to get sustainable agriculture production under normal as well stress condition. This review highlights an overview of the beneficial effect of microbes for enhancing sustainable agricultural production.

Keywords Soil microbes · Rhizobacteria · Sustainability · Soil health
Mycorrhizae

A. Kumar · J. P. Verma (✉)
Institute of Environment and Sustainable Development,
Banaras Hindu University, Varanasi 221005, UP, India
e-mail: jpv.iesd@bhu.ac.in

A. Kumar
e-mail: verma_bhu@yahoo.co.in

V. Achal and A. Mukherjee (eds.), *Ecological Wisdom Inspired Restoration Engineering*, EcoWISE, https://doi.org/10.1007/978-981-13-0149-0_14

1 Introduction

Due to continuous increase in global human population, urbanization and industrialization have reduced the arable agricultural land. These changes have been severely affected by the demand of food and supply ratio, whereas continuous increasing world population needs a substantial increase in agriculture produce to fulfill the present demand (Tilman et al. 2011). However, the conventional agriculture is facing reduction in production and increased in cost of input. In addition, loss of agriculture productivity due to natural and anthropogenic activity leads that land degradation and reduced crop yield. Land use pattern shift varies frequently due to modernization and urbanization, hence reduces arable land. Farmers are also leaving this practice because of low-cost benefits and introduction of different variety of seed and technology. The higher cost of agriculture input affects farmers' interest and takes them away from agriculture. Both of these factors severely affect the agriculture practices and produce since agriculture provides major share in our country income and grass domestic production (GDP). Above fifty percent Indian population depends on agriculture. Further, intensive use of chemical fertilizers and pesticides for higher crop production is also detrimental for soil and food quality (Kang et al. 2016). In spite of anthropogenic causes, climate change is another cause for crop productivity. Plants are exposed to various abiotic and biotic changes. As world's human population continues to increase, the demand for agriculture supply will be a challenging for agrarian country, including India. In order to fulfill the demand of population, farmers use inorganic chemical and pesticides to enhance yield. The successful implementation of green revolution in India has been result food security and plays a significant role in Indian economy. The chemical fertilizers have played a significant place in increasing agriculture productivity. However, these chemicals and pesticides have detrimental impact on soil fertility (Singh 2015). The major challenge in agriculture science is to develop a technique which increases crop yield sustainably. In recent years, due to the major negative consequences of inorganic fertilizers and pesticides, many researchers are focusing their attention on finding solution for reducing these effects. And this is where concept of "ecological wisdom" comes to guide us to look for alternative to improve crop production and soil health. Using an ever-increasing information base and understanding human "needs" and interactions between humans and nature, some planners, managers, and scientists have developed guidance for managing social–ecological systems using the concept of ecological wisdom (Patten 2016). Ecological wisdom can guide adaptive management, which then is applied to restoration or management of soil ecosystem. In ecologically meaningful engineering, projects are developed in harmony with the existing ecosystems for overall environmental benefit (Achal et al. 2016). Similarly, microbial-based fertilizers can be developed for the benefits of soils while maintaining sustainability.

Application of beneficial microbes may be a potential alternative to harmful chemical fertilizers and pesticides. Microbes stand an important role in improving crop productivity and soil management. Plant-associated soil microbes play a crucial role in plant growth and development such as nutrient cycling and crop productivity (Yan et al. 2015). Soil microbial dynamics determine the potentiality of soil crop productivity. While the interaction of plant and microbes is major factor for controlling ecosystem functioning, these plant–microbes interactions vary greatly and depend upon availability of nutrient and. Plant growth promoting microbes mostly used for plant growth promotion through various means such as plant growth regulation and nitrogen fixation (Ahmad et al. 2008). In addition, it has also been reported that PGP microbes induce plant growth under stress condition such as drought and salinity. These microbes induce plant growth through regulating cell division, enlargement, and differentiation. However, complex interactions at different level in plant involve such as genetics, physiological, ecological, and morphological events (Islam et al. 2014). These interactions vary and depend upon plant and microbes. Nowadays to make agriculture more sustainable and efficient, an alternative method is needed to introduce in agriculture system. To feed the increasing world population novel and potential approach should need to boot agriculture productivity sustainably. Food security is one of the serious challenges due to indiscriminate increase of human population worldwide. In spite, the advanced technology in agriculture practice and use of fertilizers and pesticides causes degradation of soil fertility. In addition, the pathogens cause detrimental impact on crop productivity. Therefore, now a biological control major is needed to advance our agriculture system to overcome or minimize above consequences. It is both sustainable and cost-effective approach for future. Improvement in agriculture is done through the application of Plant growth promoting bacteria which is sustainable and beneficial for soil health and farmers.

2 Beneficial Effects of Rhizobacteria

Sustainable agriculture is essential today's world to fulfill the agriculture need and future food security. Since our tradition agriculture method is unable to do so because of various concerns. We have an urgent need to develop sustainable and effective mechanism to do same. Sustainable agriculture has potential to meet our agriculture need that our convention methods were unable to do so. This types of agriculture practice use special farming technique wherein environmental resource fully utilized without compromising it. Biological methods, a component of special forming may be an important alternative to replenish gap created by traditional method. This type of agriculture is beneficial, and they use natural resources without harming future generation. Diversity of dense population of microbes including bacteria, fungi, and Actinomycetes colonizes the root of plants. These microorganisms are group of naturally occurring beneficial microbe applied as inoculant to enhance plant growth and development (Ahmad et al. 2008). These groups of microbes have several properties which attract modern scientist and policy makers. In addition, these microbial communities improve soil quality, soil health and crop quality. Organic matters, in form of root exudate, attract numerous microbes and habitat for variety of microbes. Rhizobacteria in response to root exudates by chemotactic mechanism and competent rhizobacteria reside the rhizospheric zone of plant root. Some microbes reside in close vicinity with plants and communicate through different method (Singh et al. 2011). The communication occurs at molecular level through particular signaling molecule. Depending on compatibility of plant and microbes, it may for root nodules which provide favorable condition for microbes to fix atmospheric molecular nitrogen (Masciarelli et al. 2014). However, microbial community consists of selected groups of microbes which have plant growth promoting attributes. These microbes may reside in rhizosphere and promote plant growth. Soil microorganism also contributes to a wide range of application in sustainability of all ecosystems. These microbes regulate nutrient cycling, regulation of dynamic of soil organic matter, and enhance efficiency of nutrient acquisition. The symbionts of microbes enhance the efficiency of nutrient acquisition of nutrient and water by plants. Decomposition, mineralization, and nutrient flow are also regulated by these microbial associations.

3 Soil Health and Crop Productivity

Earth crust is an important biological component for microbial activity. It is a natural habitat for diversity of microbes, and it is estimated that one gram of soil contains up to ten billion bacterial cells. Decline in soil fertility is major concern for food security. Soil microbes contribute to a wide range of function in controlling soil health and crop productivity (Sahoo et al. 2015). It managed soil fertility through the modification of soil properties either directly or indirectly. Plant–microbe interaction is one of the important aspects for agriculture system. This association may help to achieve goal of future sustainable agriculture. Soil held variety of microbial species such as bacteria, fungi, mosses, and liverwort. Bacteria fungi and Actinomycetes are three groups of microbes that form major soil biomass of microbes. Among these, rhizospheric bacteria are to establish relationship with plants and promote nutrient uptake, water supply, and ameliorate various types of abiotic and biotic stresses. The presence of microbes is indicator of soil biological activities and regulates physical and chemical properties of soil. Microorganism is fundamental component of soil for all nutrient cycles and plant nutrient. Variation in temperature, low water content, anthropogenic, and grazing causes detrimental impact on microbial diversity and soil process. Soil–root–microbes form a comparatively stable and beneficial association. Some microbes have negative impact also in rhizosphere zone and harmful for plant growth and development (Ahmad et al. 2008). Due to intensive cropping and unhealthy effect of fertilizers, this relation declines soil microbial diversity.

4 Microbial Mechanism of Plant Growth Promotion

Microbes can we use in different way for plant growth promotion required plant growth promoting rhizobacteria (PGPR). Two major ways to promote plant growth are direct and indirect mechanism. The indirect mechanism includes reduction of some of the negative impact of pathogens by different mechanism. For example, siderophore reduces availability of iron for pathogens and reduces their growth. In addition, it may increase resistance through mechanism called induced systemic resistance and provide protection against pathogens. Direct mechanism is providing by availability of different beneficial compound synthesized by microbes. Similarly, they facilitate host plant growth by solubilizing minerals, availability of nutrient, production of phytohormones and secreting siderophore (Złoch et al. 2016). Plant growth promoting microbes that enhance plant growth process includes (Table 1): (1) increase availability of nutrient, (2) production of plant growth regulator (Auxin, Cytokine, and Gibberellins), (3) metabolites such as hydrogen cyanide (HCN), 1-amino cyclopropane-1-carboxylate (ACC) deaminase, siderophore, and antibiotics, (5) induction of systemic resistance (Duan et al. 2013; Contesto et al. 2010). For example, *Pseudomonas aeruginosa* LES4 strain of Tomato rhizosphere has plant growth promoting properties such as production of siderophore, hydrogen cyanide, indole acetic acid production, and phosphate solubilization (Kumar et al. 2009). Despite these properties, plant growth promoting microbes also enhance plant growth through production of ACC deaminase enzyme, exopolysaccharides, and antibiotic. ACC deaminase enzyme produced by microbes converts ethylene it into ammonia and α-ketobutyrate and protects host plant from deleterious effect of ethylene (Glick 2014; Rashid et al. 2012; Bal et al. 2013). Exopolysaccharides protect host plant through biofilm formation and reduce uptake of unwanted elements.

Table 1 Plant growth promoting substances released by some microbes and their role in plant growth and development

Plant growth promoting microbes	Sources/ Plants	Plant growth regulation	References
Burkholderia gladioli and *Penicillium aculeatum*	Oil Palm	Convert insoluble form of phosphorous to an available form of peat soil	Istina et al. (2015)
Erwinia species and P. *chlororaphis*	*Coffea arabica L*	Efficient uptake of insoluble phosphate from soil	Muleta et al. (2013)
Pseudomonas aeruginosa FP6	Chili	Siderophore produced by biocontrol strain for *Rhizoctonia solani* and *Colletotrichum gloeosporioides*	Sasirekha and Srividya (2016)
Bacillus amyloliquefaciens 5113 and *Azospirillum brasilense* NO 40	Wheat	Promote plant growth under drought condition, increase enzyme activity in wheat plant	Kasim et al. (2013)

(continued)

Table 1 (continued)

Plant growth promoting microbes	Sources/ Plants	Plant growth regulation	References
Bacillus mucilaginosus, Bacillus edaphicus, and *Bacillus circulans*	Panax	Solubilizes the insoluble potassium (K) to soluble forms of K for plant growth through acidolysis, chelation, exchange reactions, complexolysis and production of organic acid	Meena et al. (2014)
Bacillus amyloliquefaciens HK34		Induction of systemic resistance against *Phytophthora cactorum*	Lee et al. (2015)
Bacillus licheniformis K11	Pepper	Higher expression of stress genes, HSPs, and CaPR-10	Lim and Kim (2013)
Trichoderma harzianum Tr6 and *Pseudomonas* sp. Ps14	Cucumber and *Arabidopsis thaliana*	Induced systemic resistance	Alizadeh et al. (2013)
Bacillus thuringiensis AZP2	Wheat	Decrease volatile emissions and increase	Timmusk et al. (2014)
		Photosynthesis	
B. *thuringiensis*	*Lavandula dentate*	Higher content of potassium, proline and decrease the ascorbate and glutathione reductase due to IAA	Armada et al. (2014)
Bacillus thuringiensis GDB-1		Enhanced phytoremediation of heavy metals (Pb, Zn, As, Cd, etc.)	Babu et al. (2013)
P. fluorescens CHA0 and Pf1		Enhance nutrient content and increase growth and yield	Kavino et al. (2010)
Mesorhizobium spp.	Chickpea	Increase nodulation, Enhance and uptake of nutrient yield	Verma et al. (2013)
Pseudomonas putida H-2-3	Soybean	Improve plant growth under saline and drought condition. Increase leaf length and chlorophyll content	Kang et al. (2014)
		Ameliorate low-temperature stress	Praveen Kumar et al. (2014)
Serratia nematodiphila PEJ1011	*Capsicum annuum* L	Plant growth enhancement and biocontrol management to control plant diseases	Praveen Kumar et al. (2014)
Bacillus and *Pseudomonas* spp.			

4.1 Direct Mechanism

4.1.1 Nitrogen Fixation

Nitrogen (N) is most essential for plant growth and development and constitutes 78 of atmosphere. Although it exists in molecular form and not directly available to plants use, it is available to plant through biological nitrogen fixation to in ammonia. The process of nitrogen fixation carried out by oxygen sensitive, nitrogenase enzyme complex. Nitrogenase enzymes convert atmospheric nitrogen into ammonia using ATP as energy source. Biological nitrogen fixation is an important growth parameter which affects plant growth and yield. Different microbial species from various genera have ability of nitrogen fixation such as *Bacillus, Azospirillum, Pseudomonas, Enterobacter, Flavobacterium, Erwinia,* and *Rhizobium* (Silva et al. 2016). Nitrogen-fixing microbes are categorized into (1) symbiotic nitrogen fixating bacteria and (2) non-symbiotic nitrogen-fixing bacteria. The symbiotic nitrogen-fixing microbes include members of rhizobia which form symbiotic association with leguminous plants. The non-symbiotic bacteria include *Azospirillum, Azotobacter, Gluconacetobacter diazotrophicus,* and cyanobacteria. Most of atmospheric nitrogen is fixed by symbiotic association while non-symbiotic contributes only little. Some of these microbes for root nodule with their respective host plant root product play a significant role in communication between plant and microbes (Yan et al. 2008). Various stress conditions such as drought and salinity inhibit nodulation, hence nitrogen fixation. Microbes are more tolerant then host plant for drought and salinity. Nitrogenase enzymes responsible for nitrogen fixation are highly sensitive to salinity and its activity highly reduced under stress condition. The microorganism residing soil environment can cause a dramatic change in plant growth and production through various regulators. However, various free-living considers as plant growth promoting properties. Biological nitrogen fixation is environmentally sound and economically viable alternative of chemical fertilizers.

4.1.2 Phosphate Solubilization

Phosphorus is essential micronutrient and most vital element for plant growth and development. Plants acquire phosphorus in the form of phosphate ions from soil. Plant growth promoting rhizobacteria (PGPR) having phosphate solubilization ability, available phosphorus to plant through mineralization, and solubilization. Phosphorus solubilizing microbes govern biogeochemical cycle in natural agriculture system. It is responsible for normal functioning of living organism. However, most of phosphorus present in soil in insoluble form and unavailable to plant. It plays a major role in sugar transport and stimulates root development and physiological process of plants and animals. The bioavailability of phosphate to plant depends on plants, microbes, and surrounding environment. The association of plant microbes could enhance the mobilization of phosphorus in soil and available to plants (Mehta et al. 2013). The phosphate solubilizing rhizobacteria are ubiquitous in nature and their numbers vary from soil to soil. Phosphate solubilizing microbes (PSM) such as bacteria and fungi mobilizes it by producing organic acid and phosphatase. Many genera of bacteria and fungi are described as phosphate solubilizing microbes (Yadav et al. 2014). The phosphate solubilizing bacteria are belonging to PGPR and have wide implication in plant growth and development (Naseem and Bano 2014). The solubilization of phosphate may result due to either decrease in P^H or cations chelation. The effects of phosphate solubilizing microbes vary according to soil properties. Majority of plant associated with rhizobacteria under phosphorus deficient condition improve phosphorus uptake from soil. Phosphate solubilizing microbes (PSM) as biofertilizer enhancing plant growth and provide phosphorus in sustainable way to plants (Meena et al. 2015; Naseem and Bano 2014).The principle mechanism involved in this includes lowering $p^{H,}$ production of organic acids acid phosphatase. The release of organic acids from PSB decreases the surrounding pH and release of phosphate ions from H^+. Oxalic acid, succinic acid, malic acid, etc., are important acid produced by plants. These organic acids compete with binding sites on soil and available phosphorus to plants. 2-ketogluconic acid is the most potent phosphate solubilizing acid produced by phosphate solubilizing microbes. A numbers of microbes have capacity to solubilize phosphate (Istina et al. 2015). Therefore, use of phosphate solubilizing microbes in agriculture system is may be cost-effective and sustainable approach.

4.1.3 Siderophore Production

Iron (Fe) is a principle nutrient for almost all life form crucial for respiration, photosynthesis, nitrogen fixation, and DNA biosynthesis. Iron generally inaccessible to plants because of very low solubility Fe^{3+}, and it forms hydroxides and oxyhydroxides. Almost all living organism require iron for physiological and enzymatic activity. Bacteria obtain iron secretion of iron chelating compound siderophore which available to microbes (Schalk et al. 2011). Siderophore is water soluble and has high oscillation constant for complexing with iron. Siderophore can be divided into two group extracellular and intracellular siderophore. Most of iron acquisition by plants occurs through microbial siderophore. The production of siderophore by microbes depends on immediate surrounding environment, and microbes may modulate the environment. Siderophore has higher affinity for iron (Mishra et al. 2011). Siderophore easily binds with iron and transport inside bacterial cell. The transport of iron begins through the binding of siderophore–iron complex on material membrane. The transport of iron from through bacterial membrane involves different transporter. Once inside, the iron–siderophore complex reduced, ligand exchange to convert iron ($Fe^{3+)}$ complex on bacterial membrane reduced into soluble Fe^{2+} species (Rajkumar et al. 2010). ABC transporter (ATP-binding cassette) is involved in transport of siderophore–iron complex (Braun and Hantke, 2011) Plant incorporates iron from the bacterial siderophore through different means either through direct uptake of iron–siderophore or ligand change reaction. The major effect of siderophore–iron includes chlorophyll a and chlorophyll b content of plant leaves. Pseudochelin A, a new siderophore of *Pseudoalteromonas piscicida* S2040, shows siderophore activity against *Aspergillus fumigatus.*

4.1.4 Phytohormones Production

Abiotic stresses cause change in balance of phytohormones in plants which result growth inhibition. This change in hormones concentration also leads to susceptibility to disease occurrences. A major auxin, indole acetic acid (IAA), controls various physiological aspects such as plant development, cell signaling, and induction of disease resistance (Sukumar et al. 2013). Indole acetic acid acts as signaling molecule and affects expression of gene under different conditions (Egamberdieva et al. 2015, Cassán et al. 2014; Li et al. 2016). The IAA production varies and depends on microbes and environmental conditions. The growth of plants depends on concentration gradient of IAA. This results in inhibition, stimulation, and tissue differentiation. The low level of IAA induces root elongation while higher level of causes laterals and adventitious root formation (Ghosh et al. 2013; Tognetti et al. 2012). Phytohormones associated root colonizing rhizobacteria effectively colonized and supply additional IAA for plant growth and development (Sukumar et al. 2013). Ethylene is another ubiquitous phytohormones and plays a crucial role in biotic and abiotic stress tolerance. It causes cell elongation, root initiation and nodulation, leaf senescence, and fruit ripening. Under stress condition, ethylene causes inhibition of root elongation and root hair formation. To overcome this, negative impact bacterial ACC deaminase hydrolyzes it into a-ketobutyrate and ammonia. It may be a source of nitrogen for microbes. In addition, physiological process and stress tolerance especially drought are regulated by another plant hormones abscisic acid. PGPR enhances the concentration of abscisic acid during drought stress and makes plant to tolerant stress (Meena et al. 2015; Porcel et al. 2014;). From the above discussion, it is clear that the presence of phytohormones affects the endogenous mechanism of plants and stimulates plant growth. The major effects of these phytohormones include stimulation of lateral root and root hair formations which facilitate nutrient acquisition as well as water absorption (Belimov et al. 2014).

4.2 Indirect Mechanism of Plant Growth Promotion

The indirect effect of plant growth promoting bacteria occurs when they either decrease or inhibit deleterious consequence of pathogen through various mechanisms. The application of microorganism to control disease in plants is known as biocontrol (Kumar et al. 2014). It is an environmentally sustainable and cost-effective approach to ameliorate deleterious effect of pathogens. Various types of antifungal and antimicrobial agent secreted by rhizobacteria help in disease resistance. The major indirect includes competition for nutrient, antifungal metabolites, induced systemic resistance(ISR) and niche exclusion. Rhizobacteria have been reported to secrete many metabolites such as 2,4-diacetylphloroglucinol, HCN, phenazines, pyrrolnitrin are important biocontroling agents.

5 Biotic Stress

Plants have been evolved various sophisticated (physical and chemical) method to defense against pathogens. Chemical mechanism involved various chemical interactions at different steps. However, some plants lake have weak response to pathogen due to certain reasons. However, PGPR colonizes root surface and has been showing significant beneficial effect to plants (Salas-Marina et al. 2011). These PGPR activates plant defense response against pathogen and reduces crop losses. Plant growth promoting rhizobacteria possess various mechanism of defense against biotic stress. The mechanism defense against pathogen is trigger through chemical messengers. The induction of resistance against pathogen is generally of two kinds: systemic acquired resistance (SAR) and induced systemic resistance (ISR) (Fig. 1). SAR is induced by pathogenic bacteria and ISR by reducing disease rate (Nawrocka and Małolepsza 2013; Choudhary and Johri 2009). Both SAR and ISR adopted different signaling mechanism to induced resistance. SAR involves accumulation of salicylic acid and pathogen-related (PR) protein while ISR consists of ethylene and jasmonic acid (Dimkpa et al. 2009). The synergistic effect of pathogen-related (PR) protein and oxidative enzyme causes high protection against pathogens. Salicylic acid is an essential component for development of systemic which acquires resistance. Non-pathogenic bacteria induced systemic resistance in plants and phenotypically similar to systemic acquired resistance (Wani et al. 2016). These colonizing bacteria cause an increase in concentration of Chitinase and peroxidase plants leaves and root.

The effectiveness of SAR and ISR is wide spectrum and includes bacterial, fungal, viral pathogens. The protection of plant to different pathogens includes inhibition of pathogen growth through competition for nutrient and reduction in disease symptoms (Ghazalibiglar et al. 2016). It is a highly beneficial, cost-effective, and sustainable approach to protect plant from pathogens. Once this system of defense activated increase the activity of enzyme such as peroxidase, catalase, superoxide dismutase and guaiacol, etc. These enzymes protect cells from oxidative damage due to reactive oxygen species and pathogen. In addition, several potential components of bacteria also induced ISR membrane lipopolysaccharides and iron-regulated siderophore. This defense mechanism provides resistance to pathogen above-ground part of plants. ISR induced various changes in plant parts including (1) change in epidermal and cortical cell wall (2) level of enzyme

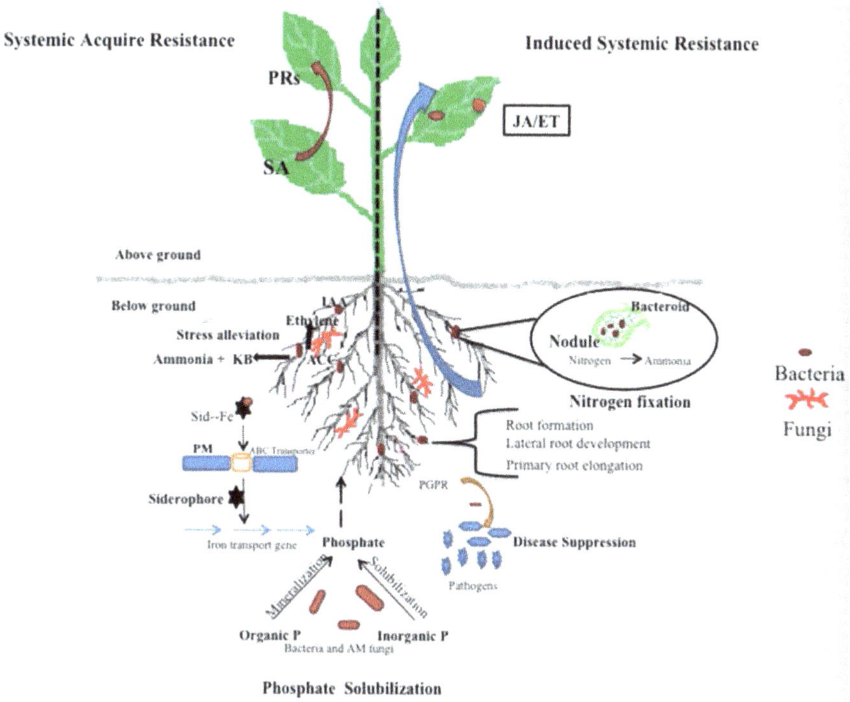

Fig. 1 Induction of resistance against pathogen by PGPR in plants

peroxidase, chitinase and phenylalanine ammonia lyase, and (3) expression of stress-related gene (Choudhary and Johri). The effect of ISR and SAR varies from plant to plant and depends on level of plant and microbes interaction. The combined effect of ISR and SAR enhances the defense level of protection against specific pathogens. Chithrashree et al. (2011) reported that PGPR mediate-induced systemic resistance in rice against bacterial leaf blight caused by *Xanthomonas oryzae* pv. *oryzae*. PGPR causes suppression of disease in plants by various mechanisms such as competition for space and nutrient, antagonism and induced systemic resistance (Beneduzi et al. 2012). *Pseudomonas* and *Bacillus* are acted as biocontroling agent and well known for their antagonistic and induced systemic properties. Bacteriocins, siderophore, and antibiotics are potential molecule and have antagonistic properties. Application of PGPR to seed and seedling also results induced systemic resistance (ISR). So plant growth promoting rhizobacteria play an important role in defense against various disease-causing agents (Choudhary and johri 2009). It is a sustainable approach to overcome disease-causing agents in economic and sustainable way.

6 Plant Growth Stress Condition

Global climate change is a major concern for agriculture productivity worldwide. Climate change will cause detrimental impact on agriculture system. Stress causes direct or indirect negative impacts on plants. For example, drought stresses condition increasing ethylene production, inhibit photosynthesis and damage photosynthesis apparatus and decrease chlorophyll content. Similarly, salinity increases ion concentration especially ions which cause injurious effect on plant growth and development. In addition, the general impact of bacteria causes hormonal and nutritional imbalance and changes in physiology of plants. Many microbial genera have been reported that has ability to survive under stress condition and improve crop productivity. Soil is a complex and dynamic system which provide habitat for microbes. In soil, the growth of microbes depends on soil environmental and their impact on plants. Under environmental stress condition such as drought, salinity heavy metal, and heat stress, etc, PGPM is effective in alleviating these negative stresses. These microbes use number of mechanisms to alleviate negative impact. The application of PGPR reduced level of ethylene overcomes the ion toxicity which caused due to salinity.

7 PGPR and Sustainable Agriculture

A fundamental shift will need to shift our conventional agriculture in toward sustainable agriculture. Plant growth promoting (PGP) microbes regarded as alternative to replace conventional agriculture practice. Plant growth promoting rhizobacteria Plant growth promoting rhizobacteria (PGPR) were first defined by Kloepper and Schroth (1978) for bacteria that colonize roots of plant after inoculation and enhance plant growth and development. PGPR includes wide range of microbes, including nitrogen fixation (Peix et al. 2015) and phosphate solubilizing bacteria useful soil microbiota important for plant growth and development. These microbes form association with microbes such as nodule formation. These microbes enhance plant growth and development through different activity, for example, enhance availability of soil nutrients, production of plant growth regulator, controlling plant growth properties, and maintain soil health. The effect of biofertilizer on growing crop in different climate region varies from neutral, positive, and negative depending upon plant microbes' interaction. The successful colonization microbes facilitate plant growth by providing major and minor nutrient to plant. This association also involves biological nitrogen fixation, phosphate solubilization, and potassium, which help in plant growth and development. Plants required phytohormones for root morphogenesis and shoot development, such as phytohormones are indole acetic acid (IAA), gibberellin, and cytokine. These hormones play an accountable role in different stage of plant during growth and development.

For example, IAA affects the root development, tissues differentiation in plants. IAA-induced root growth and enhanced root hair formation help in water and nutrient absorption due to increase surface area. This result profound growth and high yield of crop. Gibberellins are an important plant growth regulator which affects seed germination, stem elongation and development, flowering and fruit setting of plants. It also control ethylene level during stress condition by 1-aminocyclopropane-1-carboxylate (ACC) deaminase synthesized by PGPR and reduce the ethylene production (Saleem et al. 2007; Glick 2014).

8 Mycorrhizae

Mycorrhizae are mutual association between root of higher plant and fungi. Arbuscular mycorrhizae (AM) are most common fungi present in agriculture field. About 80 percent of higher plant form association with fungi. It plays in crucial role in nutrient cycling in agriculture field. Mycorrhizae are also important for improving physical properties of soil and maintain nutrient availability to plants and help in aggregation of soil (Garg and Pandey 2016; Miransari 2011; Chen et al. 2013). Mycorrhizae produce various metabolites under stress condition such as amino acid, vitamins, phytohormones and mineralization. The association with PGPR and mycorrhizae fungi supposed to cause effect on rhizospheric microbes, soil types, nutrient, moisture content, and temperature. The cumulative effects of this association also include assistance in plant growth and development in normal as well as under stress condition (Calvo-Polanco et al. 2016; Kohler et al. 2010). In addition, some study reported the mycorrhizal fungi also affect physiology process of plant through osmoregulator under stress condition. This process includes osmoregulation, such as increase proline, efficiency of water use and carbon dioxide exchange. It also enhances sugar and electrolyte concentration under drought condition. The function of mycorrhizae varies on types of stress. For example, nitrogen availability under droughts and variation in Na^+ and K^+ uptake under salinity stress (Hashem et al. 2016).

9 Conclusion

An ideal agriculture system maintains and improved soil health, benefit producer and consumer, protect human health and environment. In addition, produce enough food to feed world population and food security. Sustainable agriculture approach may fulfill these aspirations of farmers at low cost. Due to high cost of chemical fertilizers and pesticides, farmers are severely affected and compel to leave these practices. Application of biological approach decreases dependency on pest and fertilizers. This method is sustainable, cost-effective, and increase crop yield.

The application of microbial consortium reduced depletion of soil organic material and environment pollution. So plant growth promoting rhizobacteria offers an environmentally sustainable and cost-effective technology for increasing crop production. A number of studies have reported the effectiveness of PGM rhizobacteria under normal as well as in stress condition. In spite of the better performance achieved through dual inoculation of PGP rhizobacteria and mycorrhizae. PGP rhizobacteria and mycorrhizae show different response under different environmental condition. Such properties enable them to be applying as a potential alternative to traditional agriculture practices.

References

Achal V, Mukherjee A, Zhang Q (2016) Unearthing ecological wisdom from natural habitats and its ramifications on development of biocement and sustainable cities. Landsc Urban Plann 155:61–68

Ahmad F, Ahmad I, Khan MS (2008) Screening of free-living rhizospheric bacteria for their multiple plant growth promoting activities. Microbiol Res 163(2):173–181

Alizadeh H, Behboudi K, Ahmadzadeh M, Javan-Nikkhah M, Zamioudis C, Pieterse CM, Bakker PA (2013) Induced systemic resistance in cucumber and *Arabidopsis thaliana* by the combination of *Trichoderma harzianum* Tr6 and *Pseudomonas* sp. Ps14. Biol Control 65(1): 14–23

Armada E, Roldan A, Azcon R (2014) Differential activity of autochthonous bacteria in controlling drought stress in native *Lavandula* and *Salvia* plants species under drought conditions in natural arid soil. Microb Ecol 67:410–420

Babu AG, Kim JD, Oh BT (2013) Enhancement of heavy metal phytoremediation by Alnus firma with endophytic *Bacillus thuringiensis* GDB-1. J Hazard Mater 250:477–483

Bal HB, Nayak L, Das S, Adhya TK (2013) Isolation of ACC deaminase producing PGPR from rice rhizosphere and evaluating their plant growth promoting activity under salt stress. Plant Soil 366(1–2):93–105

Belimov AA, Dodd IC, Safronova VI, Dumova VA, Shaposhnikov AI, Ladatko AG, Davies WJ (2014) Abscisic acid metabolizing rhizobacteria decrease ABA concentrations in planta and alter plant growth. Plant Physiol Biochem 74:84–91

Beneduzi A, Ambrosini A, Passaglia LM (2012) Plant growth-promoting rhizobacteria (PGPR): their potential as antagonists and biocontrol agents. Genet Mol Biol 35(4):1044–1051

Braun V, Hantke K (2011) Recent insights into iron import by bacteria. Curr Opin Chem Biol 15(2):328–334

Calvo-Polanco M, Sánchez-Romera B, Aroca R, Asins MJ, Declerck S, Dodd IC, Martínez-Andújar C, Albacete A, Ruiz-Lozano JM (2016) Exploring the use of recombinant inbred lines in combination with beneficial microbial inoculants (AM fungus and PGPR) to improve drought stress tolerance in tomato. Environ Exp Bot 131:47–57

Cassán F, Vanderleyden J, Spaepen S (2014) Physiological and agronomical aspects of phytohormone production by model plant-growth-promoting rhizobacteria (PGPR) belonging to the genus *Azospirillum*. J Plant Growth Regul 33(2):440–459

Chen S, Jin W, Liu A, Zhang S, Liu D, Wang F, Lin X, He C (2013) Arbuscular mycorrhizal fungi (AMF) increase growth and secondary metabolism in cucumber subjected to low temperature stress. Sci Hortic 160:222–229

Choudhary DK, Johri BN (2009) Interactions of *Bacillus* spp. and plants–with special reference to induced systemic resistance (ISR). Microbiol Res 164(5):493–513

Contesto C, Milesi S, Mantelin S, Zancarini A, Desbrosses G, Varoquaux F, Touraine B (2010) The auxin-signaling pathway is required for the lateral root response of *Arabidopsis* to the rhizobacterium *Phyllobacterium brassicacearum*. Planta 232(6):1455–1470

Dimkpa C, Weinand T, Asch F (2009) Plant–rhizobacteria interactions alleviate abiotic stress conditions. Plant Cell Environ 32(12):1682–1694

Duan J, Jiang W, Cheng Z, Heikkila JJ, Glick BR (2013) The complete genome sequence of the plant growth-promoting bacterium *Pseudomonas* sp. UW4. PLoS ONE 8(3):e58640

Egamberdieva D, Jabborova D, Hashem A (2015) *Pseudomonas* induces salinity tolerance in cotton (*Gossypium hirsutum*) and resistance to *Fusarium* root rot through the modulation of indole-3-acetic acid. Saudi J Biol Sci 22(6):773–779

Garg N, Pandey R (2016) High effectiveness of exotic arbuscular mycorrhizal fungi is reflected in improved rhizobial symbiosis and trehalose turnover in *Cajanus cajan* genotypes grown under salinity stress. Fungal Ecol 21:57–67

Ghazalibiglar H, Hampton JG, de Jong EVZ, Holyoake A (2016) Is induced systemic resistance the mechanism for control of black rot in *Brassica oleracea* by a *Paenibacillus* sp.? Biol Control 92:195–201

Ghosh PK, Saha P, Mayilraj S, Maiti TK (2013) Role of IAA metabolizing enzymes on production of IAA in root, nodule of *Cajanus cajan* and its PGP *Rhizobium* sp. Biocatal Agric Biotechnol 2(3):234–239

Glick BR (2014) Bacteria with ACC deaminase can promote plant growth and help to feed the world. Microbiol Res 169(1):30–39

Hashem A, Abd-Allah EF, Alqarawi AA, Al Huqail AA, Egamberdieva D, Wirth S (2016) Alleviation of cadmium stress in *Solanum lycopersicum* L. by arbuscular mycorrhizal fungi via induction of acquired systemic tolerance. Saudi J Biol Sci 23(2):272–281

Islam F, Yasmeen T, Ali Q, Ali S, Arif MS, Hussain S, Rizvi H (2014) Influence of *Pseudomonas aeruginosa* as PGPR on oxidative stress tolerance in wheat under Zn stress. Ecotoxicol Environ Saf 104:285–293

Istina IN, Widiastuti H, Joy B, Antralina M (2015) Phosphate-solubilizing microbe from saprists peat soil and their potency to enhance oil palm growth and p uptake. Proc Food Sci 3:426–435

Kang SM, Radhakrishnan R, Khan AL, Kim MJ, Park JM, Kim BR, Shin DH, Lee IJ (2014) Gibberellin secreting rhizobacterium, Pseudomonas putida H-2-3 modulates the hormonal and stress physiology of soybean to improve the plant growth under saline and drought conditions. Plant Physiol Biochem 84:115–124

Kang Y, Hao Y, Shen M, Zhao Q, Li Q, Hu J (2016) Impacts of supplementing chemical fertilizers with organic fertilizers manufactured using pig manure as a substrate on the spread of tetracycline resistance genes in soil. Ecotoxicol Environ Saf 130:279–288

Kasim WA, Osman ME, Omar MN, Abd El-Daim IA, Bejai S, Meijer J (2013) Control of drought stress in wheat using plant-growth-promoting bacteria. J Plant Growth Regul:1–9

Kavino M, Harish S, Kumar N, Saravanakumar D, Samiyappan R (2010) Effect of chitinolytic PGPR on growth, yield and physiological attributes of banana (Musa spp.) under field conditions. Appl Soil Ecol 45:71–77

Kohler J, Caravaca F, Roldán A (2010) An AM fungus and a PGPR intensify the adverse effects of salinity on the stability of rhizosphere soil aggregates of *Lactuca sativa*. Soil Biol Biochem 42(3):429–434

Kumar S, Pandey P, Maheshwari DK (2009) Reduction in dose of chemical fertilizers and growth enhancement of sesame (*Sesamum indicum* L.) with application of rhizospheric competent *Pseudomonas aeruginosa* LES4. Eur J Soil Biol 45(4):334–340

Lee BD, Dutta S, Ryu H, Yoo SJ, Suh DS, Park K (2015) Induction of systemic resistance in Panax ginseng against *Phytophthora cactorum* by native *Bacillus amyloliquefaciens* HK34. J Ginseng Res 39(3):213–220

Li W, Wang D, Hu F, Li H, Ma L, Xu L (2016) Exogenous IAA treatment enhances phytoremediation of soil contaminated with phenanthrene by promoting soil enzyme activity and increasing microbial biomass. Environ Sci Pollut Res 23(11):10656–10664

Lim JH, Kim SD (2013) Induction of Drought Stress Resistance by Multi-Functional PGPR *Bacillus licheniformis* K11 in Pepper. 29(2):201–208

Masciarelli O, Llanes A, Luna V (2014) A new PGPR co-inoculated with *Bradyrhizobium japonicum* enhances soybean nodulation. Microbiol Res 169(7):609–615

Meena RK, Singh RK, Singh NP, Meena SK, Meena VS (2015) Isolation of low temperature surviving plant growth–promoting rhizobacteria (PGPR) from pea (*Pisum sativum* L.) and documentation of their plant growth promoting traits. Biocatal Agr Biotechnol 4(4):806–811

Meena VS, Maurya BR, Verma JP (2014) Does a rhizospheric microorganism enhance K + availability in agricultural soils? Microbiol Res 169(5):337–347

Mehta P, Walia A, Chauhan A, Kulshrestha S, Shirkot CK (2013) Phosphate solubilisation and plant growth promoting potential by stress tolerant *Bacillus* sp. isolated from rhizosphere of apple orchards in trans Himalayan region of Himachal Pradesh. Ann Appl Biol 163(3): 430–443

Miransari M (2011) Hyperaccumulators, arbuscular mycorrhizal fungi and stress of heavy metals. Biotechnol Adv 29(6):645–653

Mishra PK, Bisht SC, Ruwari P, Joshi GK, Singh G, Bisht JK, Bhatt JC (2011) Bioassociative effect of cold tolerant *Pseudomonas* spp. and *Rhizobium leguminosarum*-PR1 on iron acquisition, nutrient uptake and growth of lentil (*Lens culinaris* L.). Eur J Soil Biol 47(1): 35–43

Muleta D, Assefa F, Börjesson E, Granhall U (2013) Phosphate-solubilising rhizobacteria associated with Coffea arabica L. in natural coffee forests of southwestern Ethiopia. J Saudi Soc Agric Sci 12(1):73–84

Naseem H, Bano A (2014) Role of plant growth-promoting rhizobacteria and their exopolysaccharide in drought tolerance of maize. J Plant Interact 9(1):689–701

Nawrocka J, Małolepsza U (2013) Diversity in plant systemic resistance induced by *Trichoderma*. Biol Control 67(2):149–156

Patten DT (2016) The role of ecological wisdom in managing for sustainable interdependent urban and natural ecosystems. Landsc Urban Plann 155:3–10

Peix A, Ramírez-Bahena MH, Velázquez E, Bedmar EJ (2015) Bacterial associations with legumes. Crit Rev Plant Sci 34(1–3):17–42

Porcel R, Zamarreño ÁM, García-Mina JM, Aroca R (2014) Involvement of plant endogenous ABA in Bacillus megaterium PGPR activity in tomato plants. BMC Plant Biol 14(1):36

Praveen Kumar G, Mir Hassan Ahmed SK, Desai S, Leo Daniel Amalraj E, Rasul A (2014) In vitro screening for abiotic stress tolerance in potent biocontrol and plant growth promoting strains of *Pseudomonas* and *Bacillus* spp. Int J Bacteriol

Rajkumar M, Ae N, Prasad MNV, Freitas H (2010) Potential of siderophore-producing bacteria for improving heavy metal phytoextraction. Trends Biotechnol 28(3):142–149

Rashid S, Charles TC, Glick BR (2012) Isolation and characterization of new plant growth-promoting bacterial endophytes. Appl Soil Ecol 61:217–224

Sahoo RK, Ansari MW, Tuteja R, Tuteja N (2015) Salt tolerant SUV3 overexpressing transgenic rice plants conserve physicochemical properties and microbial communities of rhizosphere. Chemosphere 119:1040–1047

Salas-Marina MA, Silva-Flores MA, Uresti-Rivera EE, Castro-Longoria E, Herrera-Estrella A, Casas-Flores S (2011) Colonization of Arabidopsis roots by *Trichoderma atroviride* promotes growth and enhances systemic disease resistance through jasmonic acid/ethylene and salicylic acid pathways. Eur J Plant Pathol 131(1):15–26

Saleem M, Arshad M, Hussain S, Bhatti AS (2007) Perspective of plant growth promoting rhizobacteria (PGPR) containing ACC deaminase in stress agriculture. J Ind Microbiol Biotechnol 34(10):635–648

Sasirekha B, Srividya S (2016) Siderophore production by *Pseudomonas aeruginosa* FP6, a biocontrol strain for *Rhizoctonia solani* and *Colletotrichum gloeosporioides* causing diseases in chilli. Agric Nat Resour 50(4):250–256

Schalk IJ, Hannauer M, Braud A (2011) New roles for bacterial siderophores in metal transport and tolerance. Environ Microbiol 13(11):2844–2854

Silva WO, Stamford NP, Silva EV, Santos CE, Freitas ADS, Silva MV (2016) The impact of biofertilizers with diazotrophic bacteria and fungi chitosan on melon characteristics and nutrient uptake as an alternative for conventional fertilizers. Sci Hortic 209:236–240

Singh JS (2015) Microbes: the chief ecological engineers in reinstating equilibrium in degraded ecosystems. Agr Ecosyst Environ 203:80–82

Singh JS, Pandey VC, Singh DP (2011) Efficient soil microorganisms: a new dimension for sustainable agriculture and environmental development. Agr Ecosyst Environ 140(3):339–353

Sukumar P, Legue V, Vayssieres A, Martin F, Tuskan GA, Kalluri UC (2013) Involvement of auxin pathways in modulating root architecture during beneficial plant–microorganism interactions. Plant Cell Environ 36(5):909–919

Tilman D, Balzer C, Hill J , Befort BL (2011) Global food demand and the sustainable intensification of agriculture. PNAS 108(50):20260–20264

Timmusk S, Islam A, Abd El D, Lucian C, Tanilas T, Kännaste A, Behers L, Nevo E, Seisenbaeva G, Stenström E, Niinemets Ü (2014) Drought-tolerance of wheat improved by rhizosphere bacteria from harsh environments: enhanced biomass production and reduced emissions of stress volatiles. PLoS One 9:1–13

Tognetti VB, Mühlenbock PER, Van Breusegem F (2012) Stress homeostasis–the redox and auxin perspective. Plant Cell Environ 35(2):321–333

Verma JP, Yadav J, Tiwari KN, Kumar A (2013) Effect of indigenous *Mesorhizobium* spp. and plant growth promoting rhizobacteria on yields and nutrients uptake of chickpea (*Cicer arietinum* L.) under sustainable agriculture. Ecol Eng 51:282–286

Wani SH, Kumar V, Shriram V, Sah SK (2016) Phytohormones and their metabolic engineering for abiotic stress tolerance in crop plants. Crop J 4(3):162–176

Yadav J, Verma JP, Jaiswal DK, Kumar A (2014) Evaluation of PGPR and different concentration of phosphorus level on plant growth, yield and nutrient content of rice (*Oryza sativa*). Ecol Eng 62:123–128

Yan N, Marschner P, Cao W, Zuo C, Qin W (2015) Influence of salinity and water content on soil microorganisms. Int Soil Water Conserv Res 3(4):316–323

Yan Y, Yang J, Dou Y, Chen M, Ping S, Peng J, Lu W, Zhang W, Yao Z, Li H, Liu W (2008) Nitrogen fixation island and rhizosphere competence traits in the genome of root-associated *Pseudomonas stutzeri* A1501. Proc Nat Acad Sci 105(21):7564–7569

Złoch M, Thiem D, Gadzała-Kopciuch R, Hrynkiewicz K (2016) Synthesis of siderophores by plant-associated metallotolerant bacteria under exposure to Cd^{2+}. Chemosphere 156:312–325

Concluding Remarks

Ecological Wisdom Inspired Science and Engineering (Eco-WISE) is an emerging area of scientific investigation aiming to bring about or enhance effectiveness to ecological sustainability. It provides an overarching framework for a critical and fundamental assessment or reassessment of the theoretical basis of a sustainability initiative. Owing to its recency; however, little is known about how this wisdom-based approach can be incorporated into restoration engineering. This book is a step towards bridging this knowledge gap. It offers a unique perspective on how ecological wisdom principles can be integrated into engineering and practices for solutions related to restoration in building materials, structures, soil, water, urban and agriculture systems. Now it is time to move with the available methods or technologies and put knowledge and wisdom in restoration engineering for a better planet and ecosystem. Ecological wisdom derived from the nature can provide a powerful tool in inspiring various projects not only in theory or research but also in practice in order to restore a healthy relationship between nature and human beings. With the concept of nature-based solutions in restoration engineering, I hope that EcoWISE will bring more prudent solutions to the practice of ecological restoration in socio-ecological systems.

Varenyam Achal

V. Achal and A. Mukherjee (eds.), *Ecological Wisdom Inspired Restoration Engineering*, EcoWISE, https://doi.org/10.1007/978-981-13-0149-0

Epilogue

Ecological Practice: Original Flaw, Wickedness, and The Beacon of Ecological Wisdom

Ecological practice is the action and process that humans involuntarily engage themselves in with the aim to bring about a secure and harmonious socio-ecological condition that serves human beings' basic need for survival and flourishing. To achieve this goal, ecological practitioners, whether in planning, design, construction, restoration or management, must simultaneously attend to two distinct yet intertwining sets of relationships—the human-nature (ecological) relationship, and the human-human (social-economic-political) relationship within the human society. In doing so, ecological practitioners find themselves trapped in the messy swamps of *the original flaw* and *wickedness* in the jungle of socio-ecological systems where the ecological practice takes place.

The Original Flaw of Ecological Practice

The human-nature relationship ecological practitioners attend to is dominated by the extraordinary power of *nature's logic*. Irrational and opaque to human beings appears this logic in its own right. Not only does it operate in its own way that is entirely beyond human disposition, but it also makes sense in a deeper way that is often beyond human understanding (Laozi 2009, p. 46; Taleb 2012, pp. 348–349; Wang 2010, p. 325). As such, the unilateral dominance of nature's irrational and opaque logic in the human-nature relationship, according to American statistician and essayist Nassim Nicholas Taleb, prescribes peremptorily a concomitant feature of *original flaw* for human actions and science, which is best described in his 2012 book "Antifragile: Things that gain from disorder":

V. Achal and A. Mukherjee (eds.), *Ecological Wisdom Inspired Restoration Engineering*, EcoWISE, https://doi.org/10.1007/978-981-13-0149-0

> Just as there is a dichotomy in law: *innocent until proven guilty* as opposed to *guilty until proven innocent*... [in the human-nature relationship—the author] what Mother Nature does is rigorous until proven otherwise; what humans and science do is flawed until proven otherwise. (Taleb 2012, p. 349)

As one of "what humans and science do" that is subject to the dominance of nature's irrational and opaque logic, ecological practice is inevitably a recipient of the *original flaw* designation to its ontological identity. This prescribed designation can be stated as an inference from Taleb's proposition:

> Because "what humans and science do is flawed until proven otherwise", as one of "what humans and science do", ecological practice is also flawed until proven otherwise.

As a defining characteristic of ecological practice, *original flaw* is like the sword of Damocles hanging over ecological practitioners and the people they serve. It reminds everyone in ecological practice—practitioners, scholars, stakeholders, the public, and political leaders—to act prudently.

The Wickedness in Socio-Ecological Systems

The human-human (social-economical-political) relationship ecological practitioners involuntarily attend to is overwhelmingly contaminated by *wickedness* in socio-ecological systems where ecological practice takes place.

Wickedness refers to the ubiquity of wicked problems in socio-ecological systems (Xiang 2013, p. 2; 2016, p. 58). Wicked problems are a class of intractable and often unsolvable issues pertaining to the human-human relationship, whether it is by nature social, economic, or political. They may also be triggered by issues in the human-nature arena. These problems are wicked because they "are ill-formulated, [taking place in socio-ecological systems] where the [available] information is confusing, where there are many clients and decision makers with conflicting values, and where the ramifications in the whole system are thoroughly confusing... [such that] proposed 'solutions' often turn out to be worse than the symptoms" (Churchman 1967, p. B-141, phrases are added by the author for contextual clarity). Collectively identified and articulated by American planning scholars Horst Rittel, Mel Webber, and West Churchman in the 1960s and 1970s (For a recent review on their seminal works, see Xiang 2013), wicked problems are widely recognized to be present in almost all pressing issue areas that matter to the human society today, including those that are directly related to ecological practice. These include, but are not limited to, issues pertaining to global climate change, sustainability, resource management, urbanization (Xiang 2013, p. 2), and the use of high technologies in environmental management [for example, the use of nanoparticles in groundwater remediation, Kysar (2006), p. 7].

There are two distinct yet related characteristics of *wickedness* that are especially relevant to ecological practice. The first is what I call *the conservation law of wickedness* which states that *wickedness* (not necessarily individual wicked problems) co-exists with socio-ecological systems and co-evolves (Xiang 2013, p. 2). Metaphorically speaking, the beast of *wickedness* will not extinguish but only change its appearance from one to another as the jungle of socio-ecological systems succeeds (*Ibid.*) The co-evolution part of the law highlights the second more pertinent characteristic of *wickedness*—"solutions-are-often-worse-than-the-symptoms" (Churchman 1967, p. B-141), and explains the abundance of *iatrogeneses* in ecological practice[1]—the plethora of mutated wicked problems induced by practitioners' actions to resolve existing wicked problems.

The Beacon of Ecological Wisdom

How can ecological practitioners be "muddling through" (Lindblom 1959) the messy swamps of *original flaw* and *wickedness* in their practice?

They evidently had a time-honored beacon already. *Ecological wisdom*, that is. This reverable human virtue inspired and guided generations of human ancestors throughout history to work effectively with the daunting social reality of *original flaw* and *wickedness*, and as showcased in this book series, contributed to many great achievements in ecological practice. Despite the unfortunate under appreciation it endures during the contemporary "crisis of science without wisdom" (Maxwell 2007, p. 99), once restored and illuminated, the beacon of ecological wisdom will shine again on the jungle of socio-ecological systems to guide ecological practitioners through the messy swamps.

Let the Beacon Shine, Again

To reactivate the beacon and make it shine again, an important task ecological practitioners and scholars in the EcoWISE enterprise need to undertake immediately is to "unearth the treasure of ecological wisdom", as the late American planning scholar Robert Young once said in 2014 at *the First International Symposium on Ecological Wisdom for Urban Sustainability* in Chongqing, China,[2] and echoed by many scholars from around the world (For example, Dubos 1973; Liao and Chan 2016; Wang et al. 2016; Wang and Xiang 2016; Xiang 2014).

---

[1]In medicine, iatrogenisis (医源性疾病, in Chinese) refers to an illness or symptom induced inadvertently by a physician or surgeon or by medical treatment or diagnostic procedures. Etymologically the English word is from the Greek standing for a disease that is "brought forth by the healer". In the arena of social welfare, including ecological practice, it refers to a problem induced by the means of resolving a problem but ascribed to the continuing natural development of the problem being resolved. For more discussions, see Taleb (2012), pp. 110–133.

[2]A special ecological wisdom issue was developed by Wang and Xiang (2016) on the basis of this symposium and published in the journal *Landscape and Urban Planning*, Volume 155, 2016.

Integral to this task are the inquiries about *ecological wisdom principles*, the moral tenets for human action and scientific endeavor in ecological practice:

> *What* are the time-honored ecological wisdom principles in ecological practice? *How* are they kept up-to-date and effective in contemporary ecological practice?

More specifically, not only does this endeavor aim to revive ecological wisdom principles that had guided human ancestors through the messy swamps of *original flaw* and *wickedness*, but it also strives to keep these principles up-to-date and effective in the even messier modern-day ecological practice. With the rapid and massive increase in the human capability of "wrecking havoc" on the Earth (Maxwell 2007, p. 99), this task of unearthing ecological wisdom presents an exciting yet shrinking window of opportunity for ecological practitioners and scholars in the EcoWISE enterprise to make a difference in ecological practice. Actions, prudent but immediate, are therefore requested.

Holly Hill Farm, Charlotte, NC, USA
April 15th, 2018

Wei-Ning Xiang
University of North Carolina at Charlotte,
Charlotte, USA

and

Tongji University, Shanghai 200092 China
wxiang@uncc.edu; wnxiang@tongji.edu.cn

References

Churchman CW (1967) Wicked problems. Manag Sci 14(4):B141–B142

Dubos RJ (1973) Humanizing the Earth. Science 179(4075):769–772

Kysar DA (2006) It might have been: Risk, precaution, and opportunity costs. Cornell Legal Studies Research Paper No. 06-023 https://ssrn.com/abstract=927995 or http://dx.doi.org/10.2139/ssrn.927995

Laozi (2009) Tao te ching on the art of harmony. A translation of Dao De Jing (《道德经》) by Chad Hansen. New York: Metro Books.

Liao KH, Chan JKH (2016) What is ecological wisdom and how does it relate to ecological knowledge? Landsc Urban Plann 155:111–113

Lindblom CE (1959) The science of "muddling through". Public Adm Rev 19 (2):79–88

Maxwell N (2007) From knowledge to wisdom: the need for an academic revolution. Lond Rev Educ 5(2):97–115

Taleb NN (2012) Antifragile: things that gain from disorder. Random House, New York

Wang M(王蒙) (2010) The ease of Zhuangzi (《庄子的享受》), Hefei, China: Anhui education press (合肥,安徽教育出版社)

Wang X, Palazzo D, Carper M (2016) Ecological wisdom as an emerging field of scholarly inquiry in urban planning and design. Landsc Urban Plann 155:100–107

Wang X, Xiang W-N (2016) Ecological wisdom for urban sustainability: doing real and permanent good in ecological practice. Landsc Urban Plann 155:1–2

Xiang W-N (2013) Working with wicked problems in socio-ecological systems: awareness, acceptance, and adaptation. Landsc Urban Plann 110(1):1–4

Xiang W-N (2014) Doing real and permanent good in landscape and urban planning: ecological wisdom for urban sustainability. Landsc Urban Plann 121:65–69

Xiang W-N (2016) *Ecophronesis*: The ecological practical wisdom for and from ecological practice. Landsc Urban Plann 155:53–60

Author's Biography

Mona Wells is an Associate Professor at Department of Environmental Sciences, Xi'an-Jiaotong Liverpool University, China. She got her Ph.D. in Chemistry from Texas A&M University in 1999. She worked as Researcher at University of Otago, while Assistant Professor at Tennessee Technological University and University of New York-Oneonta. She is interested in areas of research that are tied to promoting environmental sustainability by linking science, technology, and innovation.

E-mail: mona.wells@xjtlu.edu.cn

Yusuf Çağatay Erşan conducted his Ph.D. study as a Marie Curie research fellow in European Union Seventh Framework Programme [FP7/2007–2013] and obtained his Ph.D. degree (2016) in Bioscience Engineering from Ghent University, Belgium. Since 2017 he works as an assistant professor in Department of Civil Engineering at Abdullah Gul University, Turkey. He has expertise in microbial ecology and cementitious composites. Currently, he works on microbial self-healing concrete. His research interests cover construction biotechnology, biogranulation, microbe-material interactions, biomineralization and bio-based economy. His research led to several SCI indexed journal publications and was presented in many international conferences.

Email: yusuf.ersan@agu.edu.tr

V. Achal and A. Mukherjee (eds.), *Ecological Wisdom Inspired Restoration Engineering*, EcoWISE, https://doi.org/10.1007/978-981-13-0149-0

Liang Cheng Ph.D. and Professor, graduated from Murdoch University (Australia), majored in Environmental Engineering. He has over ten years' experience in the field of biotechnology, environmental geotechnics, and bio-electrochemistry with particular emphasis on bio-mediated ground improvement, wastewater treatment, and microbial fuel cell based biosensor. He received 2014 R M Quigley Awards (honorable mention), for the best paper published in the Canadian Geotechnical Journal in 2013. His present research focuses on bio-mineralization technology for coastline protection, bio-electrochemical process leading to novel biosensor for low concentration of organic carbon detection, pretreatment of activated sludge by electrolysis process, and energy efficient wastewater treatment process.

Email: lcheng@ntu.edu.sg

I am Dr. G.G.N.N. Amarakoon and obtained my Ph. D. from Hokkaido University, Japan in 2016 with the thesis entitled as "Development of sustainable ground improvement methods using biochemical techniques". I completed my master of engineering (M. Eng.) in Foundation engineering and earth retaining systems from the University of Moratuwa, Sri Lanka in 2013. Currently, I am working as a senior geotechnical engineer in the overseas business division, Nittoc construction Co. Ltd, Japan which is a specialized contractor for soil related works such as Disaster prevention and environmental conservation, maintenance, renovation and urban regeneration. I am an associate member in Institution of Engineers, Sri Lanka (IESL).

E-mail: nishanirmi@gmail.com

I am Nadeeka Dilrukshi and obtained my Ph.D. from Hokkaido University, Japan with the thesis entitled as "Low environmental impact type ground improvement methods using biomineral precipitation". I completed my master of engineering (M.Eng.) in Foundation engineering and earth retaining systems from the University of Moratuwa, Sri Lanka. Currently, I am working as a senior lecturer in Northshore College of Business and Technology, Sri Lanka which is affiliated with a prestigious University of the West of England (UWE). I am an associate member in Institution of Engineers, Sri Lanka (IESL).

E-mail: nadilrukshi@yahoo.com

Professor Chee Seong Chin, is the Chair Professor of Civil Engineering as well as Dean of Learning and Teaching, supporting Xi'an Jiaotong-Liverpool University in the operational delivery of its academic strategy and the ongoing co-ordination of learning and teaching. He is the founding director of the Institute for Sustainable Materials and Environment and founding Head of the XJTLU-Innostrux Centre for Advanced Fibre Reinforced Concrete Technology. His primary research interests are fibre-reinforced cementitious composites and sustainable construction materials. He has acquired over 15 million RMB of research grants, produced more than 55 publications and supervised 12 Ph.D. students for their dissertations.

Email: Chee.Chin@xjtlu.edu.cn

Dr. Nadeem Akhtar has completed his Ph.D. in Biotechnology from Thapar University, India where his major area of research was focused on biodegradation of different agro-industrial waste for bioethanol and other value-added products. After graduation, he joined University of Guelph, Canada as a Postdoctoral Researcher in the Department of Animal Biosciences. Dr. Akhtar has a very board interest in improving animal production and disease resistance using biotechnological approach, with major on microbial production of antimicrobial peptides (AMPs), and testing their efficacy on animal model(s). Improving nutritional value of alternative feed stuffs for well-being of poultry and swine is another area of his current research. He has published over 15 research papers in peer-reviewed journals, contributed 6 book chapters, and published a patent in India. He has also been serving as a reviewer of many reputed journals such as Scientific Reports, Environmental Progress and Sustainable Energy, Journal of Taiwan Institute of Chemical Engineers, etc.

Email: nadeembiotech@gmail.com

Eva Stache is a practicing architect in the Amsterdam area, a member of the Royal Institute of Dutch Architects (BNA) and Ph.D. researcher at the Delft University of Technology in The Netherlands. Her main research is focused on the climatological role of urban ecosystem services in the ongoing major urbanization processes. As an architect and researcher she's been working on environmental policy studies, implementation strategies of sustainability in the architectural practice and developing new sustainability measures in the urban and architectural planning since 1998. She has won the Novem award in 2003 for the development of a sustainable dwelling concept, built in various forms in The Netherlands.

E-mail: postmaster@stache-architect.nl

Paul Bois is Assistant Professor at the National Water and Environmental Engineering School of Strasbourg —ENGEES—and at ICube Laboratory (Strasbourg—FR). His research interests include urban ecological infrastructures and their impacts on the urban water cycle (storm- and wastewater), as part of the ecosystem services these social-ecological systems provide. He studies the impact of climatic, biological, physico-chemical and hydrological drivers on the behavior of these systems. Part of this research also includes the study of how civil society action can be a driver of change in urban water cycle. He was Vice Coordinator of the Strasbourg-LTSER from 2013 to 2016.

Email: p.bois@unistra.fr

Deepika Kumari is an Associate Professor at College of Architecture and Urban Planning, Tongji University, Shanghai, China. During her Ph.D., she worked at Directorate of Mushroom Research, India and awarded with Ph.D. degree from Thapar University, India. Her major is Biotechnology. She did her postdoctoral research at Chinese Academy of Sciences, Urumqi. She is interested in studying microbiomes, environmental microbiology and ecological restoration technology.

E-mail: deepikakumari@tongji.edu.cn

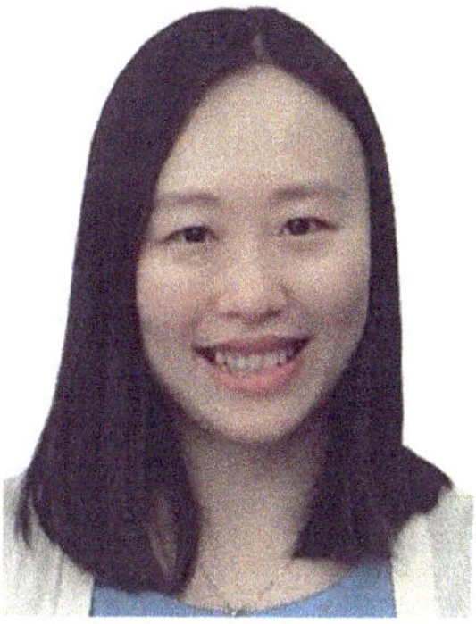

Core Member and Leading Author: Zhang Qiuzhuo

Zhang Qiuzhuo (Ph.D., Professor) got her Ph.D. from Harbin Institute of Technology (China) in 2008. Her major is Environmental Science and Technology. She is interested in ecological engineering, low carbon technology and microbiology. A lot of ecological engineering projects have been completed successfully by her efforts. She is now the Deputy Dean of School of Ecological and Environmental Sciences, East China Normal University. She is also the committee member of environmental geosciences sub-committee in Chinese Society for Environmental Sciences.

Email: qzhzhang@des.ecnu.edu.cn

MA Limin, Ph.D and Professor of School of Environmental Science & Engineering in Tongji University, China. He got his Ph.D from Chinese Academy of Sciences with major of biogeochemistry in 2001. He was as a Senior Visiting Scholar in Pisa University in Italy from 2009 to 2010 and in Karlsruhe Institute of Technology in German from 2017 to 2018. His recent academic interests include theory and technology of restoration of degraded ecological ecosystem, wetland waste water treatment and environmental behavior of organic pollutants. He has led and conducted over forty research projects, including projects sponsored by the National Natural Science Foundation, Ministry of Science and Technology of China. He has published over 120 research papers, in Chinese and English and has over 10 authorized patents. He is also a consulting expert of projects funded by the MOST of China, Miur of Italy and WWF.

E-mail: lmma@tongji.edu.cn

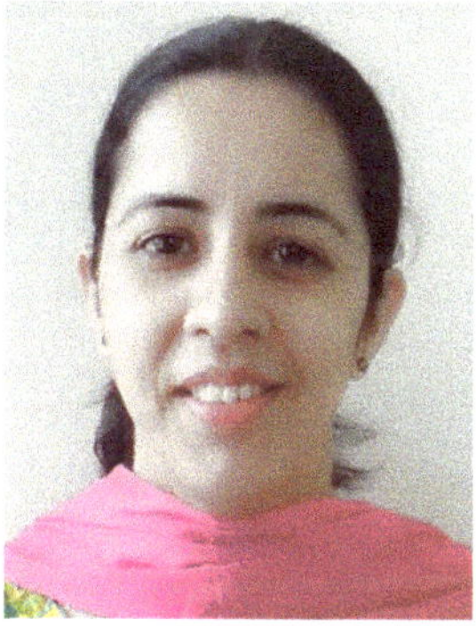

Dr. Neetu Sharma is working as an Assistant Professor in the Department of Biotechnology, GGDSD College, Chandigarh, India from last 8 years. She got her Ph.D. degree in the field of Environment Biotechnology from Thapar University, Patiala. She has worked as Research Fellow at BARC, Mumbai, on project entitled "Studies on selenium and its speciation in biological and soil samples from seleniferous Regions of Punjab". She has also worked as Research fellow in DRDO sponsored project entitled "Formation of Selenium Nanoparticle using bacterial system". Dr. Sharma has a research experience spanning over 8 years and has published a number of research and review articles in reputed national and international journals. She is also a life member of Indian Association of Nuclear Chemists and Allied Scientists (IANCAS), BARC, Mumbai. Her current research emphasis is on the biological synthesis of nanoparticles for confronting the environmental problems like water disinfection and their potential antimicrobial properties against pathogenic microbes to tackle increasing antibiotic resistance against available drugs.

Email: neetusharma4@gmail.com

Dr. Jay Prakash Verma did his Ph.D. from Banaras Hindu University (BHU) and working as a Senior Assistant Professor in IESD, Banaras Hindu University, India since 2011. He is a Visiting fellow at HIE, Western Sydney University, Australia for research on core microbiome of cotton plants since 1 Dec, 2017 to 31 July, 2018. His expertise more than 11-year on microbial consortium, PGPR, Phytomicobiome, Sustainable Agriculture, Plant-Soil-Microbe Interaction, microbial bioremediation, and bioethanol production. He has published more than 70 articles that have been cited more than 1920 times (Google citations) and has a cumulative impact factor of more than 125. He has more than 8 research project related to plant-microbe interactions. He has visited Spain, Paris and Shanghai and Australia countries. He got Indo-Australia Carrier boosting gold fellowship in 2017 and 2-time young scientist project award 2013 and 2018. He is serving as member of Editorial board for several international and national Journals. He acts as a reviewer for various peer reviewed national and international journals.

Email: jpv.iesd@bhu.ac.in

Index

V. Achal and A. Mukherjee (eds.), *Ecological Wisdom Inspired Restoration Engineering*, EcoWISE, https://doi.org/10.1007/978-981-13-0149-0

Lightning Source UK Ltd.
Milton Keynes UK
UKHW02n1328280818
327921UK00001B/24/P

9 789811 301483